AF863312

Proceedings of the Conference

ELECTRON AND ION TRANSFER IN CONDENSED MEDIA

Theoretical Physics for Reaction Kinetics

INTERNATIONAL CENTRE FOR THEORETICAL PHYSICS

UNITED NATIONS EDUCATIONAL, SCIENTIFIC AND CULTURAL ORGANIZATION

INTERNATIONAL ATOMIC ENERGY AGENCY

Proceedings of the Conference

ELECTRON AND ION TRANSFER IN CONDENSED MEDIA

Theoretical Physics for Reaction Kinetics

ICTP, Trieste, Italy 15–19 July 1996

Editors

A. A. Kornyshev
Forschungszentrum Jülich GmbH, Germany

M. Tosi
Scuola Normale di Pisa, Italy

J. Ulstrup
The Technical University of Denmark, Denmark

World Scientific
Singapore • New Jersey • London • Hong Kong

Published by

World Scientific Publishing Co. Pte. Ltd.
P O Box 128, Farrer Road, Singapore 912805
USA office: Suite 1B, 1060 Main Street, River Edge, NJ 07661
UK office: 57 Shelton Street, Covent Garden, London WC2H 9HE

British Library Cataloguing-in-Publication Data
A catalogue record for this book is available from the British Library.

ELECTRON AND ION TRANSFER IN CONDENSED MEDIA

Copyright © 1997 by International Centre for Theoretical Physics

ISBN 981-02-2929-1

This book is printed on acid-free paper.

Printed in Singapore by Uto-Print

PREFACE

Elementary charge transfer (CT), be it electron or atom group transfer, is the single most central molecular event in an extremely broad variety of physical, chemical, and biological phenomena. CT reactions are encountered in radiationless and radiative processes in the solid state, in simple and composite chemical redox processes in aqueous and other liquid solutions, and in the intriguing nonuniform and anisotropic environments of biological electron transport chains and redox enzyme chemistry. Common denominators in all these areas are ET between *localised* molecular centres, *strong* interactions between the reaction centres, and the importance of environmental configurational *fluctuations* which induce the fundamental CT events. The widely different specific environments and the increasing detail needed in the understanding of the nature of the transitions and the coupling to the environment have, however, warranted major new lines in the theoretical approaches to molecular CT phenomena. Such evolution has been impressively achieved, stimulated by parallel progress in theoretical physics rooted in areas such as theories of stochastic processes, functional integration, theory of dense polar liquids, critical phenomena, and quantum molecular dynamics simulations. Development of charge transfer science has been advanced strongly, moreover, by recent impressive achievements in femtosecond laser technology and elucidation of a wealth of new, ultrafast relaxational and coherent CT phenomena in physics, chemistry, and biology.

CT science over the last decade has prompted the idea that an international meeting of experts in the interdisciplinary CT areas of physics, chemistry, and biology is highly warranted. This view has been supported by the International Centre for Theoretical Physics (ICTP) in Trieste which has always had a keen interest in following new routes where theoretical physics could have an impact on neighbouring sciences. As proposers of such a meeting we are very grateful that this meeting came to enjoy the status of an Adriatico Research Conference entitled "ELECTRON AND ION TRANSFER IN CONDENSED MEDIA" and was held at the ICTP, Trieste–Miramare, Italy (15–19 July 1996). The conference was generously funded by the ICTP and was cosponsored by the European Community Commission, the International School for Advanced Studies and the International Society of Electrochemistry.

Updated insight, overview, and highlights in the following areas of charge transfer science constituted major conference achievements:

1. Liquid structure and dynamics, and femtosecond solvation. This included ultrafast phenomena and their accordance with nonlocal dielectric views studied by experiments and computer simulations.

2. New features of the fundamental CT phenomenon, with new analytical theories of sophisticated adiabatic and diabatic CT, dissipative media, and electronic-vibrational coupling patterns in specific systems.
3. Interfacial CT, electrochemistry, electrocatalysis, and *in situ* scanning tunnel microscopy. This extended to new theories of electrochemical electron and atom transfer, warranted by new detail experimental data. Processes in "new systems" with ionic and molecular adsorbates, electrocatalysis, solvated electrons, resonance states and interfacial tunnel phenomena, and processes with participation of "hot" electrons.
4. Ion transport, with fundamental laws and models for ion transport in liquids, glasses and molten salts, and solid-state and polymer proton conductors.
5. Electron transport in organised molecular systems, with emphasis on long-range, directional electron transfer in molecular or supramolecular assemblies, synthetic molecular devices, and electron transfer in biological systems such as photosynthesis, bacteriorhodopsin function, and metalloprotein organisation.

The contributions were timely and of inter-disciplinary look, and they were supplemented by discussions on special topics which addressed the hottest topics in the field. It was a common view that the publication of the contributions would be most expedient and that a suitably coherent volume could become a landmark in CT science and wider areas if the conference style could be transferred to the written form. About 70% of speakers were able to submit their contributions, and we are pleased to acknowledge that the present volume indeed reflects the main topics discussed at the conference. We believe that the book, as a coherent collection of reports of experts in this field will have its own life and value.

The editors wish to give their sincere thanks to all the authors who found time in their pressed schedules to contribute their communications. We also wish to thank the director of the ICTP, Professor M. Virasoro, and the members of the ARC Committee — Professors Hilda Cerdeira, Stig Lundqvist and Erio Tosatti — for the support of this project and for their encouragement. The sponsors of the conference are sincerely thanked for their financial support. Special thanks are due to Mrs. Valerie Shaw, the conference secretary, and her assistant, Mrs. Milena Poropat for perfect technical work.

Alexei A. Kornyshev (*Jülich*)
Mario Tosi (*Pisa*)
Jens Ulstrup (*Copenhagen*)

CONTENTS

SOME HISTORICAL NOTES ON CHEMICAL CHARGE TRANSFER

N.S. HUSH
School of Chemistry, The University of Sydney
NSW 2006, Australia

J. ULSTRUP
Department of Chemistry,
Technical University of Denmark
DK-2800 Lyngby, Denmark

We discuss briefly some important scientific discoveries, which were to be later adopted as foundations of chemical charge transfer science. The quantum mechanical tunnel effect, the Franck Condon principle and the Born-Oppenheimer approximations, concepts relating to chemical proton transfer and acid-base reactions, tunnelling in chemistry, and the nature of electronic transitions are early discoveries with much later strong impact. To this can be added early stochastic chemical rate theory. The postwar introduction of radioisotopes marked another major step forward and led to the notion of outer and inner sphere electron transfer, recognition of electron tunnelling in chemistry, and the importance of electronic-vibrational interaction. By this time the latter two areas had already found a detailed quantum mechanical basis from parallel but apparently quite independent efforts in solid-state physics.

The theories of R.A. Marcus and N.S. Hush, rooted in classical transition state theory, and of Levich, Dogonadze and their associates, rooted in solid-state electronic processes initiated present-day chemical charge transfer theory. The early form of these theories also mark a provisional halt to the historical perspective and later achievements are conveniently presented systematically rather than chronologically. We conclude with a brief outline of some of the later achievements and future perspectives.

1 Introduction

Molecules in solution and local centres in solids exchange elementary charges in the form of electrons and protons[1-8]. Although fundamentally different, electron and proton transfer are also in important respects formally and physically isomorphous. In both *light* particles are transferred between much heavier molecular donor and acceptor cores. Both are also in crucial respects *quantum* particles, subjected to tunnelling and other non-classical behaviour. *Electron* transfer (ET) is the simplest conceivable chemical process where no chemical bonds are broken or formed. ET can also be brought to occur over distances which significantly exceed the structural extension of the donor and acceptor. This is also two major differences from *proton* transfer (PT). PT is composed of synchronous bond breaking and formation, and PT is only possible when the actual fluctuation induced proton hopping distance is short, say a fraction of an Å or so, even though the equilibrium

PT distances are larger. A consequence is that donor-acceptor interactions in PT are much stronger than in ET, even to an extent that the quantum mechanical proton behaviour become concealed. PT is therefore overall a more composite molecular event than ET.

ET and PT are broadly encountered in the chemical, physical, and biological sciences. Simple chemical ET between molecules in solution is the most central event in inorganic and organic redox chemistry. "Isolated" ET events are processes such as

$$[MnO_4]^{2-} + [^*MnO_4]^- \rightleftarrows [MnO_4]^- + [^*MnO_4]^{2-} \quad (1)$$

$$[Fe(CN)_6]^{4-} + [IrCl_6]^{2-} \rightarrow [Fe(CN)_6]^{3-} + [IrCl_6]^{3-} \quad (2)$$

also encountered in electrochemistry, and in inorganic and organic photochemistry. Chemical proton transfer is the key "elementary" reaction step in acid-base reactions and broadly in physical organic chemistry where general acid-base catalysis is a dominating feature in hydrolytic, acid-base, isomerization processes etc. Charge transfer chemistry extends to the biological realm, summarized in the overall equation of aerobic metabolism

$$6CO_2 + 6H_2O \rightleftarrows C_6H_{12}O_6 + 6O_2 \quad (3)$$

In this way ET and PT become fully integrated elements not only in photosynthetic and respiratory *electron transport* but also in the mechanisms of a wealth of redox and hydrolytic, metallo- and non-metalloenzymes.

In physical terms ET and PT bear formal close resemblance to other, physical electronic processes. These are, for example: (a) ionization and electron capture of solid-state impurities[9,10]. We shall see below that this is in fact where contemporary quantum mechanical ET/PT theory *began*; (b) non-radiative relaxation of excited electronic states in molecules and solid-state dopants by internal conversion and intersystem crossing[11-13]; (c) electronic energy transfer[14]; (d) absorption and emission of electromagnetic radiation in molecules[15,16], and (e) physical processes such as impurity atom diffusion in metals[17] or dipolar relaxation in solid host lattices[18]. In broader terms, ET/PT is, at least conceptually related to even wider areas. Metal and semiconductor conductivity, and conductivity in low-dimensional materials involve ET with all grades of electron localization ranging from fully delocalized metallic conduction to electron hopping between states confined within the the spatial extension of the molecular groups involved. ET systems such as quantum well structures, solid-state tunnel junctions, surface modified electrochemical electrodes, and the world of scanning tunnel microscopy are other areas where the central physical event is ET controlled by electronic-vibrational interaction in much the same ways as in classical chemical redox chemistry[19].

The science areas where ET and PT are strongly conspicuous are multifarious and comprehensive. In this short introduction we therefore have to be selective. What we shall do is to illuminate, first some early scientific achievements from which ET and PT science enjoyed crucial impact, either at the time or later. This will be followed by a brief overview of the transfer of conceptual views and formalism from solid-state physics which marked the initiation of present day ET and PT science. We conclude with a brief overview of some of the broad perspectives which contemporary charge transfer science has now come to hold and which are addressed in the following chapters.

2 Some Early Views on Physical and Chemical Particle Transfer

The history of chemical charge transfer began early. Mixed-valence compounds such as Prussian Bue $Fe_4^{III}[Fe^{II}(CN_6)]_3$ which we now know owe their deep colours to photoinduced ET, were for example known already in the 18th century[20,21]. Kinetics of reactions accompanied by valence changes such as[21,22]

$$2FeCl_3 + SnCl_2 \rightarrow 2FeCl_2 + SnCl_4 \tag{4}$$

were investigated towards the end of the 19th century based on views rooted in the valence theory of the time. In a similar sense electrochemical charge flow was recognized from the times of Volta, Davy, and Faraday in the 18th and 19th centuries while proper current-voltage relations are first ascribed to Tafel in 1905[23]. Early discovery of acid catalysis are Wilhelmy's exploration of the conversion of sugar cane (1850)[24] and parallel developments of classical electrolyte theory towards the end of the 19th century where acid-base catalysis was in focus to monitor the state of the electrolyte solutions[24,25].

Charge transfer (i.e. ET and PT) as we view it today, i.e. as electron or proton transfer between independent molecular entities is, however, rooted in early views on quantum mechanical particle tunnelling and on acid-base concepts associated with proton transfer. These concepts date from the 1920's and were, interestingly, at first developed in separate worlds of physics and chemistry. Very notable achievements towards the end of the 1920's followed establishing of the new quantum mechanics, and were in due time to be adopted by chemical charge transfer science. Central were:

- The discovery, and experimental substantiation of the *quantum mechanical tunnel effect*[26]. Following a mathematical procedure for solving the Schrödinger equation due to Jeffreys[27], Wentzel[28], Kramers[29], and Brillouin[30] in the same year (1926) found the way to solve the barrier penetration problem now common in all texts on quantum theory. The WKB solution was later (1954) to be bridged with the Golden Rule of quantum mechanical perturbation theory[31]. The rigorous derivation rests on mathematical subtleties and is discussed in several excellent texts[32]. The tunnel phenomenon acquired immediate acceptance. This could partly be due to the availability of good classical optical analogues (frustrated total reflection) but probably in particular because it provided excellent and lasting explanations for several at that time exciting new phenomena such as field-induced electron emission from metals[33] and radioactive particle decay from heavy atomic nuclei[34]. This was paralleled or followed shortly by introduction of the tunnel concept in chemical phenomena such as microwave spectral broadening of chemical proton and hydrogen atom transfer[35] (cf. the next Section).
- The *Franck Condon principle* had, secondly been established and the *Born-Oppenheimer approximation* introduced. Franck's first formulation took a classical and intuitively appealing view that electronic transitions in molecules are so fast compared with vibrational motion that the configuration and velocity of the latter remain unaltered immediately before and after the transition[36]. The same principle was given a quantum mechanical formulation by Condon a little later with much the same conclusions[37]. The Born-Oppenheimer approximation and other schemes for mathematical separation of the electron and nuclear motion is a different concept but reflects similar physical features in the relation between fast electronic and much slower nuclear motion[38]. The Born-Oppenheimer approximation is a scheme for solution of composite electron-nuclear Schrödinger equations where the dynamic variables are the coordinates of the fast electron system while the nuclear coordinates are regarded as slowly varying external "field" parameters. The resulting electronic energy dependence of the parametric nuclear coordinate variables are denoted as potential energy surfaces. The Franck-Condon principle and the Born-Oppenheimer approximation, accurately established by the end of the 1920's, are implicitly or explicitly the most crucial theoretical concepts of all later approaches to ET and other light particle transfer in physics, chemistry, and biology.
- As a third fundamental concept recognized in the 1920's and central in later theoretical approaches to broad areas of chemical reaction dynamics is the clarification of the acid-base concept[39] and the notion of general acid-base catalysis[40], associated with the name of J.N. Brønsted. The acid-base concept has of course a long history prior to the 1920's, in scientific terms beginning with Lavoisier and later associated with such names as Davy, Gay-Lussac, Arrhenius and Ostwald[24,25]. In the context of charge transfer science in the 1920's the concepts were rooted in the world of physical chemistry and originally quite separate from the world of quantum mechanical particle transfer. At this time H-acid and base

function was associated with electrolytic dissociation and formation of H^+ and OH^- in aqueous solution. Brønsted's definition of an acid (A) as a species "which has a tendency to loose a proton" and a base as "a species with a tendency to add a proton" converted the notion

$$A \rightleftarrows B + H^+ \text{ to } A_1 + B_2 \rightleftarrows B_1 + A_2 \tag{5}$$

where A and B are called a corresponding acid-base pair. The importance of this almost humble statement is hard to appreciate fully today. Earlier views had, however, left many unresolved problems, for example associated with acid-base behaviour in non-aqueous solutions and with bases such as amines which do not *give off* but *produce* hydroxide ions by chemical reactions with the solvent. Most of these issues were resolved by the definition above. So was the diversity of acid and base dissociation, salt hydrolysis, neutralization, buffer, and indicator processes, at the time considered as separate areas. These now found a common conceptual order. In the context of charge transfer phenomena and history it is of additional interest that in his pioneering paper Brønsted used a tutorial parallel between acid-base and redox reactions although he did not further pursue the latter.

These three early conceptual and formal theoretical frameworks are cornerstones in contemporary ET and PT science. We proceed to three other early *applications* of these concepts specifically to ET and PT. With their strong impact much later they mark in a sense the beginning of quantitative theoretical and experimental ET and PT science.

3 Initiation of Chemical Electron and Proton Transfer Theory

The first documented case of radiationless transition from a molecular excited state is associated with ammonia absorption line broadening caused by predissociation

$$NH_3 + h\nu \rightarrow {}^*NH_3 \rightarrow NH_2 + H \tag{6}$$

(Bonhoeffer and Farkas, 1927). The recognition of *chemical* ET as a quantum mechanical physical event also originated shortly after quantum mechanics itself and is due to Gurney (1931)[42]. He considered ET between a metal electrode and an ion in solution (H^+) as a quantum mechanical tunnel transition. In retrospect the foresight of this early report is remarkable by anticipating many fundamental concepts only recognized broadly much later. In addition to the fundamental electron

tunnel view, concepts inherent in Gurney's theory are:
- The importance of the continuous electronic energy spectrum in the metal electrode
- A shift of the energy levels with the electrode potential, providing a quantum mechanical rationale for the observations of Tafel, and Erdey-Gruz and Volmer[43]
- *Conditions* that ET can occur, based (in Gurney's theory qualitatively) in strong electronic-vibrational coupling. The process considered by Gurney was

$$H^+ + e \rightarrow H + H_2O \qquad (7)$$

The condition for energy matching, being

$$\Phi \leq I - R + L \qquad (8)$$

where Φ is the work function of the metal, I the ionization potential of the hydrogen atom, R the energy of interaction between H and H_2O, and L the "lattice" (solvation) free energy of H^+. Gurney also introduced the (qualitative) notion of potential energy surfaces as both R and L depend on the environmental *motion*.

This early report initiated important later development. In retrospect it lacks precision in the sense that the chemical notion of bond energetics in the depolariser molecule H_3O^+ and the nature of the reaction product as an *adsorbed* rather than a *free* hydrogen atom were not recognized. The latter was rectified by Horiuti and Polanyi (1935)[44] but may have been a reason for the unnecessary delay of the acceptance of concepts which should much later become so central in chemical and electrochemical ET theory. As a final observation it is interesting that recognition of ET as a chemical reaction step in *electrochemistry* preceded similar recognition in *homogeneous* kinetics by at least a decade. In hindsight it is understandable that this could be so as the ET phenomenon is directly recognized by the electric current flow whereas distinction from atom group transfer in homogeneous kinetics would have been much harder at the time.

As a second major, but quite independent discovery, proton transfer *kinetics* stepped forward. In addition to the acid-base *concept*, Brønsted and his coworkers forwarded two other major early achievements[40]. One was the discovery that contrary to previous beliefs proton transfer can be *slow* and even rate determining. Brønsted's first target molecule was one of the classical "pseudo-acids", namely nitramide $H_2N\text{-}NO_2$. In the presence of bases nitramide decomposes to N_2O and H_2O

$$H_2N\text{-}N^+(O^-)(=O) \rightleftarrows HN=N^+(OH)(O^-)$$

$$HN=N^+(OH)(O^-) + B \rightarrow HN=N^+(O^-)_2 + HB \qquad (9)$$

$$HN=N^+(O^-)_2 \rightarrow N_2O + OH^- \; ; \quad OH^- + HB \rightarrow H_2O + B$$

This was followed by studies of other slow proton transfer reactions such as mutarotation of glucose. The important discoveries of Brønsted and Pedersen (1924) were:

- the notion of general base catalysis. The rate constant of eq.(9) thus followed the relation $k = k_o + k_B[B]$ for a broad range of bases. It was also shown that OH^--catalysis was unimportant under the conditions used. The catalyst was therefore the base molecule and the mechanism rooted in proton transfer.
- The discovery of a linear relation between logk and ΔpK

$$\log k = \text{const.} - \beta \Delta pK \qquad (10)$$

This kind of relation is now denoted as the Brønsted relation and the coefficient β as the Brønsted coefficient. This was the first recognized relation between a *kinetic* and a *thermodynamic* quantity and implies that the driving force is a major controlling factor in chemical processes.

- The recognition that the linear relation is a limiting case of a broader class of relations rooted ultimately in the molecular properties of the reactants and products.

As for Gurney's ET theory, the early proton transfer theory in 1923/24 exhibits remarkable intuition regarding later achievements. Brønsted and his coworkers joined this development by extending the free energy relations to broad classes of catalysts of different chemical constitution. More conspicuous results followed shortly, including important contributions by associates of Brønsted (R.P. Bell). This early age of scientific prosperity included:

- The suggestion by Hund (1927) that *nuclear* tunnelling along suitable coordinates may control chemical kinetics[44]. This was followed by other similar suggestions in preciser and quantitative dressing
- The discovery of tunnel-induced vibrational level splitting in microwave spectroscopy[34]
- Introduction of potential energy surfaces to account for the proton dynamics (Horiuti and Polanyi, 1935)[45]
- Parallel but independent evolution in electrochemical science to provide a rationale for the exponential current-voltage relation (Erdey-Gruz and Volmer[43]) observed already by Tafel

- Observation of the relation between Brønsted and Pedersen's free energy relation for proton transfer in solution and the current-overvoltage relation for electrochemical dihydrogen evolution (Frumkin, 1932)[46].

A third major theoretical result of this era with direct bearing on much later ET and PT science is the theory of electronic-vibrational *transitions* between potential surfaces, by Landau, Zener, and Stückelberg (all in 1932)[47-49]. These achievements were slightly correlated. They address, first generally electronic transitions between two intersecting, otherwise classical potential surfaces, i.e. a problem crucial in ET and PT theory. They provide exact solutions to this problem valid for all degrees of interaction. The formal solution is rigorously reflected in present-day diabatic and adiabatic ET. In modern times the theory of Landau, Zener, and Stückelberg has been extended to addressing particularly multi-dimensional nuclear motion and nuclear tunnelling.

Altogether in a decade from 1925 to 1935 an impressive amount of what today is considered as the whole basis for the ubiquitous and exciting areas of ET and PT science had thus become available.

4 Intermission - Major and Minor Observations

Depending on individual views, in the context of *charge transfer science* the period from the 1930's until the postal war era is regarded as a period of either frustration or consolidation. In chemical reaction dynamics the 1930's and 1940's were entirely dominated by transition state theory (TST)[50-52]. This broad, chemical equilibrium-based theory gave of course a tremendous boost to chemical rate theory in general. In relation to chemical ET the work of Polanyi and co-workers on the "sodium flame" reaction[53,54]

$$RX + Na^{*} \rightarrow R + X^{-} \qquad (11)$$

is most interesting. This work addresses essentially dissociative ET and was carried out initially in Berlin in the 1930's and in Manchester in the 1950's. It is probably the first extensive study of chemical ET and was interpreted in terms of the TST of which Polanyi was a co-author. The close parallel with the electrode process

$$RX + e(Hg) \rightarrow R + X^{-} \qquad (12)$$

was noted later by M.G. Evans (Polanyi's pupil) and N.S. Hush[55,56]. Over three decades later dissociative ligand substitution[57] and ET[58,59] were addressed again and brought into the framework of contemporary ET theory.

Another interesting, slightly later report from this period is the multiple-transition diabatic scheme forwarded by K.J. Laidler for quenching of excited sodium, mercury, and cadmium by different atoms and molecules in the gas phase[60]. Concepts inherent in this early report are encountered again in contemporary views on vibrational coherence and vibrationally "hot" transitions for example in the rapidly expanding areas of femtosecond chemistry.

Other broad perspectives were introduced as kinetic deuterium isotope effects became accessible from the mid-1930's. TST did not, however, offer a natural frame for quantum mechanical ET and PT in solution. ET and PT involve transfer of electrically charged, light particles which are much more mobile than the majority of the environmental nuclear motion and at the same time strongly coupled to this motion. TST is also rooted in "solid" chemical conversions with no easy place for nuclear motions following widely different time scales. At least the early TST offered no precise frame for *solvent* dynamics, reducing essentially the solvent environment to different *equilibrium* solvation in the reactants' equilibrium state and the transition state. As we know now, this view is inadequate as *non-equilibrium* solvation is exactly what controls chemical light particle transfer. The distinction between classical and quantum mechanical particle dynamics (electron and nuclear tunnelling) was also inadequately recognized in the TST of the 1930's once it came to details beyond the simplest recognition of these phenomena. A representative view is expressed concisely by Schaffer (1936)[61], namely that "...ionic reactions are immeasurably rapid, ...logical to infer that ions ...require ...little energy of activation ...". Although recognizing the fundamental chemical ET event, such views could not extend to appreciate electronic-vibrational coupling nor could they clearly distinguish ET from combined electron-ion or atomic radical transfer.

Efforts in this period with later direct impact rather stem from other areas, namely solid state physics and the beginning of semiconductor technology, and from the gradual availability of nuclear radioisotopes. Based on the impacts from both of these new areas, in a decade-and-a-half in the postwar era ET and PT were completely transformed, and at least *conceptually* approaching present day levels.

Towards the end of the intermission/consolidation period still another major achievement was, however, to appear, triggering after revival in the 1970's a new avalanche in charge transfer science. Kramers's theory of stochastic chemical rate effects (1940)[62] was the first to address *non-equilibrium* solvation in chemical reactions. In Kramers's theory, and more generally in later stochastic chemical rate theory focus is on a single "reactive" nuclear coordinate. The importance of the solvent is recognized and treated as collisions, or friction as the reacting system proceeds along the reactive mode. Some system parts are thus left out and re-introduced as diffusion or damping. The random, or stochastic motion is formally recast as a Langevin equation

$$\mu\ddot{x} = -\partial U/\partial x - \mu\zeta\dot{x} + R \qquad (13)$$

where μ is the effective mass for motion along the reactive coordinate x and U the potential energy. ζ is the friction coefficient which represents the "systematic" part of the solute-solvent interaction while R is the random environmental force. Kramers's theory is conceptually a new step relative to TST by incorporation of solvation effects in non-equilibrium modes. It also provided important specific results to be later elaborated, illustrated by the following result. In the limit of strong "enough" friction Kramers's rate constant could be recast as

$$K = \kappa_{kr}k_{TST} \; ; \; \kappa_{KR} = [1 + (\zeta/2\omega_B)^2]^{1/2} - (\zeta/2\omega_B) \qquad (14)$$

where k_{TST} is the friction-free TST rate constant and κ_{KR} a correction which depends on ζ and the barrier curvature ω_B, and represents the diffusive, barrier multiple-passage character of the motion. In particular, as friction becomes "small", $\kappa_{KR} \rightarrow 1$ and the TST rate constant is recovered. At large ζ, $\kappa_{KR} \propto \zeta^{-1}$ and damping along the reaction coordinate is directly transparent. As damping is *very* small, $\kappa_{KR} \propto \zeta$ which signifies that collision with molecules in the environment now ensures that the chemical process is truely irreversible and does not merely oscilliate forwards and backwards such as it would in truely friction-less one-dimensionless motion.

Like the theory of Gurney, Kramers's theory remained for many years with little repercussion in the world of chemistry. Contemporary charge transfer science has, however, paid full tribute to both these, in later retrospect somewhat complementary pioneering results.

5 Radioisotopes, Electron Exchange, Inner and Outer Sphere ET, and Solid-State Electronic Processes

The reaction in eq.(1) is probably the first recognized case of outer sphere ET between weakly interacting molecules. This report seems, however, to have been kept classified for a decade. $KMnO_4$ was bombarded with thermal neutrons with the purpose of testing for manganate formation[63]. Mixing of aqueous solutions of the radioactive $[^*MnO_4]^-$ with $[MnO_4]^{2-}$ showed that exchange was complete before Mn(VI) and Mn(VII) could be separated. Held against the slow oxygen atom exchange with surrounding water molecules this prompted the recognition of electron tunnelling between the two manganese ions. The availability of radioisotopes initiated a new age in experimental ET science. It was discovered particularly that many ET reactions could be slow (cf. the work of Brønsted and Pedersen on slow PT) such as in the following early reported system (1948)[64]

$$[Co(en)_3]^{3+} + [^*Co(en)_3]^{2+} \rightleftarrows [Co(en)_3]^{2+} + [^*Co(en)_3]^{3+} \quad (15)$$

(en = 1,2-ethanediamine), with many others to follow shortly. This was parallelled by precise distinction between ET and redox *atom group* transfer, or outer and inner sphere ET. Binuclear bridge-group intermediate complexes in redox reactions involving for example Sb(III)/Sb(V) had been discussed since the late 1940's[65]. The direct proof of such a mechanism was provided by Taube, Myers and Rich (1953) based on reactions of which[66]

$$[Co(NH_3)_5Cl]^{2+} + [Cr(H_2O)_6]^{2+} + 5H^+ + 6H_2O \rightarrow$$
$$[Co(H_2O)_6]^{2+} + [Cr(NH_3)_5Cl]^{2+} + 5NH_4^+ \quad (16)$$

is the first prominent representative, and where the close correlation between substitution lability and redox state for the Co(III)/Co(II) and Cr(III)/Cr(II) couples could be exploited.

This new era of experimental ET science was accompanied by the beginning of a more precise description of the role of electronic-vibrational coupling extending to both the molecular structure and the solvent environment. The recognition of the nuclear Franck-Condon effects in chemical ET is commonly ascribed to W.F. Libby (1952)[67]. Some indications in Libby's own report and in a following report by Platzman and Franck[68] suggest, however, that Franck had already recognized particularly the Franck-Condon factors associated with the solvent. These views do not, however, seem to have been published in any other way. The early 1950's also saw the first quantitative theories of chemical ET. The tunnel concept in bimolecular ET was introduced in a lucid report by Marcus (R.J.), Zwolinskij and Eyring (1955)[69], and the distinction between the *electronic* and *nuclear* solvent polarization was recognized (Weiss, 1954)[70].

Meanwhile but long unnoticed by the chemical communities, by the early 1950's a parallel development had grown out of solid state *physics*, namely electronic processes in solid materials. Like the advent of radioisotopes, this had been advanced by Second World War efforts. Central notions were *localized electronic states*, *guest-host interactions*, and "deep" semiconductor *impurity levels*. As a result, the physical nature of both radiationless processes (thermal ionization and electron capture) and optical absorption and emission bandshapes had largely become understood[9,10,71-74]. So had the notion of electron self-trapping by strong electronic-vibrational interaction[75,76], and of local molecular and bulk polarization modes. Last but not least, the treatment of both electronic and nuclear dynamics was fully quantum mechanical with recognized correspondence to the appropriate classical (high-temperature) limits.

The formal theory is broadly discussed in the important paper by Kubo and Toyozawa (1955)[10], in Holstein's paper on small polaron theory (1959)[77] and in the comprehensive review by J.J. Markham (1959)[78]. The following observations for the transition probability, broadly used in chemical and biological ET theory also in the 1980's and 1990's, are for example reported already by Kubo and Toyozawa[10]

$$W(\Delta G^o \text{ or } \nu) = (2\pi/\hbar)(V_{fi})^2 \sum_{v,w} p(v)\, S_{fi}^{wv}\, \delta(\epsilon_{fw} - \epsilon_{iv}) \tag{17}$$

where V_{fi} is the electronic factor coupling the initial (i) and final (f) Born-Oppenheimer electronic states, and S_{fi}^{wv} the nuclear Franck-Condon overlap factor coupling the nuclear states v and w, of energies ϵ_{iv} and ϵ_{fw} in the initial and final electronic states, respectively. p(v) is the population of the iv state, and $2\pi\hbar$ Planck's constant. For a pair of displaced nuclear harmonic oscillators of frequency ω and displacement Δ, $W(\Delta G^o$ or $\nu)$ takes the particular form

$$W(\Delta G^o \text{ or } \nu) = (2\pi/\hbar^2\omega)(V_{fi})^2 \exp[-S(2n+1)] \cdot$$

$$\cdot\ [(n+1)/n]^p\, I_p\, \{2S[n(n+1)]^{1/2}\}\ ; \quad S = \tfrac{1}{2}\Delta^2 \tag{18}$$

$$n = [\exp(\hbar\omega/k_BT) - 1]^{-1}\ ; \quad p = |\ \Delta G^o/\hbar\omega\ |\ \text{ or }\ |\ \Delta G^o - h\nu\ |\ /\hbar\omega$$

where k_B is Boltzmann's constant and T the temperature. $I_p(z)$ is the modified Bessel function of order p

$$I_p(z) = (\tfrac{1}{2}z)^p \sum_{k=0}^{\infty} (\tfrac{1}{2}z)^{2k}/[k!(p+k)!] \tag{19}$$

Eqs.(17)-(19) give in the low- and high-T limits, respectively

$$W_{low\text{-}T} = (2\pi/\hbar^2\omega)(V_{fi})^2 \exp(-S)\, S^p/p!$$

$$W_{high\text{-}T} = (V_{fi})^2 \sqrt{\pi/S\hbar\omega k_BT}\, \exp[-(S\hbar\omega + p\hbar\omega)^2/4S\hbar\omega k_BT] \tag{20}$$

corresponding to ET dynamics by nuclear tunnelling and classical thermal activation, respectively.

A further important development of the early 1950's was introduction of the *crystal* (and later *ligand*) field theory of transition metal ions[79], deriving from the earlier solid state physics work of Bethe[80]. This made possible calculation of many properties of transition metal ions in condensed phases, such as, geometries and vibrational frequencies of ion-water complexes[81]. The importance of the *crystal field potential* for transition ion transfer dynamics soon emerged.

By the early 1950's the nature of chemical ET as an electron tunnel event and the qualitative Franck-Condon role of the solvent in chemical redox reactions were thus recognized. The nature and data for an increasing number of chemical ET systems were beginning to accumulate. The electronic structure of transition metal complexes was well understood. Correct and detailed quantum mechanical theories for solid-state electronic processes were finally, and independently available. The road was thus prepared for shortly forthcoming quantitative theories of chemical electron transfer.

6 Classical and Quantum Mechanical Theories of Chemical Electron Transfer

During the 1950's such theoretical frames appeared, following different major lines of concept and formalism. These frames have had the greatest impact on all later ET research. The theory of Marcus[82-84] rooted in *classical* TST was the first to introduce correctly solvent *dynamics* in chemical ET. While Libby had a "vertical" electronic transition in mind as in spectroscopy, it was shown by Marcus and a little later by Hush[57,85-87] that solvent polarization *fluctuations* were needed to overcome the Franck-Condon barrier. Understanding of the environmental dynamic reorganization and the transparent formal representation of this central feature was a major and lasting achievement in chemistry. The total solvent polarization holds an inertial (i.e. nuclear) and an inertia-less (electronic) component

$$\mathbf{P}_{tot} = \mathbf{P}_{in} + \mathbf{P}_{in\text{-}l} \qquad (21)$$

The non-equilibrium state refers to the inertial part only, analogous to the non-equilibrium configuration of two intersecting molecular potential surfaces. The reversible work to go from the initial *equilibrium* state polarization$\mathbf{P}_{in}^{io}(\mathbf{r})$ is

$$U_i^{\neq} = (2\pi/c) \int d\mathbf{r}\, [\mathbf{P}_{in}^{\neq}(\mathbf{r}) - \mathbf{P}_{in}^{io}(\mathbf{r})]^2 \; ; \;\; c = \epsilon_\infty^{-1} - \epsilon_s^{-1} \qquad (22)$$

where ϵ_∞ and ϵ_s are the optical and static dielectric constant of the (assumed structureless, dielectric) medium, respectively. This scheme follows closely the one previously developed for electron autolocalization in polar crystals[75]. $P^{\neq}_{in}(r)$ is, further, ultimately identified with the minimum Gibbs free energy subject to the constraint $\Delta G^o = U_f^{\neq} - U_i^{\neq}$ where ΔG^o is the reaction Gibbs free energy (driving force).

The approach of Marcus and Hush gave a simple bimolecular rate constant form which immediately reflects all the system properties

$$k = Z\exp(-\Delta G^{\neq}/k_BT) \qquad (23)$$

Z is *related* to the bimolecular collision number. The activation Gibbs free energy is

$$\Delta G^{\neq} = \alpha^2\lambda = (\lambda + \Delta G^o)^2/4\lambda \qquad (24)$$

where α is the chemical transfer coefficient, and λ the solvent reorganization Gibbs free energy

$$\alpha = \tfrac{1}{2} + \Delta G^o/2\lambda \qquad (25)$$

λ holds a contribution from the solvent, i.e.

$$\lambda_s = (2\pi/c) \int dr\, [P^{fo}_{in}(r) - P^{io}_{in}(r)]^2 = (c/8\pi) \int dr\, [D_f(r) - D_i(r)]^2 \qquad (26)$$

$D_i(r)$ and $D_f(r)$ being the electric induction vectors in the initial and final state, respectively. However, it was recognized that it is necessary to consider also the contributions of local (inner) classical harmonic, λ_i and the total reorganization energy then becomes $\lambda = \lambda_s + \lambda_i$.

These achievements aroused prompt attention, their attractive formal transparency making them well suited for experimental investigations. In the following years these took particularly the form of free energy relations (cf. the theory of Brønsted and Pedersen[40]), cross relations (see for example ref.88), solvent variation, and correlations between ET in homogeneous solution and electrochemical ET. Altogether simple and extremely useful systematic data frames had thus been provided, maintaining much of their merit until the present day.

Regarding the inherent importance of the solvent dynamics the theory of Hush[57,84-86] follows the theory of Marcus. The adiabatic theory proposed by Hush, first for electrode processes[57,84] and later for homogeneous ET[85,86] was, however, rooted in quantum mechanical mechanical concepts. A key quantity was thus *electron occupation numbers* which assume precise measurable values in contexts such as mixed-valence complexes and electrosorption valencies. This view has been expanded by Hush[89]. The concept of electron occupation numbers has, importantly, also become central in much more recent views on adiabatic electrochemical ET where the concept is now inserted in preciser frames (see Section 7).

The explicit quantum mechanical dimension was first introduced by Levich and Dogonadze (1959-62)[1,90-92]. Although three years later than the first report of Marcus the theory of Levich and Dogonadze was no less far-reaching. It was the *first* theory to view chemical ET as a quantum mechanical transition between two manifolds of electronic-vibrational levels induced perturbatively by suitable level coupling terms, similarly to other condensed phase electronic processes. By this view it was also the first theory to introduce the Born-Oppenheimer approximation and the notion of potential energy surfaces. The electronic transmission coefficient appeared here correctly for the first time. From the very beginning the theory was further, fully quantum mechanical and included nuclear tunnelling. In communications to follow shortly (Dogonadze, 1962)[92] the theory clarified, finally the question of strong and weak intermolecular interactions corresponding to the adiabatic and diabatic limits. There has been a tendency in modern times to overlook all these achievements. Proper and full scientific credit should of course be maintained to these far-reaching pioneering results in *quantum mechanical* ET theory.

The theories of Marcus and Hush, and of Levich and Dogonadze mark the initiation of contemporary chemical charge transfer theory on which a voluminous literature now rests. With these observations we also gradually step out of the *historical* sphere and into the *systematics* of contemporary charge transfer science. The Chapters in the present volume cover a wealth of exciting areas and results which show how charge transfer science has now evolved. Before concluding we shall endeavour briefly and in a highly schematic fashion into an overview of *some* of the achievements of charge transfer science since the 1960's.

7 Concluding Remarks

At least the following major areas, listed here with little comment, in focus of charge transfer science in the 1960's and 1970's, can be distinguished[1-5,8]:

7.1 Diabatic and Adiabatic ET and PT Processes

This extends particularly to multi-dimensional[93] and nuclear tunnel features[4,5,8,94].

7.2 Local Molecular and Collective Solvent Modes

Focus on nuclear tunnelling[95-97] and dielectric solvent spatial and vibrational dispersion[98-100].

7.3 Electrochemical ET Rate Theory

This covered the electronic spectrum of metal and semiconductor electrodes[3,101-103], adiabatic and diabatic electrochemical processes[104], and elastic, resonance, and inelastic mechanisms for tunnelling through thin oxide and polymer films[105].

7.4 Light-assisted Processes in Chemistry

Strong focus has been on chemical mixed-valence compounds and correlations between optical and thermal transitions[106]. These offer approaches to some of the elusive features in thermal processes such as nuclear tunnelling, the inverted free energy region, and the electronic coupling factor[107]. Particularly from the intervalence spectra of intramolecular transfer it is possible to extract details of the reorganization energy and donor-acceptor coupling[108,109].

7.5 Proton Transfer in New Perspectives

Proton tunnelling became recognized as a dominating feature of chemical proton transfer, rather than a minor correction[95,110-112]. New attention was given to the kinetic deuterium isotope effect and the electrochemical dihydrogen evolution reaction in the light of these views.

7.6 Nuclear Tunnelling at Low Temperatures and in the Inverted Gibbs Free Energy Region

Nuclear tunnelling dominates chemical processes at low temperatures[100] and is verified experimentally for hydrogen atom and heavier atom group transfer[113-115]. The inverted Gibbs free energy region, inherently recognized in solid state physics from the 1950's[10] and in chemical ET from 1960[83], is a target for nuclear tunnelling also at ambient temperature[116,117] because the barrier for the nuclear transition is much narrower than in the normal free energy region (see examples discussed in refs.5-8 and 19).

7.7 Multi-level Processes and Bridge-group Assistance

The notion of ET involving intermediate states, in addition to donor and acceptor states originates from magnetic superexchange in solids[118,119]. Such views were introduced early in chemical ET science[120-121]. Evolution of this concept towards more exact theory was initiated in the 1970's and is also presently an area of intense focus[123] (see overview in ref. 19).

7.8 Tunnelling and ET in Biological Systems

The 1960's and 1970's was, finally an age where major attention to charge transfer in biological systems began[124]. These highly interdisciplinary endeavours were stimulated by elucidation of biological structures, by some of the earliest reported low-temperature processes in bacterial photosynthesis, and in the spin conversion of the recombination of myoglobin with small ligand molecules. The fundamental concepts of nuclear tunnelling and long-range ET received notable substantiation from these efforts which also provided new substance to the otherwise elusive notion of conformational dynamics in reactive protein systems.

The most recent decade-and-a-half is too close to call for a historical perspective but this period has witnessed exciting achievements also with historical *threads*. The period has in a sense been an era of rediscovery but with results which have lifted these areas of charge transfer science to new and higher levels. This applies to superexchange and long-range electron tunnelling[19,125-127], and to stochastic chemical rate theory[128-131]. Beginning with the work of Zusman[128] the formalism of the latter has come to dominate large parts of charge transfer theory, new attention being given for example to multi-dimensional nuclear motion[132], to the interplay between friction-free and damped nuclear motion[133], and to the role of nuclear tunnelling in dissipative media[134]. It should, however, be recognized that stochastic and dynamic views on chemical solution rate processes are of course not out of entirely different worlds. Stochastic views are, first, heavily impacted by both the earlier and recent quantum mechanical views. Stochastic effects are, secondly important when the charge transfer event is fast and approaches environmental relaxational control. Stochastic chemical rate formalism refers, finally particular, arbitrary but precisely defined parts of the solute-solvent interaction to the stochastic diffusion or friction effects. The bridge between stochastics and dynamics in chemical charge transfer science not entirely settled but discussed in several recent reports[134-136].

One of the most important merits of the stochastic theories is that they attend *in principle* to details of the molecular features of environmental dynamics in the fundamental chemical charge transfer event. Formal theory has been parallelled by comprehensive computational schemes, including quantum computer simulations, addressing solvation equilibria, the solvation process, ET and ion transfer kinetics, and macromolecular dynamics. Specific target systems are, for example "simple" ET between solute transition metal ions, proton transfer, and ET in large protein systems. Reports are, for example refs. 137-140, and later chapters of the present volume. Complementary evolution has been achieved by experimental investigations of "ultrafast" charge transfer in part controlled by environmental relaxation. The fast processes extend to all the different categories noted and now cover broadly the nano- pico- and femtosecond time ranges (reviews are given in refs. 131 and 141-144).

Interfacial electrochemical ET has also been in focus and new achievements have lifted electrochemistry to levels approaching those of surface science:

- The adoption broadly of single-crystal "ultra-pure" electrode-electrolyte surfaces
- Statistical mechanical and electronic structural models of the interface, and computer simulation of the interfacial structure[145-148]
- The formal theory of *adiabatic* electrochemical ET has evolved[149-152]. Progress has been along different routes emphasizing either individual level crossing[104,153], or electron delocalization in ways similar to adsorption phenomena[149,151-154]. A key notion is, interestingly electron occupation numbers as in both adsorption and in early proposals of electrochemical ET theory[57] (cf. Section 6). In this way the whole range from the diabatic limit of weak electrode-molecule interaction to the fully adiabatic limit has become covered.
- Tunnelling and adsorbate states, pointing towards new views on electrocatalysis[155,156]
- New experimental and theoretical perspectives associated with imaging of the surface in real time and space by *in situ* scanning tunnel and atomic force microscopy[157,158]
- Construction and design of functional molecular materials and organized assemblies in which ET is the central functional element[159-162]
- Biological systems have, finally come much more in focus over the last decade-and-a-half. Fundamentally important is that structural biology has evolved immensely, with a wealth of new redox and hydrolytic enzyme structures. To this adds the parallel evolution of chemical and microbiological techniques. As a result, biological structure and charge transfer dynamics have not only added greatly to our general understanding of biological function but have themselves been among the most versatile target systems for fundamentals of charge transfer phenomena and theory[163-166].

No doubt most of these areas will continue into the 21st century. Ultrafast phenomena and spectroscopies will continue to be in focus. So will new quantum chemical calculations of large molecular fragments, to which fluctuational electronic effects and coherent multi-ET dynamics would be added. Apolar solvents, nonlinear solvation, conformational dynamics and protein folding, response functions for organized and anisotropic media, and critical phenomena are other new areas where initiated efforts are likely to continue. In electrochemistry and other areas of interfacial charge transfer structure-reactivity relations of "*ab initio*" surfaces, modified surfaces, and micro- and nanoelectrochemistry offer new perspectives. The scanning probe microscopies still hold puzzles regarding the imaging process and tunnel mechanisms, in addition to being powerful "practical" tools. On the applied side are supramolecular materials, working monolayer assemblies, solid-molecule interactions, surface attachment sites etc.

In these and in new areas of ion transfer there will be a need for understanding increasing detail of molecular and collective behaviour of new and untraditional systems, and how this is controlled by the structural, electronic, and thermodynamic system properties. Rooted in great historical discoveries, contemporary charge transfer science therefore appears to thrive most excitingly as it has done now for several decades.

Acknowledgement

Financial support from the Australian Research Council (NSH) and the Danish Natural Science Research Council (JU) is acknowledged.

References

1. V.G. Levich and R.R. Dogonadze, *Coll. Czech. Chem. Comm.* **26**, 193 (1961).
2. R.R. Dogonadze, in *Reactions of Molecules at Electrodes*, ed. N.S. Hush (Wiley, London, 1971).
3. R.R. Dogonadze and A.M. Kuznetsov, *Progr. Surf. Sci.* **6**, 1 (1975).
4. J. Ulstrup, *Charge Transfer Processes in Condensed Media* (Springer, Berlin, 1979).
5. R.A. Marcus and N. Sutin, *Biochim. Biophys. Acta* **811**, 265 (1985).
6. *Chem. Rev.* **92**, 365-490 (1992), ed. M.A. Fox.
7. *Chem. Phys.* **176**, 289-649 (1993), eds. T.J. Meyer and M.D. Newton.
8. A.M. Kuznetsov, *Charge Transfer in Physics, Chemistry and Biology* (Gordon and Breach, Reading, 1995).
9. R. Kubo, *Phys. Rev.* **86**, 929 (1952).
10. R. Kubo and Y. Toyozawa, *Progr. Theor. Phys.* **13**, 160 (1955).
11. M. Bixon and J. Jortner, *J. Chem. Phys.* **48**, 715 (1968).
12. N.R. Kestner, J. Logan and J. Jortner, *J. Phys. Chem.* **78**, 2148 (1974).
13. R. Englman, *Nonradiative Decay of Ions and Molecules* (North-Holland, Amsterdam, 1979).
14. D.L. Dexter, *J. Chem. Phys.* **21**, 836 (1953).
15. W. Heitler, *The Quantum Theory of Radiation* (Oxford, University Press,

Oxford, 1936).
16. P.A.M. Dirac, *The Principles of Quantum Mechanics(Carendon, Oxford, 1957)*.
17. C.P. Flynn, *Point Defects and Diffusion* (Oxford University Press, Oxford, 1972).
18. K.F. Freed and F.K.Fong, *J. Chem. Phys.***63**, 2890 (1975).
19. G. Iversen *et al.*, *Adv. Chem. Phys.*, in press.
20. H.M. Powell, *Proc. Chem. Soc.* 73 (1953).
21. Noted in: A. Broo, *Electron Transfer and Electronic Spectra in Metalloproteins and Model Systems*, Ph.D. Diss (University of Göteborg, 1991).
22. L. Kahlenberg, *J. Am. Chem. Soc.* **16**, 314 (1894).
23. J. Tafel, *Z. Phys. Chem.* **50A**, 641 (1905).
24. R.P. Bell, *Acid-Base Catalysis* (Oxford University Press, Oxford, 1941).
25. R.P. Bell, *The Proton in Chemistry, 2nd ed.*(Chapman and Hall, London, 1973).
26. R.P. Bell, *The Tunnel Effect in Chemistry* (Chapman and Hall, London, 1980).
27. H. Jeffreys, *Proc. London Math. Soc.* **23**, 428 (1923).
28. G. Wentzel, *Z. Phys.* **38**, 518 (1926).
29. H.A. Kramers, *Z. Phys.* **39**, 828 (1926).
30. L. Brillouin, *J. Physique* **7**, 353 (1926).
31. E. Fermi, *Nøtes on Quantum Mechanics* (The Chicago University Press, Chicago, 1954).
32. L.D. Landau and E.M. Lifshitz, *Quantum Mechanics, 2nd ed.* (Pergamon, Oxford, 1965).
33. R.H. Fowler and L. Nordheim, *Proc. Roy. Soc. A* **119**, 173 (1928).
34. G. Gamow, *Z. Phys.* 204 (1928).
35. C.E. Cleeton and N.H. Williams, *Phys. Rev.* **45**, 234 (1934).
36. J. Franck, *Trans. Faraday Soc.* **21**, 536 (1925).
37. E.U. Condon, *Phys. Rev.* **32**, 858 (1928).
38. M. Born and J.R. Oppenheimer, *Ann. Phys.* **84**, 457 (1927).
39. J.N. Brønsted, *Rec. Trav. Chim. Pays-Bas* **42**, 718 (1923).
40. J.N. Brønsted and K.J. Pedersen, *Z. Phys. Chem.* **108**, 185 (1924).
41. K.F. Bonhoeffer and L. Farkas, *Z. Phys. Chem.* **134**, 337 (1928).
42. R.W. Gurney, *Proc. Roy. Soc. A* **134**, 137 (1931).
43. R. Erdey-Gruz and M. Volmer, *Z. Phys. Chem.* **150A**, 203 (1930).
44. F. Hund, *Z. Phys.* **43**, 805 (1927).
45. J. Horiuti and M. Polanyi, *Acta Physicochem. URSS* **2**, 305 (1935).
46. A.N. Frumkin, *Z. Phys. Chem.* **164A**, 121 (1932).
47. L.D. Landau, *Physik. Z. d. Sowietunion* **1**, 88 (1932).
48. C. Zener, *Proc. Roy. Soc. A* **137**, 696 (1932).
49. E.C.G. Stückelberg, *Helv. Chim. Acta* **5**, 369 (1932).
50. S. Glasstone *et al.*, *The Rheory of Rate Processes* (McGraw-Hill, New York, 1941).

51. K.J. Laidler, *Theories of Chemical Reaction Rates* (McGraw-Hill, New York, 1969).
52. D.G. Truhlar *et al.*, *J. Phys. Chem.* **87**, 2664 (1983).
53. M. Polanyi, *Atomic Reactions* (Williams and Norgate, London, 1932).
54. E. Warhust, *Quart. Rev. Chem Soc.* **5**, 44 (1951).
55. M.G. Evans and N.S. Hush, *J. Chim. Physique* **49**, 159 (1952).
56. N.S. Hush, *Z. Elektrochem.* **61**, 734 (1957).
57. R.R. Dogonadze and E.D. German, *J. Res. Inst. Catalysis. Hokkaido Univ.* **20**, 34 (1972). *Int. J. Chem. Kin.* **6**, 457, 467 (1974)
58. J. Ulstrup, *Proc. Sandbjerg Meeting on Organic Electrochemistry*, ed. H. Lund (University of Aarhus, 1982), p. 1.
59. J.-M. Savéant, *J. Am. Chem. Soc.* **109**, 6788 (1987).
60. K.J. Laidler, *J. Chem. Phys.* **10**, 34, 43 (1942).
61. P.A. Schaffer, *J. Phys. Chem.* **40**, 1021 (1936).
62. H.A. Kramers, *Physica* **7**, 284 (1940).
63. H.C. Hornig *et al.*, *J. Am. Chem. Soc.* **72**, 3808 (1950).
64. W.B. Lewis and C.D. Coryell, *Brookhaven Conf. Rept. BNL-C8*, 131 (1948).
65. J.E. Whitney and N. Davidson, *J. Am. Chem. Soc.* **69**, 3076 (1947).
66. H. Taube, H. Myers and R.L. Rich, *J. Am. Chem. Soc.* **75**, 4118 (1953).
67. W.F. Libby, *J. Phys. Chem.* **56**, 863 (1952).
68. R.L. Platzman and J. Franck, *Z. Phys.* **138**, 411 (1954).
69. R.J. Marcus *et al.*, *J. Phys. Chem.* **58**, 432 (1954).
70. J.J. Weiss, *Proc. Roy. Soc. A* **222**, 128 (1954).
71. K. Huang and A. Rhys, *Proc. Roy. Soc. A* **204**, 406 (1950).
72. M. Lax, *J. Chem. Phys.* **20**, 1752 (1952).
73. M.A. Krivoglaz, *Zhur. Teor. Eksp. Khim.* **25**, 191 (1953).
74. G. Rickayzen, *Proc. Roy. Soc. A* **241**, 480 (1957).
75. S.I. Pekar, *Untersuchungen über die Elektronentheorie der Kristalle* (Akademie Verlag, Berlin, 1954).
76. H.A. Fröhlich, *Adv. Phys.* **3**, 325 (1954).
77. T. Holstein, *Ann. Phys.* **8**, 325 (1959).
78. J.J. Markham, *Rev. Mod. Phys.* **31**, 956 (1959).
79. J.S. Griffith, *Theory of Transition Metal Ions* (Cambridge University Press, Cambridge, 1961).
80. H. Bethe, *Ann. Physik* **3**, 133 (1929).
81. N.S. Hush, *Disc. Faraday Soc.* **26**, 145 (1958).
82. R.A. Marcus, *J. Chem. Phys.* **24**, 966 (1956).
83. R.A. Marcus, *Disc. Faraday Soc.* **29**, 21 (1960).
84. R.A. Marcus, *Ann. Rev. Phys. Chem.* **15**, 155 (1964).
85. N.S. Hush, *Proc. 4th Moscow Conf. Electrochemistry* (English Transl. Consultants Bureau, 1961) **1**, 99 (1956).
86. N.S. Hush, *J. Chem. Phys.* **28**, 962 (1958).

87. N.S. Hush, *Trans. Faraday Soc.* **57**, 557 (1961).
88. P. Bernhard and A.M. Sargeson, *Inorg. Chem.* **26**, 4122 (1987).
89. N.S. Hush, in *Mixed-Valence Compounds*, ed. D.B. Brown (Reidel, Dordrecht, 1980), p. 151.
90. V.G. Levich and R.R. Dogonadze, *Dokl. Akad. Nauk SSSR, Ser. Fiz. Khim.* **124**, 123 (1959).
91. V.G. Levich and R.R. Dogonadze, *Adv. Electrochemistry Electrochem. Eng.* **4**, 149 (1966).
92. R.R. Dogonadze, *Dokl. Akad. Nauk SSSR, Ser. Fiz. Khim.* **142**, 1108 (1962).
93. R.R. Dogonadze and Z.D. Urushadze, *J. Electroanal. Chem.* **32**, 235 (1971).
94. M.S. Child, *Molecular Collision Theory* (Academic, Press, London, 1974).
95. R.R. Dogonadze and A.M. Kuznetsov, *Elektrokhimiya* **3**, 739 (1967).
96. R.R. Dogonadze, A.M. Kuznetsov and M.A. Vorotyntsev, *Phys. Stat. Sol. (b)* **54**. 125 (1972).
97. N.R. Kestner, J. Logan and J. Jortner, *J. Phys. Chem.* **78**, 2148 (1974).
98. R.R. Dogonadze and A.M. Kuznetsov, *Elektrokhimiya* **7**, 763 (1971).
99. A.A. Kornyshev, in *The Chemical Physics of Solvation. Part A. Theory of Solvation*, eds. R.R. Dogonadze *et al.* (Elsevier, Amsterdam, 1985), p. 77.
100. R.R. Dogonadze *et al.*, *J. Electroanal. Chem.* **75**, 315 (1977).
101. H. Gerischer, *Z. Phys. Chem. NF* **26**, 223, 325 (1960).
102. R.R. Dogonadze and Yu.A. Chidzmazhev, *Dokl. Akad. Nauk SSSR, Ser. Fiz. Khim.* **144**, 1077 (1962). **145**, 848 (1962).
103. R.R. Dogonadze and A.M. Kuznetsov, *Izv. Akad. Nauk SSSR, Ser. Khim.* **12**, 2140 (1964).
104. R.R. Dogonadze, A.M. Kuznetsov and M.A. Vorotyntsev, *Croat. Chim. Acta* **44**, 257 (1972).
105. For a review, see: J. Ulstrup, *Surf. Sci.* **101**, 564 (1980).
106. N.S. Hush, *Progr. Inorg. Chem.* **8**, 391 (1967).
107. E.M. Itskovitch *et al.*, in *The Chemical Physics of Solvation. Part B. Spectroscopy of Solvation*, eds. R.R. Dogonadze *et al.* (Elsevier, Amsterdam, 1986), p. 223.
108. H. Taube, *Electron Transfer Reactions in Solution* (Academic Press, New York, 1971)
109. T.J. Meyer, *Acc. Chem. Res.* **22**, 1048 (1989).
110. R.R. Dogonadze *et al.*, *Electrochim. Acta* **13**, 1025 (1968).
111. *Chem. Phys.* **136**, 153-360 (1989), eds. P.F. Barbara and H.P. Trommsdorf.
112. For an overview, see: L.I. Krishtalik, *Charge Transfer Reactions in Electrochemical and Chemical Processes* (Consultants Bureau, New York and London, 1986).
113. V.A. Benderskij *et al.*, *Chem. Phys.* **67**, 301 (1982).
114. *Chem. Phys.* **170**, 265-460 (1993), eds. V.A. Benderskij, V.I. Gol'danskij and J. Jortner.

115. V.A. Benderskij, D.E. Makarov and C.A. Wright, *Chemical Dynamics at Low Temperatures* (Wiley, Chichester, 1994).
116. R.P. Van Duyne and S.F. Fischer, *Chem. Phys.* **5**, 183 (1974).
117. J. Ulstrup and J. Jortner, *J. Chem. Phys.* **63**, 4358 (1975).
118. H.A. Kramers, *Physica* **1**, 182 (1934).
119. P.W. Anderson, *Phys. Rev.* **79**, 350 (1950).
120. P. George and J.S. Griffith, in *The Enzymes*, ed. K. Myrbäck (Academic Press, New York, 1959), p. 347.
121. J. Halpern and L.E. Orgel, *Disc. Faraday Soc.* **29**, 32 (1960).
122. H.A. McConnell, *J. Chem. Phys.* **35**, 508 (1961).
123. M.D. Newton, *Chem. Rev.* **91**, 767 (1991).
124. *Tunnelling in Biological Systems*, eds. B. Chance *et al.* (Academic Press, New York, 1979).
125. J. Reimers and N.S. Hush, *Chem. Phys.* **197**, 323 (1995), and references there.
126. P. Siddarth and R.A. Marcus, *J. Phys. Chem.* **97**, 2400 (1993), and references there.
127. J.M. Gruschus and A. Kuki, *J. Phys. Chem.* **97**, 5581 (1993).
128. L.D. Zusman, *Chem. Phys.* **49**, 295 (1980).
129. H. Frauenfelder and P.G. Wolynes, *Science* **223**, 337 (1985).
130. M. Maroncelli, *J. Mol. Liq.* **57**, 1 (1993).
131. K. Yoshihara *et al.*, *Bull. Chem. Soc. Japan* **68**, 696 (1995).
132. A.M. Berezhkovskij and V. Yu. Zitserman, *Chem. Phys.* **164**, 341 (1992).
133. H. Sumi and R.A. Marcus, *J. Chem. Phys.* **84**, 4894 (1986).
134. *Tunnelling*, eds. J. Jortner and B. Pullman (Kluwer, Dordrecht, 1986).
135. A.M. Kuznetsov, *Elektrokhimiya* **20**, 1069, 1226 (1984).
136. A.A. Belousov, A.M. Kuznetsov and J. Ulstrup, *Chem. Phys.* **129**, 311 (1989).
137. P.J. Rossky and J.D. Simon, *Nature* **370**, 263 (1994), and references there.
138. J.S. Bader, R.A. Kuharski abd D. Chandler, *J. Chem. Phys.* **93**, 230 (1990), and references there.
139. W.W. Parson, in *Protein Electron Transfer*, ed. D.S. Bendall (BIOS, Oxford, 1996), p. 125.
140. M. Marchi *et al.*, *J. Am. Chem. Soc.* **115**, 4178 (1993).
141. P.F. Barbara and W. Jarzeba, *Acc. Chem. Res.* **21**, 195 (1988).
142. R. Jimenez *et al.*, *Nature* **369**, 471 (1994), and references there.
143. *Femtochemistry. Ultrafast Chemical and Physical Processes in Molecular Systems*, ed. M. Chergui (World Scientific, Singapore, 1996).
144. *Femtosecond Chemistry*, eds. J. Manz and L. Wöske (Verlag Chemie, Weinheim, 1995).
145. S.L. Carnie and G.M. Torrie, *Adv. Chem. Phys.* **56**, 141 (1984).
146. A.A. Kornyshev, *Electrochim. Acta* **34**, 1829 (1989).

147. *Chem. Phys.* **141**, 1-188 (1990), eds. W. Schmickler and J. Ulstrup.
148. *Electrochim. Acta* **41**, 2061-2338 (1996), ed. S. Trasatti.
149. W. Schmickler, *J. Electroanal. Chem.* **204**, 31 (1986).
150. A.M. Kuznetsov, *J. Electroanal. Chem.* **241**, 45 (1988).
151. K.L. Sebastian and P. Ananthapadmanabhan, *J. Electroanal. Chem.* **230**, 43 (1987).
152. B.B. Smith and J.T. Hynes, *J. Chem. Phys.* **99**, 6517 (1993).
153. A.A. Kornyshev and W. Schmickler, *J. Electroanal. Chem.* **185 253 (1985).**
154. W. Schmickler, *Interfacial Electrochemistry* (Oxford University Press, New York and Oxford, 1996).
155. W. Schmickler, *J. Electroanal. Chem.* **137**, 189 (1982).
156. A.M. Kuznetsov and J. Ulstrup, *J. Electroanal. Chem.* **195**, 1 (1985).
157. C.J. Chen, *Introduction to Scanning Tunnel Microscopy* (Oxford University Press, New York, 1991).
158. S.N. Magonov and M.H. Whangbo, *Surface Analysis with STM and AFM* (Verlag Chemie, Weinheim, 1996).
159. *New J. Chem.* **15**, 95-229 (1991), ed. J.P. Launay.
160. *Mol. Cryst. Liq. Cryst.* **234**, 1-794 (1992), eds. C. Pecile *et al.*
161. *Nanostructured Based on Molecular Materials*, eds. W. Göpel and C. Ziegler (Verlag Chemie, Weinheim, 1992).
162. M.A. Ratner and J. Jortner, *Mol. Electronics*, in press.
163. *Metal Ions in Biological Systems*, eds. H. Sigel and A. Sigel (M. Dekker, New York, 1991), *Vol. 27.*
164. H.B. Gray and J.R. Winkler, *Ann. Rev. Biochem.* **65**, 537 (1996).
165. *Chem. Phys.* **197**, 223-472 (1995), eds. M.E. Michel-Beyerle and G.J. Small.
166. *The Reaction Center of Photosynthetic Bacteria. Structure and Dynamics*, ed. M.E. Michel-Beyerle (Springer, Berlin, 1996).

CHARGE TRANSFER PROCESSES ASSOCIATED WITH ENERGY CONVERSION

U. STIMMING and W. LEHNERT
Institute for Energy Process Engineering
Research Center Jülich
D-52425 Jülich, Germany

Fuel cells convert chemical energy directly into electricity and are thus an alternative to conventional energy conversion techniques for mobile as well as stationary applications. The energy conversion process in the fuel cell is based on charge transfer reactions at the electrode-electrolyte interfaces of the fuel cell. Detailed knowledge about the microscopic processes occurring at these interfaces allow optimization of the fuel cell performance. Most important are the charge transfer processes at the three phase boundary and the electrocatalytic properties of the materials. The influence of different catalysts on activation energy and pre-exponential factor of the oxygen reduction reaction in a high-temperature fuel cell are discussed. The dependence of the charge transfer coefficient on temperature and overpotential is also investigated for the hydrogen oxidation reaction. Model systems for solid electrolytes such as the polyhydrates $HClO_4 \cdot 5.5H_2O$ allow a better understanding of the microscopic effects relating to the liquid or the solid state of the electrolyte. For low-temperature fuel cells the influence of CO on performance is examined and the reaction mechanisms using CV, IR, and DEMS is investigated. A $Pt_{0.5}/Ru_{0.5}$ catalyst shows the highest CO tolerance in PEMFCs while $Pt_{0.9}/Ru_{0.1}$ is favored for DMFCs for low temperature operation.

1 Introduction

During the last 20 years the energy consumption increased drastically. New energy conversion techniques have been developed to decrease emissions and to increase the efficiency of the conversion of chemical energy into electricity[1]. Two principal routes for this conversion are possible. The first way is the conversion of the fuel in heat engines to mechanical energy. The second route is the fuel cell process. In a fuel cell electrochemical processes convert the chemical energy contained in the fuel directly into electricity. The electrochemical reactions take place at the boundaries of the electrodes and the electrolyte. The fuel is continuously fed to the cells and the reaction products are continuously removed.

In thermal processes the maximum efficiency is limited by the Carnot cycle:

$$\eta_c = 1 - \frac{T_1}{T_2}$$

T_1 and T_2 are the lower and the upper temperatures of the process. By contrast, the efficiency of a fuel cell does not depend on the Carnot cycle. The theoretical efficiency of the electrochemical processes is given by:

$$\eta_{th} = \frac{\Delta G}{\Delta H} = 1 - \frac{T\Delta S}{\Delta H}$$

In a fuel cell the Gibbs Free Energy ΔG can be converted directly into electricity. ΔH is the chemical energy stored in the fuel, ΔS is the Entropy. Under standard conditions at 25°C heat, the theoretical efficiency can be >80% which is much higher than the efficiency of heat engines. Another advantage of fuel cells compared to combustion engines are their lower emissions. Especially NO_x production is avoided because of the lower operating temperatures and the electrochemical process. In the following we will focus on the charge transfer process in fuel cells which have to be well understood in order to increase the efficiency of the system.

1.1 Basic principal operation of fuel cell operation

The main parts of a fuel cell are the anode, the cathode and the electrolyte. In a hydrogen fuel cell the gases fed to the anode and the cathode are hydrogen and oxygen, respectively. The hydrogen is oxidized at the anode, while oxygen is reduced at the cathode. For these reactions catalysts are necessary which are placed at the boundaries of the electrolyte and the electrodes. The electrodes must have good electronic conductivity, whereas the electrolyte has to be a good ionic conductor and should not have any electronic conductivity. The electrodes have to be permeable to gas, the electrolyte has to be gas-tight. For a hydrogen fuel cell with a proton conducting electrolyte, the reactions at the anode and the cathode can be written as follows:

Anode:	H_2	→	$2H^+ + 2e^-$
Cathode:	$\frac{1}{2} O_2 + 2H^+ + 2e^-$	→	H_2O
overall reaction:	$H_2 + \frac{1}{2} O_2$	→	H_2O

When the electrodes are fed with their respective gases, the reactions described above result in an open cell voltage (OCV) of 1.2 V. When a current is delivered by the fuel cell, the cell voltage drops due to various losses in the cell. The difference between the OCV and the cell voltage at a given current denote the losses. They are composed out of the overpotentials at each electrode and the ohmic loss in the electrolyte.The overpotential depends on the kinetics of the electrode reactions and the mass transport limitations in the cell. The importance of the electrode kinetics for energy conversion will be shown in the next sections.

2 High Temperature Fuel Cells

2.1 Introduction

The operating temperature of solid oxide fuel cells (SOFC) is about 950 °C. Because of the high operating temperature this type of fuel cell will be most suitable for stationary applications like power plants in the range of kW up to MW. For larger power plants the waste heat can be used for producing additional electricity in a conventional power plant with gas and steam turbines (combined cycle power plants). One advantage of Solid Oxide Fuel Cells compared to low temperature fuel cells is that they can not only convert hydrogen into electricity, but also carbon monoxide. Furthermore, the fuel cell can be fed with methane. Due to the high temperature, the methane can be converted to H_2 and CO in a steam reforming process within the cell. The main parts of the SOFC are the ceramic electrodes and the ceramic electrolyte (fig. 1). The anode is a porous nickel cermet. A typical material for the cathode is $La_{.84}Sr_{.16}MnO_3$ (LSM). The standard material for the electrolyte is YSZ8 (yttria-stabilized zirconia) which is an oxygen-ion conductor. The reactions at the anode and the cathode are:

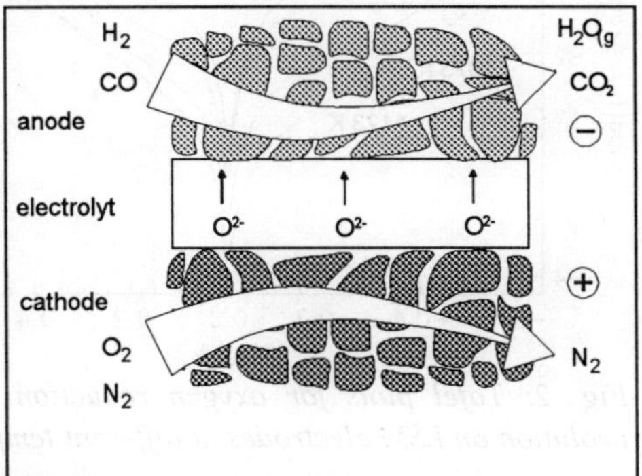

Fig. 1. Operation of a SOFC

	cathode:	$O_2 + 4e^-$	$\rightarrow$	$O^{2-} + O^{2-}$
	anode:	$H_2 + O^{2-}$	$\rightarrow$	$H_2O + 2e^-$
or:	anode:	$CO + O^{2-}$	$\rightarrow$	$CO_2 + 2e^-$
	overall reactions:	$H_2 + 1/2O_2$	$\rightarrow$	$H_2O + 2e^-$
or		$CO + 1/2O_2$	$\rightarrow$	$CO_2 + 2e^-$

Two different microscopic mechanisms relevant for the improvement of the SOFC will be described in the next sections: The effect of noble metal catalysts at the cathode on the electrochemical performance and the kinetics of the anode reaction.

2.2 Kinetics of the oxygen reduction

It is an important goal of the SOFC development to reduce the high operating temperature of approximately 950°C. Several factors contribute currently to losses of operation at e.g. 700 - 800 °C such as ohmic losses in the electrolyte and high overpotentials for the electrode reactions. While the former can be addressed by e.g. changing the cell design to the substrate concept[2], the latter requires improved reaction kinetics. An important aspect is the high activation energy of the oxygen reduction reaction.

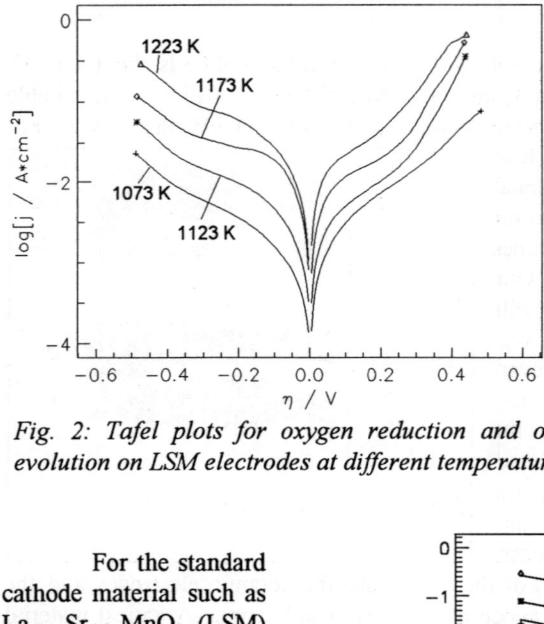

Fig. 2: Tafel plots for oxygen reduction and oxygen evolution on LSM electrodes at different temperatures.

For the standard cathode material such as $La_{.84}$ $Sr_{.16}$ MnO_3 (LSM) the activation energy is about 200 kJ/mol. In order to determine that activation energy and the pre-exponential factor in detail, current-potential curves were measured in the temperature range 1073 - 1223 K. While the activation energy of the reaction is related to the

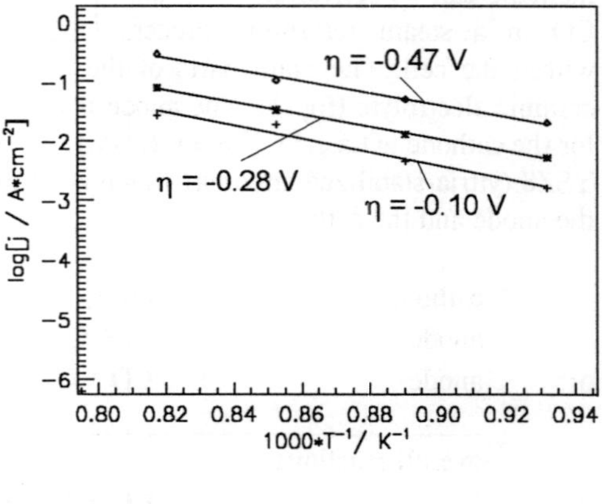

Fig. 3: Arrhenius plots for three different overpotentials.

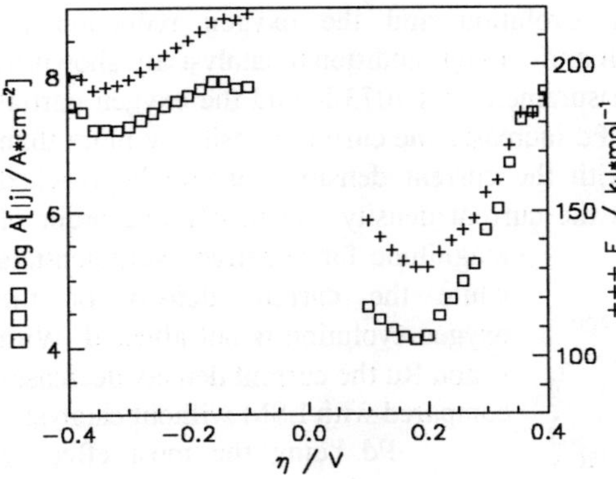

Fig. 4: Activation energy and pre-exponential factor as a function of overpotential for a $L_{a.84}Sr_{.16}MnO_3$ electrode.

electrocatalytic properties of the cathode, the pre-exponential factor is proportional to the length of the reaction zone and in this way it reflects the microstructure of the cathode.

Commercially available yttria-stabilized zirconia foils (YSZ8) were used as electrolyte. The cathodes were prepared by screen printing. Potentiodynamic current-potential curves were recorded for different cathodes. The temperature dependence of the Tafel plot for LSM is shown in fig. 2. For oxygen reduction (at $\eta<0$) the increase of the current density with temperature is more pronounced than for oxygen evolution ($\eta>0$). Arrhenius plots at three different negative overpotentials are shown in fig. 3. From the slope and the intercept of the Arrhenius plots the activation energies and the pre-exponential factors are calculated. For oxygen reduction, activation energies of more than 200 kJ/mol are obtained. At overpotentials up to 200 mV the activation energy is about 150 kJ/mol and at overpotentials of more than 200 mV the activation energy and the pre-exponential factor increases with increasing overpotential (fig. 4). In order to reduce this high activation energy, noble metal catalysts such as Pt, Pd, Ir and Ru were investigated regarding their potential ability to enhance catalysis.[3,4] The finely dispersed catalyst particles were located at the interface between electrode and electrolyte.

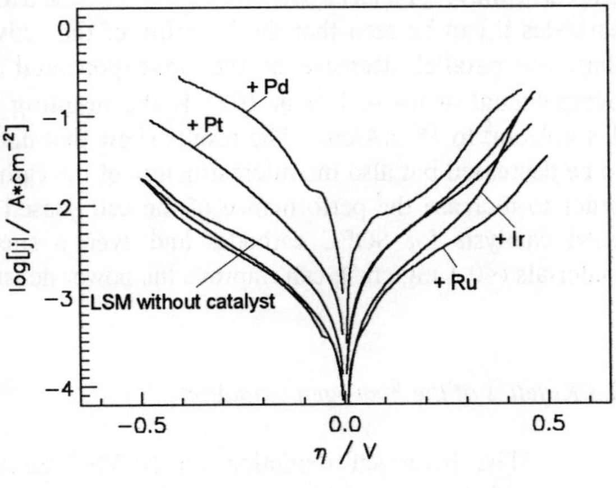

Fig. 5: Tafel plots for oxygen reduction and oxygen evolution on LSM electrodes. Comparison of electrodes without additional catalyst and with different additional metal catalysts.

Tafel plots for the oxygen evolution and the oxygen reduction on $La_{.84}Sr_{.16}MnO_3$ (LSM) electrodes with and without additional catalyst are shown in fig. 5. The temperature during the measurement was 1073 K and the oxygen partial pressure was 20 kPa. The presence of Pd increases the current density by more than one order of magnitude compared with the current density that can be reached without catalyst. Adding Pt increases the current density by nearly one order of magnitude for negative overpotentials while the current density of the oxygen evolution is not affected. With Ir and Ru the current density decreases compared with LSM without catalyst.

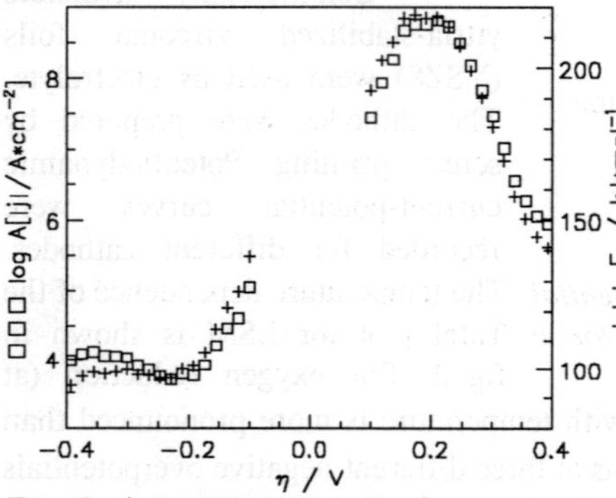

Fig. 6: Activation energy and pre-exponential factor as a function of overpotential for $La_{.84}Sr_{.16}MnO_3$ *+ Pd electrode.*

Pd being the most effective catalyst, the resulting activation energies and pre-exponential factors are shown in Fig. 6. The effect of decreased activation energy compared to LSM without noble metal catalyst is partially compensated by a decrease of the pre-exponential factor. Comparing the pre-exponential factors and the activation energies of the different cathodes it can be seen that the lowering of the activation energy is more effective than the parallel decrease of the pre-exponential factor. At a typical cathodic overpotential of η= -0.1 V at 1073 K the resulting current density increases from 1.8 mA/cm^2 to 18 mA/cm^2. The results show that not only the activation energy has to be decreased but also the microstructure of the cathodes have to be investigated in order to increase the performance of the cell. Based on the results, palladium is a good catalysts for SOFC cathodes and even a small addition of these catalyst-materials (<0.1 mg/cm^2), can improve the power density of SOFCs significantly.

2.3 Kinetics of the hydrogen oxidation

The hydrogen oxidation on Ni/YSZ electrodes was investigated as a function of potential and temperature in order to obtain information about kinetic parameters such as the activation energy, the pre-exponential factor and their possible potential dependence. The studies were conducted using screen printed Ni/YSZ cermet electrodes.[5] The current overpotential curves of a Ni/YSZ electrode in the temperature range 725 - 950 °C are shown in fig. 7. An Arrhenius analysis of the current densities at constant overpotentials shows that two temperature regimes

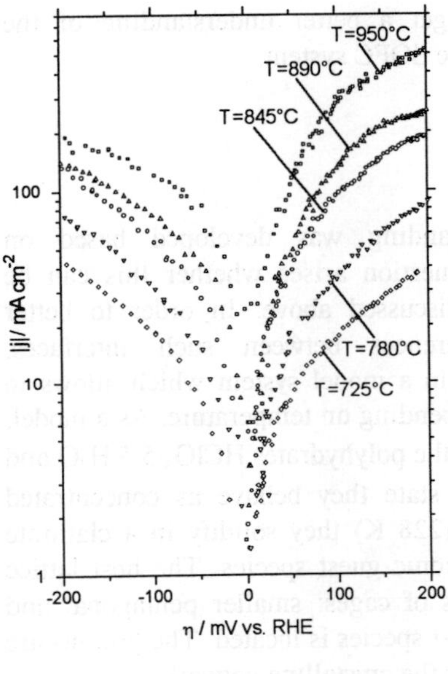

Fig. 7: Temperature dependent current-overpotential curves.

have to be distinguished, one at 725 - 845 °C and the other at 845 - 950 °C. The pre-exponential factors and activation energies as a function of the anodic overpotential are shown in fig. 8. It can be seen that in the low temperature region the activation energy decreases linearly and the pre-exponential factor is constant (full symbols). In the high temperature regime (open symbols) the pre-exponential factor and the activation energy are independent of potential at overpotentials $\eta > 140$ mV. From the dependence of the activation energy and the pre-exponential factor on the overpotential in the low temperature range, the enthalpic (α_H) and the entropic (α_S) contribution to the charge transfer coefficient (α) can be determined.[6] The charge transfer coefficients in the low temperature region are: $\alpha_H = 0.7$, $\alpha_S = 0$. The resulting charge transfer coefficient $\alpha = \alpha_H + T\ \alpha_S$ is therefore independent of temperature and has the same value as α_H. These results are in agreement with a determination of the charge transfer coefficients from current-overpotential curves in this temperature region. From the results it can be concluded that at low temperatures a charge transfer step is rate limiting. In the high temperature region this conclusion is not valid.

The experiments show that in the two temperature regimes investigated, different rate limiting steps are present and more detailed investigations of the hydrogen

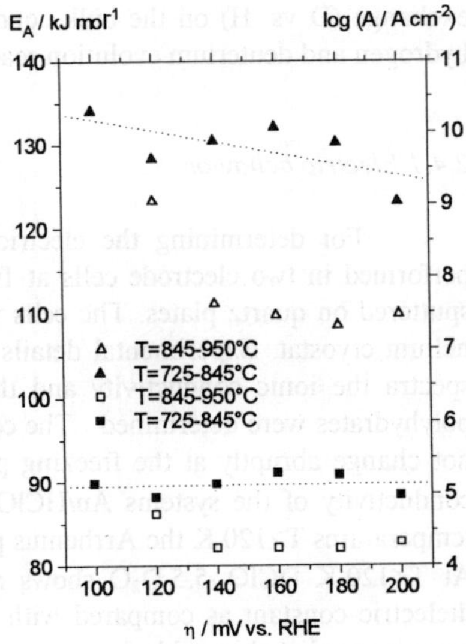

Fig. 8: Overpotential dependence of the activation energy.

oxidation reaction is necessary in order to get a better understanding of the microscopic effects in order to improve the whole SOFC system.

2.4 A model system for solid electrolytes

Since most electrochemical understanding was developed based on experiments with solid-liquid interfaces the question arises whether this can be applied also to the solid-solid interface as discussed above. In order to better understand the similarities and the differences between such interfaces, electrochemical investigations were performed in a model system which allows to have a one-phase solid or a liquid electrolyte depending on temperature. As a model, e.g. gold or silver were chosen as electrode, and the polyhydrates $HClO_4{\cdot}5.5\ H_2O$ and $DClO_4{\cdot}5.5\ D_2O$ as electrolyte.[7] In the liquid state they behave as concentrated electrolyte solutions, below the freezing point (228 K) they solidify in a clathrate structure consisting of a host lattice and an anionic guest species. The host lattice consists of water molecules forming two kinds of cages: smaller pentagonal and larger hexagonal cages in which the anionic guest species is located. The protons are part of the water structure and are delocated over the crystalline network.

Some aspects will be described in the next sections: The effect of isotopic exchange (D vs. H) on the bulk conductivity and static dielectric constant and the hydrogen and deuterium evolution reaction.

2.4.1 Electric behavior

For determining the electrical properties impedance measurements were performed in two electrode cells at 10 K - 300 K. The electrodes were gold films sputtered on quartz plates. The cells were mounted in a liquid nitrogen or a liquid helium cryostat. Experimental details can be found elsewhere.[7,8,9] From impedance spectra the ionic conductivity and the static dielectric constant of the solid acid polyhydrates were determined . The conductivity of the polyhydrate of the acid does not change abruptly at the freezing point. Fig. 9 shows the Arrhenius plot of the conductivity of the systems $Au/HClO_4{\cdot}5.5\ H_2O/Au$ and $Au/DClO_4{\cdot}5.5\ D_2O/Au$. At temperatures $T>120$ K the Arrhenius plots of both isotopic compounds are identical. At $T<120$ K $DClO_4{\cdot}5.5\ D_2O$ shows a lower conductivity and also a lower static dielectric constant as compared with the protonated form (fig. 10). The activation energies are listed in table 1.

Table 1: Activation energies in different temperature regions

Polyacid	E_a/eV above 165 K	E_a/eV between 120 and 165 K	E_a/eV below 120 K
$HClO_4 \cdot 5.5\ H_2O$	0.35	0.48	0.25
$DClO_4 \cdot 5.5\ D_2O$	0.35	0.5	0.2

At 163 K a second order phase transition is most likely responsible for the change of the activation energy. At temperatures below 120 K a deviation from the Arrhenius behavior can be observed. This behavior can be explained by a change from a thermally activated process to a proton tunneling process which can be described on the basis of quantum theory.[8] The dielectric constant ϵ (fig. 10) shows a strong isotope effect: At very low temperatures a limiting value of ϵ can be observed which is $\epsilon=10$ for $HClO_4 \cdot 5.5\ H_2O$. For $DClO_4 \cdot 5.5\ D_2O$, however, ϵ is only 5 which is higher than the dielectric constant of pure ice ($\epsilon=3$).

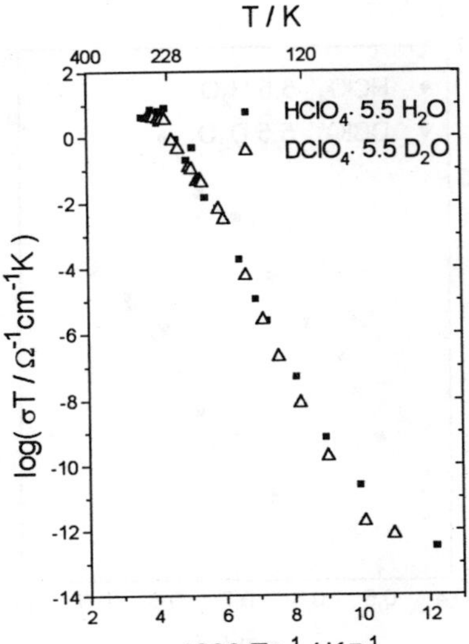

Fig. 9: Arrhenius plot of the conductivity.[11]

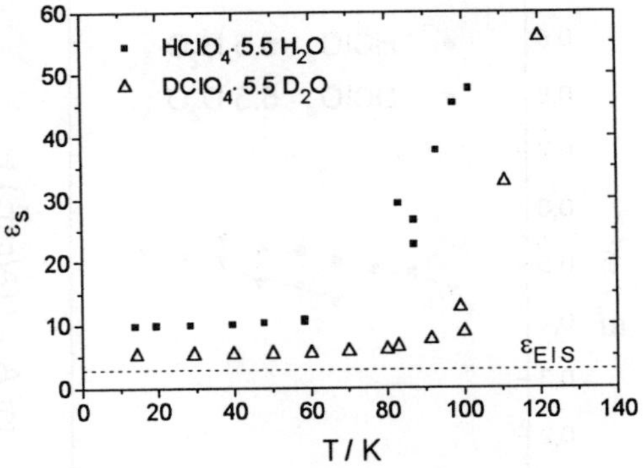

Fig. 10: Temperature dependence of the static dielectric constant.

2.4.2 Electrochemical behavior

Hydrogen and deuterium evolution at silver electrodes of $HClO_4{\cdot}5.5\ H_2O$ and $DClO_4{\cdot}5.5\ D_2O$ was investigated in the temperature region of 133 K - 300 K with cyclic voltammetry.[9] The Tafel plot of the current density vs. overpotential for hydrogen and deuterium evolution shows a linear relationship of ln |j| vs. η at all temperatures investigated. A comparison of the current density of the hydrogen and deuterium evolution exhibits a pronounced isotope effect.

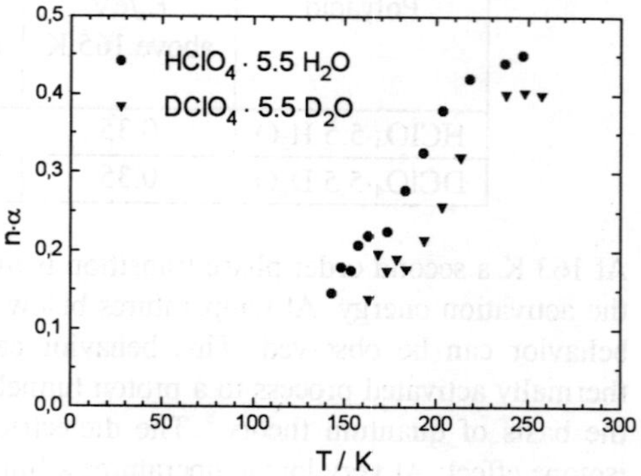

Fig. 11: Charge transfer coefficient as a function of temperature.

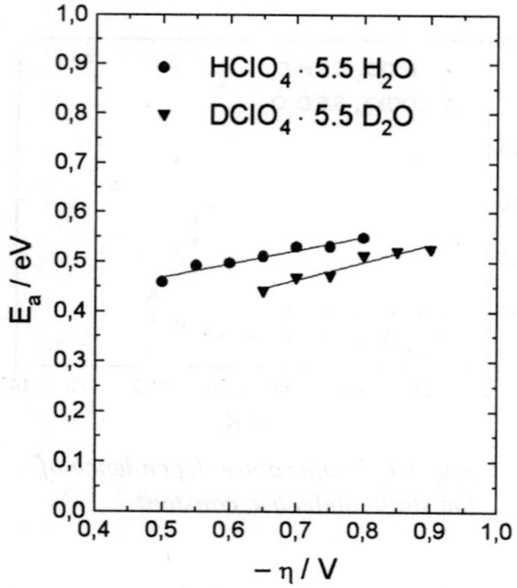

Fig. 12: Activation energies as a function of overpotential.

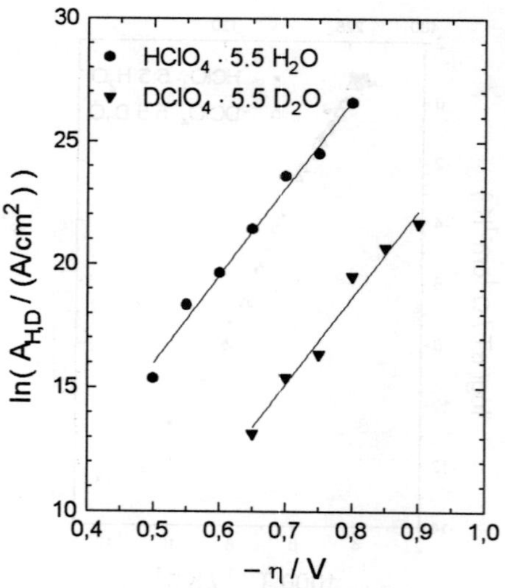

Fig. 13: pre-exponential factors as a function of overpotential.

At the same temperature and overpotential the current density of the hydrogen evolution is about one order of magnitude higher than for deuterium evolution (T=203 K, η=0.5 V). The effect becomes more pronounced with decreasing temperature. The resulting charge transfer coefficient $n\alpha$ as a function of temperature is shown in fig. 11: The charge transfer coefficient for hydrogen evolution decreases with decreasing temperature from 0.44 down to 0.15 in the temperature region of T=250 K to T=140 K. In the same temperature region the charge transfer coefficient for deuterium evolution decreases from 0.4 down to 0.1. Arrhenius plots of the current density at constant overpotentials show nearly parallel slopes. The resulting activation energies are almost equal and only weakly potential dependent (fig. 12). The pre-exponential factors as functions of overpotential are shown in fig. 13. The strong potential dependence of the pre-exponential factor results in the potential dependence of the current.

The results suggest that the reaction of hydrogen evolution on silver electrodes occurs in a two-step mechanism with charge transfer in both steps. The first step is the discharge of protons or deuterons. The second step includes the recombination reaction with the simultaneous discharge of a proton/deuteron and the recombination with an adsorbed hydrogen/deuterium atom similar to the Heyrovski reaction. The second step is considered the rate-determining step of the mechanism, leading to adsorbed H or D.

3 Low Temperature Fuel Cells

3.1 Introduction

Proton Exchange Membrane Fuel Cells (PEMFC) using hydrogen as fuel are regarded as highly attractive for mobile application due to their high power density at temperatures of 60°C - 100°C. Methanol as energy carrier in a fuel cell powered vehicle exhibits advantages regarding energy density compared to compressed or liquid hydrogen. In a fuel cell system using methanol, a reformer converts methanol into a hydrogen-rich fuel gas. This gas contains about 75% H_2, 24% CO_2 and 1% CO. The electrochemical reactions in PEMFCs have to be catalyzed in order to achieve high power densities. Pt is the most effective catalyst for hydrogen oxidation but problems arise when the fuel gas contains CO. Even a few ppm CO produce a substantial degradation of the fuel cell performance. This degradation is associated with CO adsorption on active Pt sites at the anode thus inhibiting the hydrogen oxidation. There are two possible ways to solve this problem: A gas treatment can be used to remove the CO from the fuel gas. The second possibility is to increase the CO-tolerance of the anode or a combination of both. The most promising electrocatalysts for the oxidation of CO containing hydrogen are Pt/Ru alloys. Studies in aqueous solutions at room temperature have shown that the oxidation of an adsorbed CO- monolayer is shifted to more negative potentials compared to pure Pt. A composition of $Pt_{0.5}/Ru_{0.5}$ is the best composition regarding CO-tolerance in comparison to pure Pt.[10,11]

Another way to use methanol as fuel is the direct oxidation of methanol in a direct methanol fuel cell (DMFC). Using a DMFC, the reforming of methanol to a hydrogen rich gas is no longer necessary. The whole fuel processing with its additional weight, volume, and costs can be avoided. The basic idea is known for many years but the realization has only been demonstrated on a laboratory scale. The direct electrochemical methanol oxidation in a DMFC also requires the development of optimized catalysts since the oxidation of methanol on Pt leads to the formation of poisoning species such as CO, COH and CHO on the electrode.[12,13]
Both types of fuel cells (PEMFC, DMFC) will be described in the following sections.

3.2 Principle of a PEMFC

The basic components of a PEMFC are the electrodes, the electrolyte and the catalyst. Fig. 14 shows a single cell with planar geometry. The porous gas diffusion electrodes with the carbon bond catalyst are pressed onto both sides of the ion exchange membrane. Titanium bipolar plates serve as current collectors and as distributors for the reaction gases. Proton conducting polymer membranes are used as electrolyte. Hydrogen fuel gas is fed to the anode, oxygen or air to the cathode. The experiments were done with a single laboratory cell with an active area of $4cm^2$. Details of the experimental setup can be found elsewhere.[14]

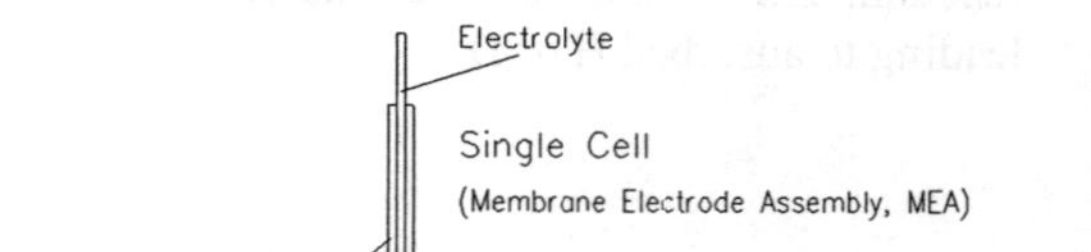

Fig. 14: Principle of a PEMFC stack.

3.3 Influence of CO in the fuel on the power density

Fig. 15 shows the influence of CO on the cell performance with a Pt catalyst. For different CO concentrations the steady state current-voltage- curves are plotted. In all curves an open cell voltage of 1.05 V was observed, independent of the CO concentration in the fuel. The open cell voltage is obviously predominantly determined by the H_2 /H^+ equilibrium. With increasing CO concentration the cell voltage drops to lower values with increasing current density compared to pure hydrogen as fuel gas. With 25 ppm CO in the fuel gas the stationary current-voltage-curve is linear and the slope of the curve is more negative compared to the one measured in pure hydrogen. This effect is due to a higher overvoltage at the anode. With 100 ppm and 250 ppm CO the current-voltage- curves have two distinct slopes. The lower slope of the curve can be explained by the adsorption and oxidation kinetics of H_2/CO at the anode. With increasing current densities the potential at the anode increases to values where adsorbed CO can be oxidized. An example of the influence of Ru as catalyst is shown in fig. 16. It can clearly

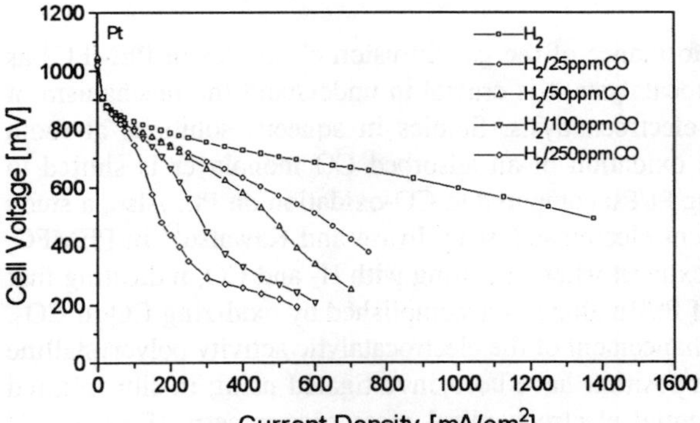

Fig. 15: Cell voltage vs. Current density plots of a PEM single cell using H_2 and H_2/CO. Anode Pt.

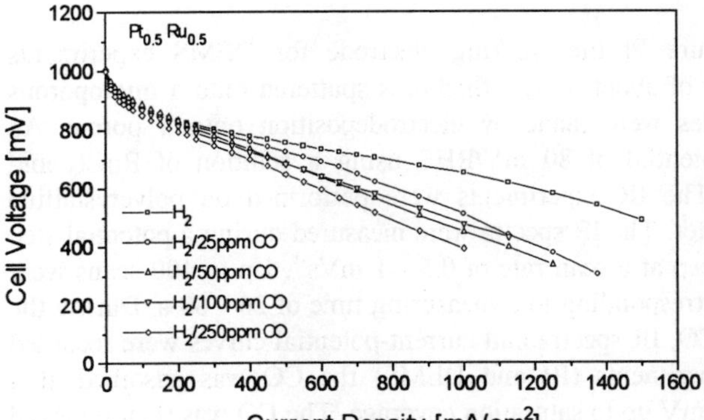

Fig. 16: Cell voltage vs. Current density plots of a PEM single cell using H_2 and H_2/CO. Anode $Pt_{0.5}RU_{0.5}$.

be seen that the cell voltage with $Pt_{0.5}/Ru_{0.5}$ catalyst is significantly higher for all gas compositions than for pure Pt catalyst.

3.4 Investigation of the mechanism of CO-oxidation on polycrystalline Pt/Ru alloys

To optimize the performance of the gas diffusion electrodes in PEMFCs as well as to develop new electrocatalysts it is crucial to understand the mechanism of CO-oxidation on bimetallic electrocatalysts. Studies in aqueous solutions at room temperature showed that the oxidation of an adsorbed CO monolayer is shifted to more negative potentials using Pt/Ru compared to CO-oxidation on Pt.. Also, a study on the performance of different electrocatalyst by Iwase and Kawatsu[15] in PEMFC, CO_2 was found in the anode exhaust when operating with H_2 and CO, indicating that the increased CO-tolerance of Pt/Ru alloys is accomplished by oxidizing CO to CO_2. In order to understand the enhancement of the electrocatalytic activity polycrystalline Pt/Ru, alloys of different composition have been investigated using in-situ infrared spectroscopy (IR) and differential electrochemical mass spectrometry (DEMS).[10,16] Using DEMS an analysis of the volatile products formed during the CO-oxidation was performed. In this way ion current-potential curves can be correlated with faradaic current-potential curves. For the identification of adsorbed CO and of reaction intermediates in-situ IR spectroscopy was used.

3.4.1 Experimental

In the case of pure Pt the working electrode for DEMS experiments consisted of a porous layer of about 50 μm thickness sputtered onto a microporous membrane. Pt/Ru electrodes were made by electrodeposition onto a porous Au substrate at a constant potential of 80 mV/RHE using a solution of $RuCl_3$ and H_2PtCl_6 in 1M $HClO_4$. The IR experiments were performed on polycrystalline electrodes of 10 mm diameter. The IR spectra were measured during a potential step or during a continuous sweep at a scan rate of 0.5 - 1 mVs^{-1}. Up to 400 scans were taken for each spectrum corresponding to a measuring time of 20 - 60 s. During the oxidation of the adsorbed CO, IR spectra and current-potential curves were recorded simultaneously. In all experiments (IR and DEMS) the CO was adsorbed at a constant potential of U=70 mV up to saturation coverage. The CO was then removed from the solution by purging with Ar while the potential was kept constant. The measurements were performed at room temperature in 0.1 M and in 1 M $HClO_4$, respectively.[10]

3.4.2 Results and discussion

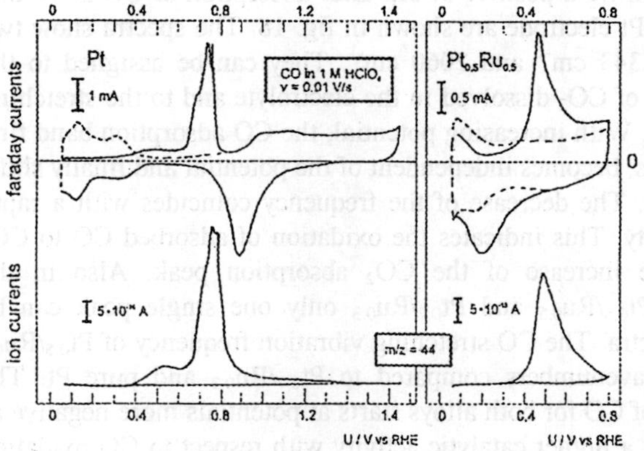

Fig. 17: Oxidation of adsorbed CO on a porous Pt electrode and on a porous $Pt_{0.5}Ru_{0.5}$ electrode, respectively. Dashed line: second potential scan.

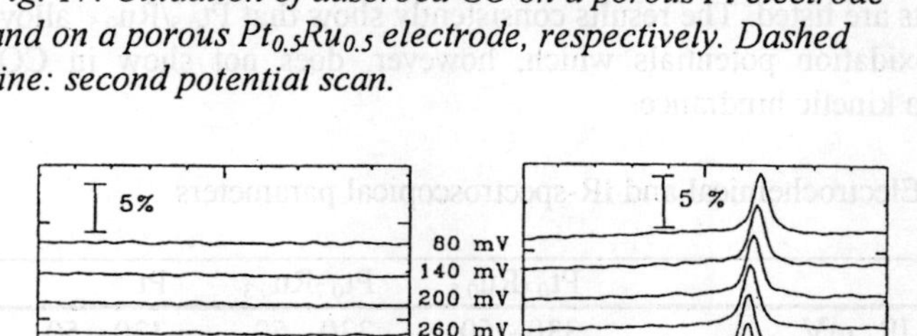

Fig. 18: In situ IR spectra taken in 0.1M $HClO_4$ on a polycrystalline Pt electrode.

Fig. 17 shows the cyclic voltammogram (CV) for the oxidation of adsorbed CO on a porous Pt electrode and the corresponding mass spectrometric cyclic voltammogram (MSCV) for the CO_2 formation. In the CV a broad oxidation prewave peak at 400 mV and a sharp oxidation peak at 640 mV can be observed. The oxidation peaks in the CV correspond to the beginning of the CO_2 production as can be seen in the MSCV since a signal appears at m/z = 44. The profile of the MSCV follows exactly the oxidation peak in the CV with a delay time of about 0.5 s which is due to the diffusion process through the porous electrode.

The corresponding measurements with a porous $Pt_{0.5}/Ru_{0.5}$ electrode are also shown in fig. 17. Only one oxidation peak can be observed at U=470 mV. This is a shift of

170 mV to lower potentials compared to the main oxidation peak of pure Pt. The delay time of about 4 s is due to the longer diffusion pathway in the more densely packed electrode.

IR spectra taken during a positive sweep after adsorption of CO at 70 mV on a smooth polycrystalline Pt electrode are shown in fig. 18. The spectra show two absorption bands at about 2343 cm^{-1} and 2060 cm^{-1}. They can be assigned to the asymmetric stretching mode of CO_2 dissolved in the electrolyte and to the stretching mode of linearly bonded CO. With increasing potential, the CO adsorption band first shifts to higher wavenumbers, becomes independent of the potential and finally shifts back to lower wavenumbers. The decrease of the frequency coincides with a rapid decrease of the band intensity. This indicates the oxidation of adsorbed CO to CO_2 which can be seen by the increase of the CO_2 absorption peak. Also in the corresponding spectra for $Pt_{0.5}/Ru_{0.5}$ and $Pt_{0.7}/Ru_{0.3}$ only one single peak can be observed in the infrared spectra. The CO stretching vibration frequency of $Pt_{0.5}/Ru_{0.5}$ is shifted towards lower wavenumbers compared to $Pt_{0.7}/Ru_{0.3}$ and pure Pt. The decrease of signal intensity of CO for both alloys starts at potentials more negative as compared with Pt indicating a higher catalytic activity with respect to CO-oxidation for the Pt/Ru alloys. In table 2 the experimental data obtained from DEMS as well as from IR measurements are listed. The results consistently show that $Pt_{0.5}/Ru_{0.5}$ alloys exhibit the lowest oxidation potentials which, however, does not show in CO_2 formation indicating a kinetic hindrance.

Table 2: Electrochemical and IR-spectroscopical parameters.

	$Pt_{0.5}Ru_{0.5}$	$Pt_{0.7}Ru_{0.3}$	Pt
CO_2 formation onset, IR, mV	330± 50	330± 50	330± 50
CO_2 formation onset, DEMS, mV	300	-	300
CO-oxidation peak potential, IR, mV	390	480	570
CO-oxidation peak potential, DEMS, mV	470	-	640

It is known that in order to oxidize an adsorbed CO molecule an oxygen containing adsorbate site is needed. It has been suggested e.g. by Gasteiger et al[17,18] that the adsorbed oxygen-containing species has to be in the vicinity of CO. The increased reactivity with respect to CO oxidation observed on Pt/Ru alloys can thus be understood in terms of the lower potential for OH-adsorption on Ru atoms compared to Pt. $Pt_{0.5}/Ru_{0.5}$ shows the lowest CO-oxidation potential because the alloy combines the adsorption of OH at low potential with the high activity of Pt for this reaction. This has been denoted bifunctional mechanism[19] and consists of the following steps:

$$Ru + H_2O \rightarrow Ru\text{-}OH + H^+ + e^-$$

$$Ru\text{-}Oh + Pt\text{-}CO \rightarrow Ru + Pt + CO_2 + H^+ + e^-$$

As a consequence, the maximum oxidation rate would be expected if both components are present in equal concentrations. This is in accordance with the observations but surface processes influencing the kinetics have to be taken into account as will be shown in the next section.

3.5 CO adsorption and oxidation on modified Pt(111) surfaces

As was mentioned earlier, the understanding of CO adsorption and electro-oxidation on platinum-based and particularly on Pt/Ru electrodes is of importance for the development of CO-resistant PEMFC anodes. The identification of surface adsorbates and the study of their potential-dependent surface concentrations on well-defined electrode surfaces should provide a deeper insight into the mechanism of CO adsorption and oxidation on a molecular level.

Using IR-spectroscopy in combination with cyclic voltammetry and scanning tunneling microscopy (STM) on Pt(111) surfaces modified with nanometer sized Ru deposits were investigated.[20,21] This system may serve as a model system bridging the gap between the behavior of pure Pt and Pt/Ru alloys and allowing the separation of the electrochemical properties of Ru and Pt.

3.5.1 Results and discussion

In fig. 19 an STM image is shown for a Pt(111) electrode with a Ru coverage of Θ=0.25 according to an XPS coverage determination. The images show mono-atomic islands of about 3 nm in diameter on a Pt(111) surface which are attributed to Ru deposits on the Pt substrate as found from in-situ XRSD measurements.[22] In fig. 20 the cyclic voltammograms of the CO electro-oxidation on a clean and a Ru-modified Pt(111) surface with a coverage of ≈ 0.4 (XPS) is shown. In order to enable a direct comparison of the voltammograms and the IR-spectra the voltammograms were performed with a scan rate of 1 mVs^{-1}. The CO electro-oxidation peak can be observed at a potential of U_{ox}=0.63 $V/_{RHE}$

Fig. 19: In-situ STM image of a Pt(111) electrode with a ruthenium coverage of 0.25

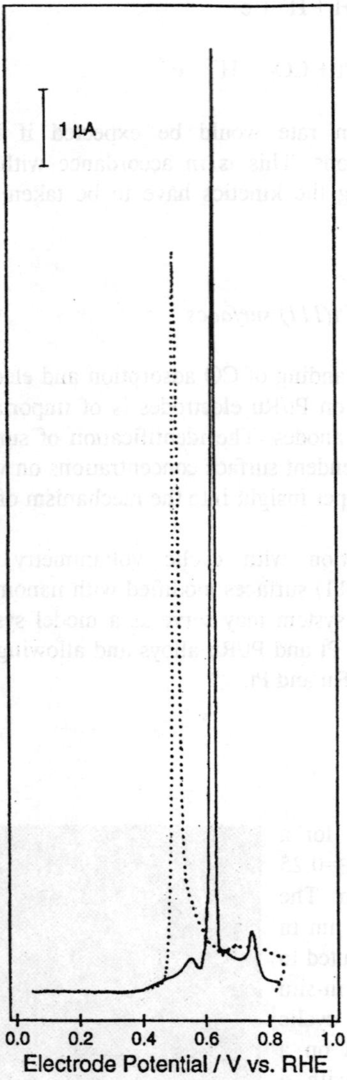

Fig. 20: Cyclic voltammograms for electro-oxidation of a CO adsorbate layer on Pt(111) (solid curve) and on Ru-modified Pt(111) (dotted curve) in 0.1 M $HClO_4$.

on a pure Pt(111) electrode whereas on a Ru-modified Pt(111) surface the oxidation potential is 0.5 $V/_{RHE}$. The Ru-modified surfaces exhibit a negative shift of 130 mV compared to pure Pt(111), this shift being smaller than that observed for Pt/Ru alloys (250 mV). However, this shift can be determined much more accurately on the single-crystal surfaces since the alloys show very broad current waves. Fig. 21 shows the corresponding IR absorption spectra on Ru-modified Pt(111). Three potential dependent absorption bands around 2070 cm^{-1}, 2010 cm^{-1} and 1790 cm^{-1} are found. The bands around 2070 cm^{-1} and 1790 cm^{-1} can be assigned to CO in a linear-bonded (on-top) and a threefold hollow coordination, respectively.[23] The third band around 2000 cm^{-1} has not been observed on pure Pt(111) surfaces and is associated with CO adsorbed on Ru.

In contrast to this result, IR adsorption spectra of CO on Pt/Ru alloys show only one vibrational band shifted to lower frequencies compared to polycrystalline Pt. This has been interpreted in terms of a homogeneous distribution of Ru and Pt atoms at the surface.[16] The observation of two distinct vibrational bands for CO adsorbed on Pt and Ru indicates that sufficiently large areas with Ru or Pt properties are present.

Despite the spatially separated Pt and Ru areas on the electrode, the Ru-modified electrodes show a clear promotion effect for CO electro-oxidation as can be seen in the CV (fig. 20).

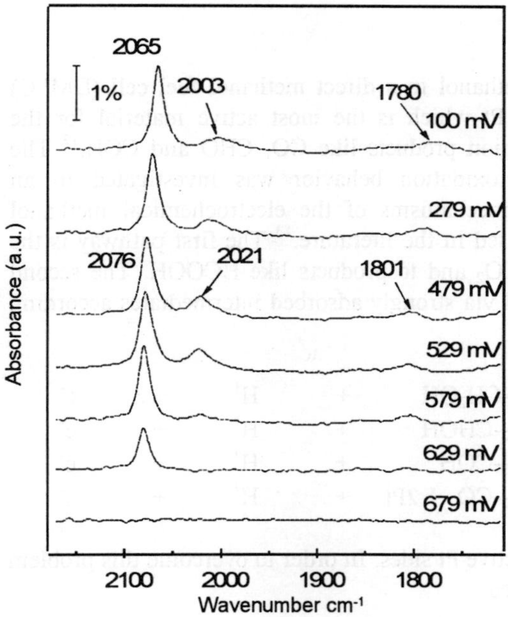

Fig. 21: Potential dependent IR adsorption spectra for Ru-modified Pt(111).

Fig. 22 shows the integrated band intensities of CO adsorbed on-top of Pt, of CO adsorbed on top on Ru and of the CO_2 asymmetric stretch vibration as a function of potential. It can be seen that CO adsorbed on Ru and on Pt coexist in a wide potential range between 50 mV and 550 mV with a constant ratio of the integrated band intensities of the CO-Ru and the CO-Pt bands. The intensities remain stable up to a potential of about 550 mV and diminish simultaneously and uniformly. This is a surprising observation since the bifunctional mechanism (see section 3.4.2) postulates OH adsorption on Ru at lower potentials than on Pt. Therefore, on a heterogeneous surface like the Ru-modified Pt(111) surface seen in fig. 19, one would expect a faster decrease of the CO-band on Ru. The observation of the coupling of the reaction on the whole surface in combination with the observation of enhancement in reactivity, requires a modification of the bifunctional mechanism: CO electro-oxidation starts at the boundary of the Ru islands, because at these sites the combined properties of Ru and Pt are present. Furthermore, we have to assume a sufficiently high mobility of CO on the surface which is supported by current transient experiments.

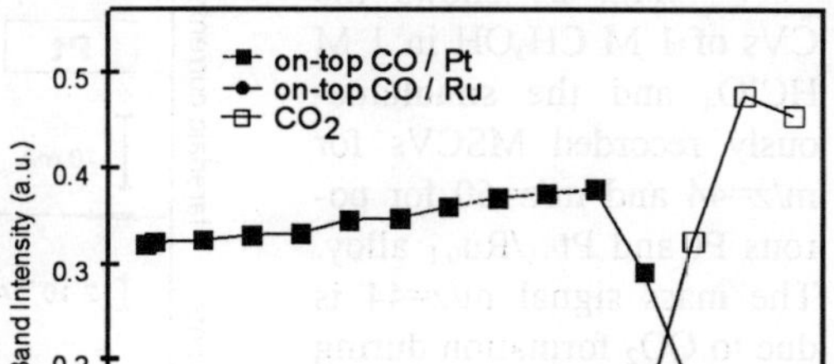

Fig. 22: Integrated band intensities as a function of potential.

3.6 Direct methanol fuel cell

The electrochemical oxidation of methanol in a direct methanol fuel cell (DMFC) requires optimized anode catalysts. Pt which is the most active material for the oxidation will be poisoned by oxidation products like CO, CHO and COH.[12] The influence of Pt/Ru alloys on the oxidation behavior was investigated in an electrochemical cell. Two parallel mechanisms of the electrochemical methanol oxidation on Pt electrodes are discussed in the literature.[13] The first pathway is the oxidation of adsorbed methanol to CO_2 and to products like HCOOH. The second pathway is the oxidation of methanol via strongly adsorbed intermediates according to the following reaction:

$$Pt + CH_3OH \rightarrow Pt\text{-}CH_2OH + H^+ + e^-$$
$$Pt + Pt\text{-}CH_2OH \rightarrow Pt_2\text{-}CHOH + H^+ + e^-$$
$$Pt + Pt_2\text{-}CHOH \rightarrow Pt_3\text{-}COH + H^+ + e^-$$
$$Pt_3\text{-}COH \rightarrow Pt_2\text{-}CO + 2Pt + H^+ + e^-$$

The intermediates are blocking the active Pt sides. In order to overcome this problem different Pt/Ru alloys were investigated.

3.7 Results and discussion

Fig. 23 shows the CVs of 1 M CH_3OH in 1 M $HClO_4$ and the simultaneously recorded MSCVs for m/z=44 and m/z=60 for porous Pt and $Pt_{0.9}/Ru_{0.1}$ alloy. The mass signal m/z=44 is due to CO_2 formation during methanol oxidation. The mass signal m/z=60 is assigned to the radical cation of $CHOOCH_3$ which indicates the formation of HCOOH as an oxidation product of methanol. The ester formation can be ex-

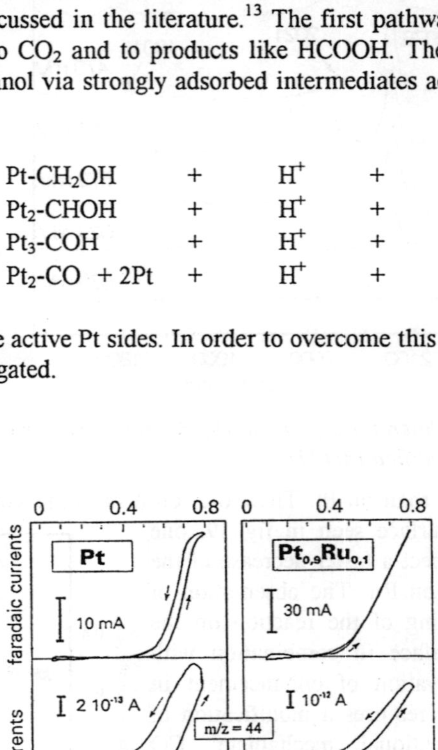

Fig. 23: Potentiodynamic profiles for electro-oxidation of 1 M CH_3OH in 1 M $HClO_4$ porous $Pt_{0.9}/Ru_{0.1}$ and Pt electrodes.

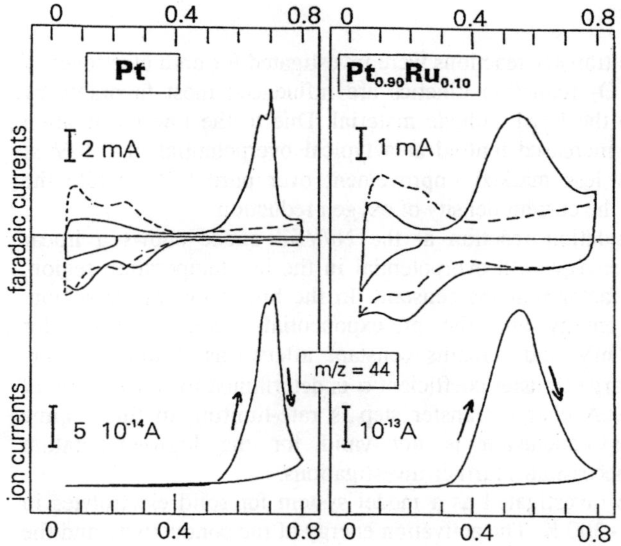

Fig. 24: Oxidation of methanol on $Pt_{0.9}Ru_{0.1}$ and Pt electrodes in 1 M $HClO_4$ at room temperature.

plained by a chemical reaction of HCOOH with CH_3OH in the solution. Comparing the CVs and the MSCVs for Pt and $Pt_{0.9}/Ru_{0.1}$ it can be seen that the onset potential for the methanol oxidation is shifted by 0.15 V to negative potentials. An integration of the ion current shows that the ratio $Q_{(m/z=44)}/Q_{(m/z=60)}$ changes in dependence of the alloy composition. The ratio has a maximum at $Pt_{0.9}/Ru_{0.1}$ indicating that the complete oxidation to CO_2 is favoured at this alloy composition (second pathway). Results of oxidation experiments of methanol oxidation on Pt and on $Pt_{0.9}/Ru_{0.1}$ in a methanol free solution are shown in fig. 24. In these experiments the oxidation potential was also found to be shifted by 0.15 V to negative potentials on the alloy compared to pure Pt. This is the same shift as could be observed for the oxidation of dissolved CH_3OH. This observation gives evidence that Ru promotes the oxidation of the poisoning intermediates during the oxidation of methanol. With the bifunctional mechanism described for the CO-oxidation the promoting effect of Ru can be explained:

$$Ru + H_2O \rightarrow Ru\text{-}OH + H^+ + e^-$$

$$Ru\text{-}OH + Pt\text{-}CO \rightarrow Ru + Pt + CO_2 + H^+ + e^-$$

or

$$Ru\text{-}OH + Pt\text{-}COH \rightarrow Ru + Pt + CO_2 + H^+ + e^-$$

The results show that Pt/Ru catalysts can improve the cell performance compared to pure Pt catalysts. Best results could be achieved for $Pt_{0.9}/Ru_{0.1}$ catalysts working at room temperature.

4 Conclusions

Kinetics of charge transfer reactions were investigated for both electrodes of a Solid Oxide Fuel Cell. O_2 reduction kinetics are influenced most favorably by adding Pd as a catalyst to the LSM cathode material. Due to the lower activation energy, current density is increased tenfold at a typical overpotential of 0.1 V at 1073 K. Pt catalysts show less marked improvement over pure LSM, while the addition of Ir or Ru lowers the current density of oxygen reduction.

The hydrogen oxidation reaction at the Ni/YSZ anode shows a linear decrease of the activation energy with overpotential in the low-temperature region, while the pre-exponential factor remains constant. In the high-temperature region, however, the activation energy and the pre-exponential factor decreases for overpotentials up to 140 mV and remains constant afterwards. From the low-temperature results, the charge transfer coefficient α is determined to consist only of the enthalpic contribution. A charge transfer step is rate-limiting in this regime (T<850 °C), whereas this conclusion is not valid for the high-temperature region(T>850 °C), which necessitates further investigations.

Polyhydrates were investigated as a model system for solid electrolytes in the temperature range 100 - 300 K. The activation energy of the conductivity and the static dielectric constant show a strong dependence on temperature. Hydrogen evolution in the systems $Ag/HClO_4{\cdot}5.5\ H_2O$ and $Ag/DClO_4{\cdot}5.5\ D_2O$ occurs in a two-step reaction. The second step, a recombination reaction associated with charge transfer, seems to be rate-limiting with a potential dependent pre-exponential factor.

For low-temperature Proton Exchange Membrane Fuel Cells, the influence of CO on power density was investigated using different catalyst materials. The observations support the bifunctional mechanism for the electro-oxidation of CO. The CO oxidation is initiated around the Ru islands. These islands act as "sinks" for the CO which is adsorbed on Pt and has a high mobility on the Pt surface. The experiments show that $Pt_{05}/Ru_{0.5}$ exhibits the best CO tolerance due to the lowest oxidation potential.

For Direct Methanol Fuel Cells, a $Pt_{0.9}/Ru_{0.1}$ catalyst is preferable at room temperature. Here Ru promotes the oxidation of poisoning reaction intermediates.

References

1. P. Biedermann, B. Höhlein, B. Sackmann and U. Stimming, *Electrochem. Eng. and Energy*, ed. F. Lapicque et al. (Plenum Press, New York, 1995)
2. H.P. Buchkremer, U. Diekmann, L.G.J. de Haart, H. Kabs, U. Stimming and D. Stöver, *1996 Fuel Cell Seminar Abstracts,* 175, Courtesy Associates, Inc., Washington, D.C. (1996)
3. J. Divisek, L.G.J. de Haart, P. Holtappels, T. Lennartz, W. Mallener, U. Stimming and K. Wippermann, *J. Power Sources* **49**, 257 (1994)
4. J.W. Erning, W. Schaffrath, U. Stimming, E. Syskakis and K. Wippermann, *Proc. of the fourth int. Symp. on Solid Oxide Fuel Cells, ed M. Dokiya, O. Yamamoto, H. Tagawa and S.C. Singhal, The Electrochem. Soc., Inc., Pennington, NJ, Vol. 95-1(1995)*
5. J. Divisek, L.G.J. de Haart, P. Holtappels, U. Stimming and I.C. Vinke, *Proc. 17th int. symp. on mat. sci*., ed. F.W. Poulsen et al. , Roskilde, Denmark 1996
6. J.N. Agar, *Discuss. Faraday Soc.* **41**, 64 (1945)
7. S. Krause, *Jül-3128*, Forschungszentrum Jülich 1995
8. M. Cappadonia, A.A. Kornyshev, S. Krause, A.M. Kuznetsov and U. Stimming, *J. Chem. Phys.* **101**, 7672 (1994)
9. M. Cappadonia, S. Krause and U. Stimming, *Electrochim. Acta* **42**, 841 (1997)
10. R. Iannielle, V.M. Schmidt, U. Stimming, J. Stumper and A. Wallau, *Electrocim. Acta* **39**, 1863 (1994)
11. V.M. Schmidt, R. Ianniello, H.-F. Oetjen, H. Reger, U. Stimming and F. Trila, in *Proton Conducting Membrane Fuel Cells I*, ed. S. Gottesefeld et al. PV 95-23, The Electrochemical Society Proceedings Series, Pennington, NJ (1995)
12. T. Iwasita, F. C. Nart, B. Lopez and W. Vielstich, *Electrochim. Acta* **37**, 2361 (1992)
13. V.M. Schmidt, R. Ianniello, H.-F. Oetjen, H. Reger, U. Stimming, in *Proton Conducting Membrane Fuel Cells I*, ed. S. Gottesefeld et al. PV 95-23, The Electrochemical Society Proceedings Series, Pennington, NJ (1995)
14. H.-F. Oetjen, V.M. Schmidt, U. Stimming and F. Trila, *J. Electrochem. Soc.*, **143**, 3838 (1996)
15. M. Iwase and S. Kawatsu, in *Proton Conducting Membrane Fuel Cells I*, ed. S. Gottesfeld et al. PV 95-23, The Electrochemical Society Proceedings Series, Pennington, NJ (1995)
16. D. Olligs, U. Stimming and J. Stumper, *Proc. of the first int. Symp. on new materials for fuel cell systems,* ed. O. Savadogo et al., Montreal, Canada (1995)
17. H.A. Gasteiger, N. Markovic, P.N. Ross and E.J. Cairns, *J. Phys. Chem.*, **97**, 12020 (1993)
18. H.A. Gasteiger, N. Markovic, P.N. Ross and E.J. Cairns, *J. Phys. Chem.*, **98**, 617 (1994)
19. M. Watanabe and S. Motoo, *J. Electroanal. Chem.*, **60**, 267 (1975)

20. K.A. Friedrich, K.-P. Geyzers, U. Linke, U. Stimming and J. Stumper, *J. Electroanal. Chem.*, **402**, 123 (1976)
21. K.A. Friedrich, K.-P. Geyzers, F. Henglein, A. Marmann, U. Stimming, W. Unkauf and R. Vogel, *6th int. Symp. on elctrode processes*, ed. Itaya et al., Electrochemical Society, in press
22. M. Cappadonia, J. Schmidberger, W. Schwegle and U. Stimming, *Proc. of the Sixth Int. Symp. on Electrode Processes,* ed. K. Itaya and A. Wieckowski, (1996), submitted
23. J. Villegas and M.J. Weaver, *J. Chem. Phys.*, **101**, 1648 (1994)

THE MICROSCOPIC STRUCTURE OF WATER UP TO SUPERCRITICAL CONDITIONS: COMPARISON OF NEUTRON DIFFRACTION EXPERIMENTS AND SIMULATIONS

MARIA ANTONIETTA RICCI
Dipartimento di Fisica "E. Amaldi", Universita' degli Studi di Roma Tre,
Istituto Nazionale di Fisica della Materia, Unita' di Roma Tre.
Via della Vasca Navale 84, 00146 Roma, Italy.

The method for the experimental determination of the microscopic structure of a hydrogenated molecular fluid, such as water, by neutron diffraction with isotopic substitution is outlined. The experimental bias and their corrections are discussed, as well as the degree of reproducibility of the experiments and present performances of diffractometers. The site-site pair correlation functions of water at ten state points, including a supercritical one, are presented and discussed in comparison with Molecular Dynamics simulations performed using either pairwise additive or polarizable water models. The experimental data show clear density dependence, not reproduced by the simulations.

1 Introduction

Water is the most familiar and ubiquitous solvent in nature for both polar and non polar solutes and its properties as a solvent are of interest to many branches of science and technology. Biologists are mainly interested to ambient water properties, while for instance geologists are often involved in studying supercritical water solutions. The latter is a field of increasing technological interest, since supercritical water is an ideal medium for supercritical fluid extraction or waste treatment, as its solvent power and dielectric permittivity continuously change when the density changes from high liquid-like values down to low gas-like ones.

The chemico-physical description of the properties of water and aqueous solutions in the overall thermodynamic range from ambient up to supercritical states ultimately must be unique: this means that a unique model of the intermolecular forces between neighbouring molecules must be found. This attempt has been so far biased by the exaggerated interest devoted by physicists to ambient water compared to higher temperature/higher pressure states and to the consequent luck of experimental information. As a matter of fact experiments at these states deserve to solve a number of technical difficulties and the extrapolation from known ambient properties as well as the calculation of the quantities of interest via a potential model is sometimes preferred.

It is largely assumed that the key to understanding water and aqueous solutions lies in the concept of the "hydrogen bond": a strongly directional bond with intermediate strength between that of van der Waals interactions and that of covalent chemical bonds, according to the chemical literature. The steric structure of a water molecule is indeed particularly suited to engage in hydrogen bonding, with tetrahedral symmetry as the crystal structure of ice I_h vividly demonstrates. Two main questions concerning the hydrogen bonding in water are still matter of discussion. The first concerns the temperature and density evolution of the hydrogen bonding far from ambient conditions; the second concerns the availability of analytical models for the intermolecular interactions suitable to simulate water.

The structural implications of the hydrogen bonding in water can only be studied by X-ray or neutron diffraction experiments, the latter probe being more efficient. Neutrons are able indeed to interact with both oxygen and deuteron atoms, being thus sensible to the relative orientations of molecules, contrarily to X-rays that essentially probe the oxygen atom positions. Moreover the large difference in the scattering length for the interaction of a neutron with a hydrogen or a deuteron nucleus allows the experimental determination of the three site-site pair correlation functions of water. The knowledge of these three functions gives a description as complete as experimentally possible of the degree of hydrogen bonding in water, in terms of preferred atomic distances and number of atoms within a given distance from the atom sitting on the origin of the reference frame. More precisely we can address two aspects of the hydrogen bonding. First, to what extent does the OH bond on one molecule point towards the oxygen of another: this is a characteristic feature of hydrogen bonding, which implies the presence of a distinct peak (usually referred to as the H-bond peak) in the oxygen-hydrogen pair correlation function, $g_{OH}(r)$, at $r \sim 1.85\mathring{A}$. Such a peak is clearly visible in water under ambient conditions, and the number of hydrogen bonds per molecule can be defined by integrating the function $4\pi\rho r^2 g_{OH}(r)$ over the corresponding peak (ρ indicates the atomic number density). Second, we can determine to what extent the relative orientations of neighboring molecules are correlated, that is how far the local orientational order proceeds into the liquid. While the first feature probes the directionality of the hydrogen bonds, the latter aspect of the hydrogen bonding is related to the presence and extension of the so-called hydrogen bond network and mainly influence the hydrogen-hydrogen pair correlation function, $g_{HH}(r)$, and the longer range behavior of both $g_{OH}(r)$ and the oxygen-oxygen pair correlation function, $g_{OO}(r)$. The presence of peaks at

$r \sim 2.3\AA$ in $g_{HH}(r)$ and at $r \sim 4.3\AA$ in $g_{OO}(r)$, as well as the two peaks at $r \sim 5\AA$ and $r \sim 7\AA$ in $g_{OH}(r)$ are usually regarded as signatures of the tetrahedral coordination of an extended hydrogen bond network, present in ambient water. More precise geometrical and/or energetic criteria aimed to the definition of the extent of hydrogen bonding are only accessible to a Molecular Dynamics (MD) or Monte Carlo (MC) simulation, possibly combined to the experiment. Experimental studies of the site-site radial correlation functions are now available over a wide thermodynamic range, including a point at supercritical conditions[1,2,3,4,5]. These may be used as a sensible and straightforward test-bench for the interaction potential models so far proposed.

As far as the hydrogen bonding interaction is concerned quantum mechanical studies suggest that it has a cooperative character[6], that is after forming a first hydrogen bond, the charge distribution within the participating monomers changes in such a way that both the hydrogen acceptor and the hydrogen donor molecule have now an enhanced ability of forming a second bond. It is hard to account for this aspect of the hydrogen bonding interaction by classical MD. Those using computer simulation have tried to get rid of this problem implementing "effective" pairwise additive potential models, which incorporate the increased binding effect of cooperativeness. The more popular effective potential models consist of long range Coulomb's law interaction between a few strategically placed charges on the molecules, plus a Lennard-Jones interaction between the oxygen atoms[7,8]. The hydrogen bonding is then described by the strong attractive interaction between a positive charge sitting on a hydrogen site of one molecule and a negative charge usually placed in the vicinity of the oxygen site on another. Such models describe quite satisfactorily many properties of water from the supercooled states up to the boiling point[9], i.e. in those thermodynamic states where the density is similar to that of ambient water. They predict that hydrogen bonding is retained even at temperature and pressures well above the critical point[10,11], as expected for strong electrostatic interactions, and in contrast with our experimental results [1,3,5]. On the other hand the distinctive features of hydrogen bonding in ambient water have been shown to appear also in calculations performed by using a sticky hard sphere model, which is short range, for the water molecules[12]. Unluckily calculations at lower densities and higher temperatures are not available and we cannot establish whether this model works better than the other ones.

In this paper we will review the existing Neutron Diffraction with Isotopic H/D Substitution (NDIS) data for water from ambient conditions up to the supercritical state and discuss the experimental method and the accuracy of the site-site pair correlation functions obtained. This discussion and the comparison with the existing MD simulations will suggest which are the more relevant and reliable features of the site-site pair correlation functions that MD simulations should but not yet reproduce.

2 Theoretical background and relevance of NDIS experiments

According to the text books[13], the cross-section per molecule and steradian for a neutron diffraction experiment on a molecular liquid, heavy water in the present instance, can be regarded as the sum of a coherent plus an incoherent contribution:

$$\frac{d\sigma}{d\Omega}(Q) = \left[\frac{d\sigma}{d\Omega}\right]_c + \left[\frac{d\sigma}{d\Omega}\right]_i = b_O^2 H_{OO}(Q) + 4b_O b_D H_{OD}(Q) + 4b_D^2 H_{DD}(Q) + \frac{\sigma_{sc}}{4\pi} \tag{1}$$

where Q represents the exchanged wavevector, b_O and b_D are the neutron scattering lengths for oxygen and deuteron, $\frac{\sigma_{sc}}{4\pi} = \left[\frac{d\sigma}{d\Omega}\right]_i$ represent the atomic self-scattering contribution to the total cross-section and $H_{OO}(Q)$, $H_{OD}(Q)$, $H_{DD}(Q)$ are the partial structure factors (PSF). These correspond to the Fourier transforms of the site-site pair correlation functions (SSPCF) and are defined as:

$$H_{ij}(Q) = \int_{-\infty}^{E_0/\hbar} \frac{k_1}{k_0} S_{ij}(Q,\omega) d\omega \tag{2}$$

being k_0 and k_1 the moduli of the wavevectors of the incoming and outcoming neutron, E_0 and $\hbar\omega$ the initial and exchanged energy of the neutron. The integral of the dynamic structure factor, $S_{ij}(Q,\omega)$, in eq.2 must be performed over a constant Q path, corresponding to elastic neutron scattering events. The same integral is performed on the self contribution to the dynamical structure factor to get $\frac{\sigma_{sc}}{4\pi}$.

Since the scattering length of hydrogen is $b_H = -3.74 fm$ while that of deuteron is $b_D = 6.671 fm$, it is in principle straightforward to extract the three PSF from eq.1, performing three independent experiments on three isotopic mixtures of D_2O and H_2O, and derive the SSPCF. These are shown in fig.1 in the case of D_2O at ambient conditions, with the intensity with which they contribute to the so-called neutron weighted radial correlation function $g(r) = 0.090 g_{OO}(r) + 0.421 g_{OD}(r) + 0.489 g_{DD}(r)$, also reported in the figure as a dash-dotted line. We notice all the peaks characteristics of the SSPCF of ambient water already quoted in the previous section: the comparison of the SSPCF with the neutron weighted $g(r)$ stresses the relevance of extracting the SSPCF in order to better understand the hydrogen bonding in water. The neutron weighted $g(r)$ is indeed a very smooth function, due to the position and relative intensities of the sharp structures characteristics of the SSPCF, for this reason it may well be compatible with different sets of SSPCF. Moreover the weak shoulders visible between $1.5\AA$ and $3\AA$ are due to the superposition of two or more peaks of the SSPCF, and the modifications of such shoulders with the thermodynamic parameters might not have a direct interpretation.

3 From the real experiment to the site-site pair correlation functions

When we run a real experiment, instead of $\frac{d\sigma}{d\Omega}(Q)$, we measure the intensity of diffracted neutrons as a function of the diffraction angle 2ϑ, given the neutron wavelength λ_0, or of the neutron time of flight t, given the diffraction angle, depending on which neutron source is used (steady state or pulsed, respectively), and several experimental bias have to be accounted for in order to get the cross-section per molecule and steradian. First of all we notice that equations 1 and 2 have been derived in the hypothesis that each neutron undergoes a single scattering event and that it is never absorbed by the sample. This determines limitations on the amount of sample in the neutron beam, and a compromise has to be found in order to get enough statistics on the measured intensity; analytical correction of the data is however needed[14,15].

Moreover a fluid sample needs a container: this will interact with the neutrons and its contribution to the diffracted intensity must be accounted for. In order to reduce the bias from the container these are usually made of vanadium or a Ti-Zr alloy, which have an almost flat diffraction pattern. In the case of measurements performed on water at high temperature and high pressure presented here, the corrections for the container contribution are particularly severe, since the volume of container material in the beam is ten times the volume of water.

The measured intensity must then be put on an absolute scale, accounting for the incident neutron spectrum and detector efficiency: the intensity diffracted by a standard vanadium sample is used for this purpose.

At this point of the data handling we have a measured differential cross-section, Σ, which is distorted with respect to $\frac{d\sigma}{d\Omega}(Q)$, because the integral in equation 2 cannot be executed in practise on a constant Q path, but either at constant 2ϑ or at constant t, for data collected at a reactor or at a pulsed source respectively. The distortion of the measured differential cross-section is usually referred to as "inelasticity effects": these are particularly severe for hydrogenated materials and increase with increasing the diffraction angle and decreasing the energy of the incident neutrons. As a consequence time of flight

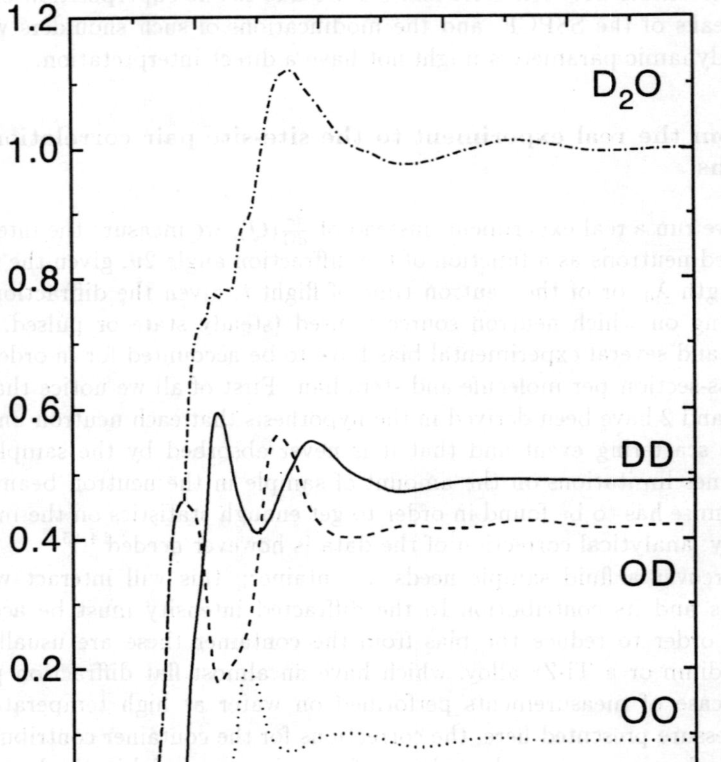

Figure 1: The site-site pair correlation functions of D_2O at ambient conditions: $g_{DD}(r)$ (continuous line), $g_{OD}(r)$ (dashed line), $g_{OO}(r)$ (dotted line), reported with the relative intensite they have in the total neutron weighted $g(r)$ (dash-dotted line).

diffractometers installed at pulsed neutron sources (as for instance that used for the experiments on water) are more suitable for structural studies of hydrogenated samples, than two axis diffractometers operating at a reactor source. Epithermal neutrons from pulsed spallation sources are indeed more energetics in the average than thermal neutrons, and the experiment can be limited to very low diffraction angles, without loosing information[16]. Yet analytical corrections need to be applied. These would require reasonably detailed knowledge of the dynamic structure factor of water, and although many approaches to this problem have been attempted[13,17,18], these are not yet accurate enough to enable a practical corrrection procedure to be implemented. Currently used procedures [2,5,19] assume that the cross-section measured at a time of flight spectrometer for a given sample α, and at an angle 2ϑ is:

$$\Sigma^{\alpha}(Q,2\vartheta) = F^{\alpha}(Q,2\vartheta) + \Sigma^{\alpha}_{self}(Q,2\vartheta) \tag{3}$$

where:

$$\Sigma^{\alpha}_{self}(Q,2\vartheta) = \int_{-\infty}^{E_0/\hbar} S_{self}(Q,\omega)W(Q,\omega)d\omega \tag{4}$$

and

$$F^{\alpha}(Q,2\vartheta) = \int_{-\infty}^{E_0/\hbar} S(Q,\omega)W(Q,\omega)d\omega \tag{5}$$

$W(Q,\omega)$ is a weighting function accounting for detector efficiency, incident and scattered flight paths and incident neutron spectrum, and the integration is performed over an integration path determined by the energy and momentum conservation laws. It is usually assumed that the distortion on the PSF is negligible and hence $F^{\alpha}(Q,2\vartheta) = F^{\alpha}(Q) = \left[\frac{d\sigma}{d\Omega}(Q)\right]_c^{\alpha}$. The atomic self scattering contribution is on the contrary affected by the inelastic terms and approaches the theoretical value of $\frac{\sigma_{sc}}{4\pi}$ only at high Q values. The correction for the inelasticity contributions is then accomplished along with the subtraction of the atomic self scattering, which does not contain information about the microscopic structure of the fluid. A polynomial fitting procedure of the profile of the $\Sigma^{\alpha}_{self}(Q,2\vartheta)$ term is usually adopted. The reproducibility of the $F^{\alpha}(Q)$ functions obtained at different diffraction angles after the subtraction of the $\Sigma^{\alpha}_{self}(Q,2\vartheta)$ terms is a test of the reliability of the above procedure.

At this point if we have done the experiments on three samples (H_2O, D_2O, and for instance an isomolar mixture of the two), we can combine the $F^{\alpha}(Q)$ functions to extract the three PSF and then Fourier transform these to r-space. This procedure was used until recently, but then it has been shown[20,21] that the high pressure data were affected by a residual systematic error, preventing the reconstruction by a Reverse Monte Carlo simulation of physical configurations for a set of water molecules compatible with the three SSPCF, and an alternative method for the extraction of the SSPCF from the $F^{\alpha}(Q)$ functions has been developed[5]. According to this method each $F^{\alpha}(Q)$ functions is written as the usual weighted sum of PSF, plus a background function, $T(Q)$, which will in general be different for different samples. Thus

$$F^{H_2O}(Q) = b_O^2 H_{OO}(Q) + 4b_O b_H H_{OH}(Q) + 4b_H^2 H_{HH}(Q) + T^{H_2O}(Q) \quad (6)$$

$$F^{D_2O}(Q) = b_O^2 H_{OO}(Q) + 4b_O b_D H_{OH}(Q) + 4b_D^2 H_{HH}(Q) + T^{D_2O}(Q) \quad (7)$$

$$F^{m}(Q) = b_O^2 H_{OO}(Q) + 4b_O b_m H_{OH}(Q) + 4b_m^2 H_{HH}(Q) + T^{m}(Q) \quad (8)$$

where $b_m = xb_H + (1-x)b_D$ (i.e. $0.5b_H + 0.5b_D$ in the case of an isomolar mixture). Each background function is then represented as a weighted sum of partial background functions, one for each PSF:

$$T^{H_2O} = b_O^2 T_{OO}(Q) + 4b_O b_H T_{OH}(Q) + 4b_H^2 T_{HH}(Q) \quad (9)$$

where

$$T_{\alpha\beta}(Q) = 4\pi\rho \int_0^\infty r^2 t_{\alpha\beta}(r) \frac{\sin Qr}{Qr} dr \quad (10)$$

and $t_{\alpha\beta}(r)$ is a background term associated with the $\alpha\beta$ site-site pair correlation function. Note however that unlike $g_{\alpha\beta}(r)$, $t_{\alpha\beta}(r)$ is allowed to take on unrestricted values, positive or negative, if necessary.

In order to define the $t_{\alpha\beta}(r)$ it is necessary to place restrictions on the $g_{\alpha\beta}(r)$. In particular it is required that

$$g_{\alpha\beta}(r) \begin{cases} \geq 0, & \text{for } r \geq r_{\alpha\beta} - \delta \,; \\ =0, & \text{for } r < r_{\alpha\beta} - \delta. \end{cases} \quad (11)$$

and

$$4\pi\rho \int_0^\infty r^2 dr(g_{\alpha\beta}(r) - 1) = \rho\chi_T k_B T - N_a \tag{12}$$

where χ_T is the isothermal compressibility and N_a the number of atoms per molecule. For the background functions the constraints are roughly the inverse of equation 11:

$$t_{\alpha\beta}(r) \begin{cases} anyvalue, & \text{for } r \leq r_{\alpha\beta} + \delta \text{ ;} \\ \begin{cases} \leq 0, & \text{if } g_{\alpha\beta} = 0, \text{ and } r > r_{\alpha\beta} + \delta; \\ = 0, & \text{if } g_{\alpha\beta} > 0, \text{ and } r > r_{\alpha\beta} + \delta. \end{cases} \end{cases} \tag{13}$$

and with no constraint on the integral of these functions. In equations 11 and 13 $r_{\alpha\beta}$ is a minimum radius to be specified for each site-site correlation function, and δ is a small region where each background function and its corresponding site-site correlation function are allowed to overlap, in order to prevent discontinuities occurring in either.

The $g_{\alpha\beta}(r)$ and $t_{\alpha\beta}(r)$ are refined against the supplied data for the $F^\alpha(Q)$ functions, plus the constraints 11, 12 and 13, using the Minimum Noise principle [22] to extract the smoothest possible set of functions. Any tendency for the data to require any of the correlation functions to violate the conditions 11 and 12 is treated as a background function, which is estimated alongside the true site-site correlation function. A satisfactory fit is deemed to have been achieved when the difference between supplied data for $F^\alpha(Q)$ and fit shows no systematic deviation outside the statistical error.

4 Experimental detailes

Table 1 lists the ten state points for which neutron diffraction data are available using the H/D isotopic substitution. All the experiments have been performed using the SANDALS time of flight diffractometer[16] installed at the spallation neutron source ISIS in the UK. In the present configuration, SANDALS has 14 detector banks at diffraction angles 2ϑ ranging from 3.8° to 31.2°. It will be noted that two of the datasets, namely run numbers 5429 and 8255, have been measured under identical sample conditions, that is close to the coexistence density at $573K$. These two experiments were however run under different experimental conditions. Specifically run 8255 was run with an ambient $(300K)$ water neutron moderator, compared to run 5429 which was achieved with a liquid methane $(109K)$ neutron moderator. The two moderators produce markedly different incident neutron spectra due to the different neutron

Table 1: Neutron diffraction experiments on the Small Angle Neutron Diffractometer for Amorphous and Liquid Samples (SANDALS), at the ISIS pulsed neutron source, for water with hydrogen/deuterium isotope substitution in the period 1992-1995. The experiments are listed as a function of increasing temperature and decreasing density. The run number (left hand column) refers to the measurement of the light water sample, and indicates the rough chronology of the measurements. For each light water run there is a corresponding run for heavy water, and a run for a mixture of heavy and light water. For completeness the critical conditions for light water are shown on the bottom line.

Run No.(H_2O)	$T(K)$	$P(bar)$	$\rho(molecules/\mathring{A}^3)$
4705	298	1	0.0334
6137	423	1900	0.0334
6132	423	100	0.0308
6142	573	2800	0.0308
8258	573	1970	0.0296
8261	573	1100	0.0278
8265	573	500	0.0260
8255	573	100	0.0240
5429	573	95	0.0240
5435	673	800	0.0221
Critical point	646	221	0.0107

thermalisation temperatures, with a corresponding substantially different inelasticity correction at low Q. If the correction procedure described in the previous section is working correctly it should nonetheless produce equivalent site-site correlations for either dataset. Thus the comparison of runs 5429 and 8255 is a direct indication of the success with which this method is working and of the accuracy of such experiments. Figure 2 compares the site-site pair correlation functions for the two runs. It can be seen that in all three cases there are differences in detail, with perhaps the HH and OO functions showing the greater discrepancies. The estimated uncertainties are clearly not as small as would be ideally necessary for a detailed comparison with the computer simulation results, and may be of the order of 14% below 3Å. However qualitative trends are certainly discernible in these data. In particular the positions of the peaks are the same for both datasets and are clearly shifted with respect to their value at ambient conditions. This suggests that MD calculations must primarily reproduce the peak positions, their intensities being less well determined.

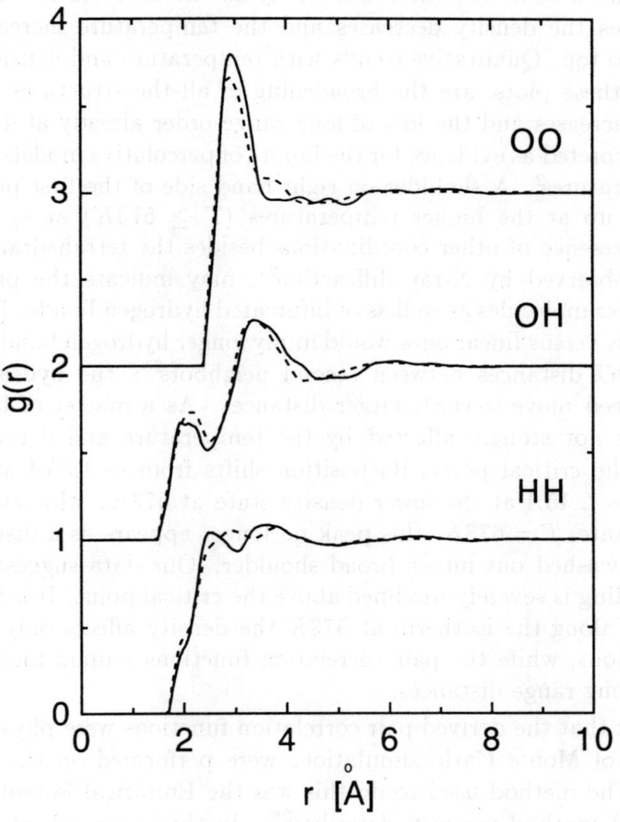

Figure 2: The site-site pair correlation functions obtained for run 5429 (continuous line) and run 8255 (dashed line).

5 Evolution of the site-site pair correlation functions with temperature and density

Figures 3, 4 and 5 show the three SSPCF of all the datasets listed in Table 1: in all figures the density decreases and the temperature increases going from bottom to top. Qualitative trends with temperature and density, clearly discernible in these plots, are the broadening of all the structures when the temperature increases and the loss of long range order already at $423K$: this result was interpreted as evidence for the failure of percolative models of water[23] at such temperatures[2]. A shoulder on right hand side of the first peak of the $g_{OO}(r)$ shows up at the higher temperatures ($T \geq 573K$) at $\sim 3.5\AA$ and suggests the presence of other coordinations besides the tetrahedral one: this feature, also observed by X-ray diffraction[24], may indicate the presence of interstitial water molecules as well as of bifurcated hydrogen bonds. Bifurcated hydrogen bonds versus linear ones would imply longer hydrogen bond distances and shorter OO distances between second neighbors[25]: the hydrogen bond peak does indeed move towards larger distances. As a matter of fact, while its intensity is not stongly affected by the temperature and density in the states below the critical point, its position shifts from $\sim 1.85\AA$ at ambient conditions to $\sim 2.15\AA$ at the lower density state at $573K$. Moreover, in the supercritical state, $T = 673K$, this peak no longer appears as a distinct peak and has been washed out into a broad shoulder. Our data suggest that the hydrogen bonding is severely modified above the critical point. It is important to notice that along the isotherm at $573K$ the density affects only the short range correlations, while the pair correlation functions remain unchanged at medium and long range distances.

As a check that the derived pair correlation functions were physically sensible, a series of Monte Carlo simulations were performed on the results of Figures 2-4. The method used to do this was the Empirical Potential Monte Carlo (EPMC) method recently described[26]. In this approach an empirical site-site potential function is derived which when applied in a MC simulation reproduces the estimated SSPCF. Unlike the previous results[1,2,3] the results of the new analysis can be adequately fit by physical distributions of water molecules, which satisfies some of the earlier criticisms[20,21]. A further advantage of this method is that once these distributions have been produced, may be used to calculate quantities otherwise not directly accessible to the experiment.

Several analyses[11,24,27] suggest for instance that the height of the OH peak is not a reliable measure of the degree of hydrogen bonding in water, and geometric or energetic or combined geometric and energetic definitions should

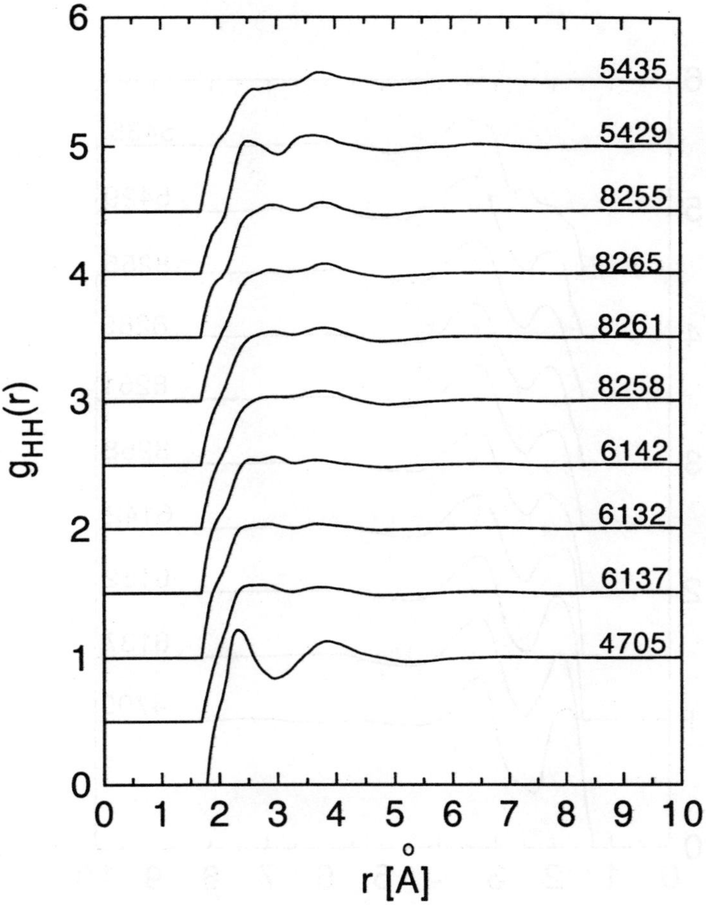

Figure 3: The HH radial correlation functions of all the dataset listed in Table I. The density decreases and the temperature increases going from bottom to top: the curve at the bottom refers to water at ambient conditions, that on the top to water at supercritical conditions. The different curves have been shifted by an arbitrary quantity.

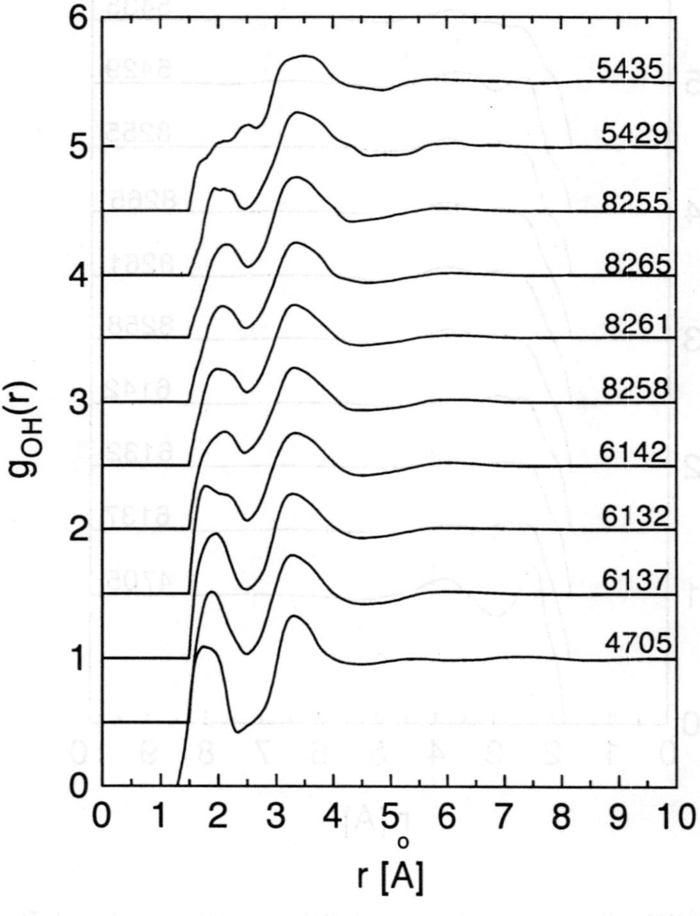

Figure 4: The same as Fig.3 for the OH radial correlation function.

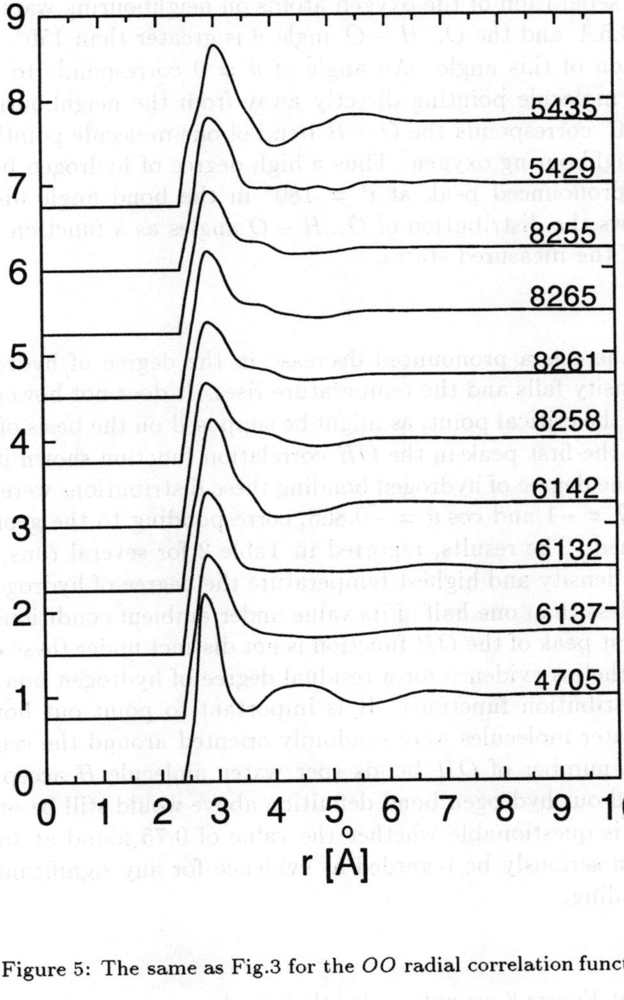

Figure 5: The same as Fig.3 for the *OO* radial correlation function.

be used. Obviously such definitions cannot be applied directly to the measured site-site correlation functions, but if the data are used to derive 3-dimensional distributions of water molecules in the manner described above then a sensible geometric definition can be applied to establish the degree of hydrogen bonding in the real liquid. In the present instance it is assumed that a hydrogen bond occurs if the separation of the oxygen atoms on neighbouring water molecules is less than 3.5Å, and the $O...H-O$ angle ϑ is greater than 150°. (See Ref.[26] for a definition of this angle. An angle of $\vartheta = 0$ corresponds to the $O-H$ bond of one molecule pointing directly away from the neighbouring oxygen, while $\vartheta = 180°$ corresponds the $O-H$ bond of one molecule pointing directly towards a neighbouring oxygen. Thus a high degree of hydrogen bonding will show up as pronounced peak at $\vartheta = 180°$ in the bond angle distribution.) Figure 6 shows the distribution of $O...H-O$ angles as a function of $\cos\vartheta$ for a selection of the measured states.

Clearly there is a pronounced decrease in the degree of hydrogen bonding as the density falls and the temperature rises. It does not however entirely vanish above the critical point, as might be supposed on the basis of the height and shape of the first peak in the OH correlation function shown in Figure 4. To estimate the degree of hydrogen bonding these distributions were integrated between $\cos\vartheta = -1$ and $\cos\vartheta = -0.866$, corresponding to the geometric definition assumed. The results, reported in Table 2 for several runs, show that at the lowest density and highest temperature the degree of hydrogen bonding has fallen to less than one half of its value under ambient conditions. Thus although the first peak of the OH function is not distinct under these conditions, there is nonetheless evidence for a residual degree of hydrogen bonding in the measured distribution functions. It is important to point out however that even if the water molecules were randomly oriented around the central water molecule, the number of OH bonds (per water molecule H-acceptor) which would satisfied our hydrogen bond definition above would still be on the order of 0.62, so it is questionable whether the value of 0.75 found at supercritical conditions can seriously be regarded as evidence for any significant degree of hydrogen bonding.

Looking at Figure 6 we notice also that the hump centered around $\cos\vartheta = 0.2$ at ambient conditions is smeared out at the higher temperatures: this is a further evidence of the loss of long range coordination as the temperature increases[2].

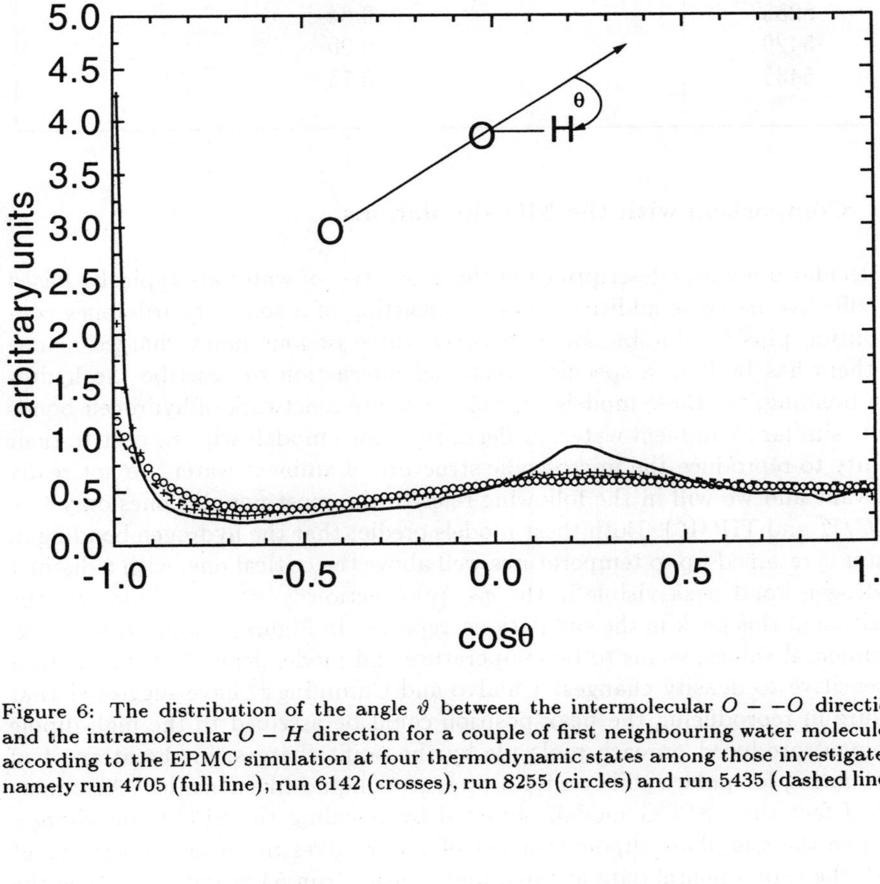

Figure 6: The distribution of the angle ϑ between the intermolecular $O - -O$ direction and the intramolecular $O - H$ direction for a couple of first neighbouring water molecules, according to the EPMC simulation at four thermodynamic states among those investigated, namely run 4705 (full line), run 6142 (crosses), run 8255 (circles) and run 5435 (dashed line).

Table 2: Number of hydrogen bonds per water molecule on the hydrogen accepting site of a water molecule.

Run No.(H_2O)	No.of hydrogen bonds (per water molecule H-acceptor)
4705	1.79
6142	1.53
8255	0.84
5429	0.90
5435	0.75

6 Comparison with the MD simulations

Molecular dynamics descriptions of the properties of water are typically based on effective pairwise additive models, consisting of a soft Lennard-Jones contribution plus Coulumbic forces between three or four point charges. None of them has built in a specific directional interaction to describe the hydrogen bonding, yet these models are able to create a network of hydrogen bonds quite similar to ambient water. Differences among models with respect to their ability to reproduce the microscopic structure of ambient water are not really relevant and we will in the following refer to the most popular ones only (i.e. SPC/E[7] and TIP4P[8]). Both these models predict that the hydrogen bonding in water is retained up to temperatures well above the critical one, with a distinct hydrogen bond peak visible in the $g_{OH}(r)$ function[10,11,27,28,29]. Moreover the position of this peak in the simulations, reported in Figure 7 along with the experimental values, seems to be temperature and model dependent, but almost insensitive to density changes. Chialvo and Cummings[27] have suggested that failure in reproducing the peak position could be ascribed to the high dipole moment produced on each molecule by the point charges, in the attempt of accounting for polarizability effects present at liquid-like densities. As a matter of fact their SPCG model, obtained by rescaling the SPC point charges to give the gas phase dipole moment of water, gives much better agreement with the experimental data at the lowest density (run 5429 and 5435) then the original model, as far as both peak position and intensity is concerned. If this is the case polarizable water models should be able to reproduce the density dependence of the position of the hydrogen bond peak, as also auspicated by ref.[30]. The polarizable water model proposed by Brodholt et al.[30] has been then used to simulate water along the isotherm at $573K$ at exactly the same state points as the experiments[31]. At the lowest density we get a real improvement in reproducing the position of the hydrogen bond peak (see Fig.8), but at the same time when the density increases we have the undesired result of modifying the $g_{OH}(r)$ function at large distances, leaving the hydrogen bond peak fixed at $r \sim 1.94 \AA$ (see Figure 9).

7 Conclusions

We have shown that it is possible to extract from NDIS experiments on water a set of physically sensible site-site pair correlation functions with a reasonable degree of confidence on the qualitative features. The quantitative accuracy, particularly as regards the heights of peaks is less certain, clear trends with temperature and density can however be identified.

First we notice the loss of long range tetrahedral coordination already at $423K$, and a broadening of all structures as the temperature increases. We also notice that the $g_{OO}(r)$ at temperatures $\geq 573K$ shows a feature centered at $\sim 3.4 \AA$ ascribable to the presence of interstitial molecules or distorted or less energetic hydrogen bonds. The number of such bonds, as evaluated by integration of the bond angle distribution obtained by the EPMC simulation, does indeed decrease as the temperature increases and the density decreases, giving at the supercritical state an estimate of 0.75 hydrogen bonds per molecule, i.e. a value not seriously higher that that expected for a randomly oriented ensamble of molecules. This result confirms our previous claim of a substantial reduction of the hydrogen bonding in water above the critical point. Last, but not least, we have evidentiated a strong density dependence of the short range site-site pair correlation functions and mainly of the position of the hydrogen bond peak. This result implies that the local minimum of the interatomic potential is changing markedly with density, an effect that can only occur if many-body interactions are an important component of the interatomic potential. In fact even in a pairwise additive system small shifts in the near neighbour peak of a liquid can occur at high densities, depending on the relative strengths of the attractive forces binding the molecules together and the repulsive forces which prevent molecular overlap. The point to be emphasized however is that this shift is not observed in the traditional pairwise additive water potentials. The fact that these potentials do not reproduce the observed peak movements suggests that they overemphasize both the strength of the attractive hydrogen bond interaction and the corresponding repulsive interaction preventing molecular overlap: changes in density at constant temperature do not then produce an observable peak shift with density change. Therefore our data indicate that for accurate predictions of water and aqueous solution

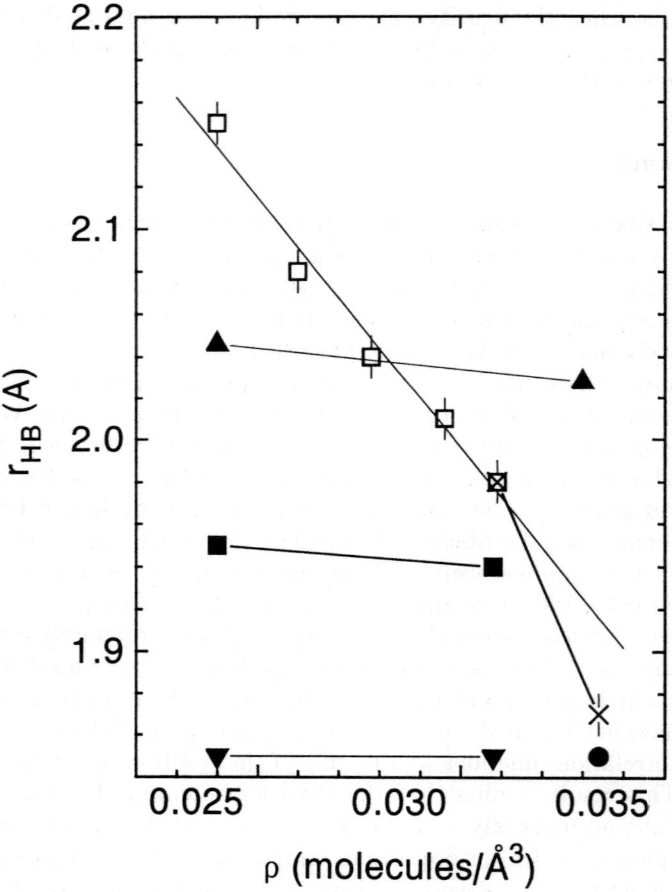

Figure 7: Position of the centroid of the H-bond peak as a function of density. The circle indicates the peak position at ambient conditions; the crosses and the open squares the experimental data at $T = 423K$ and at $T = 573K$ respectively. The up-triangles are the result of a simulation of TIP4P water at $T = 773K$ [11], while down-triangles and full squares are the results of simulations performed at the same states as the experiment [31], using the SPC/E model or a polarizable water model [30]. Solid lines are guides for the eye.

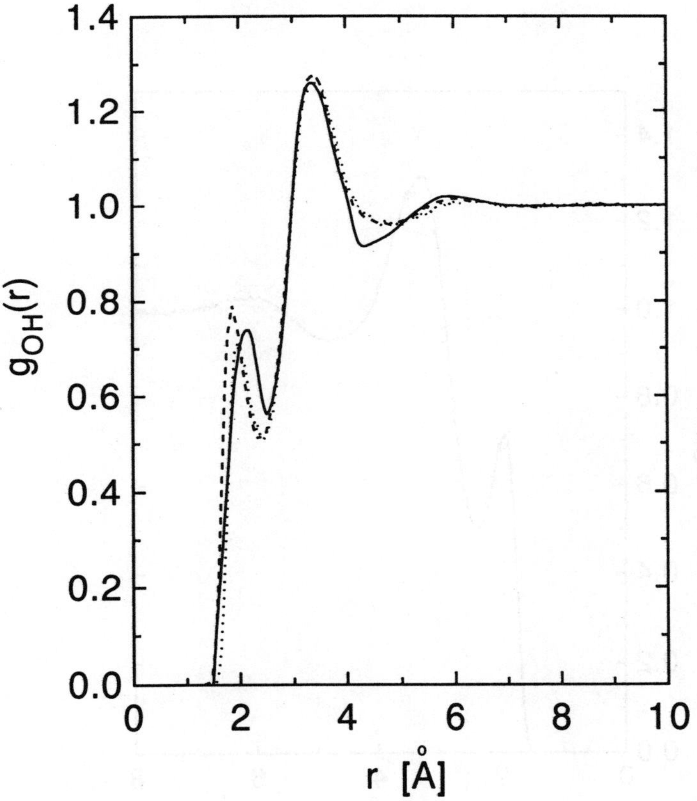

Figure 8: The $g_{OH}(r)$ function at $T = 573K$ and $\rho = 0.0240 molecules/\AA^3$: the full line represents the experimental results (run 8255), the dashed line the MD calculations performed usyng the SPC/E model and the dotted line those for the polarizable water model introduced in ref.[30]

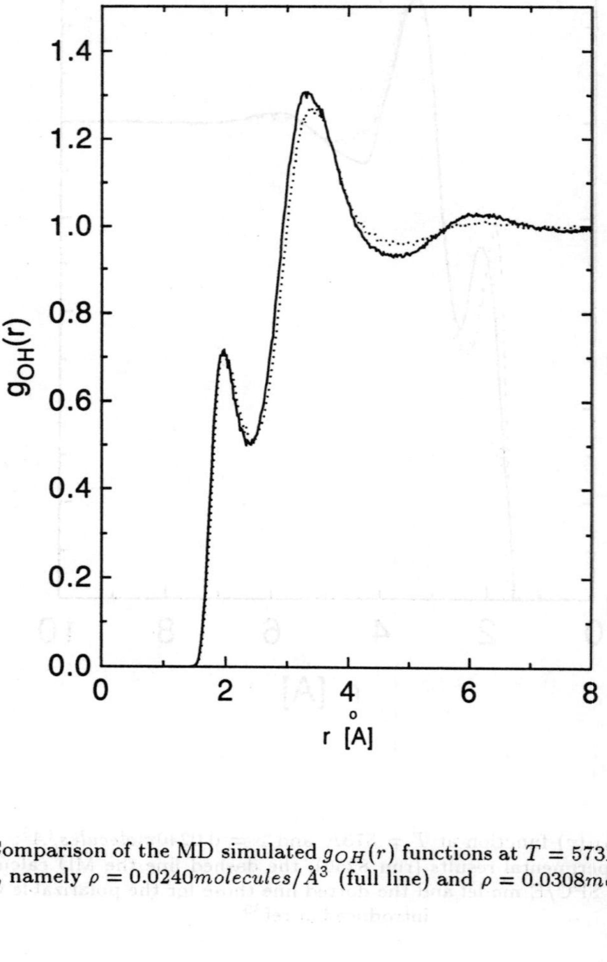

Figure 9: Comparison of the MD simulated $g_{OH}(r)$ functions at $T = 573K$ and two density states, namely $\rho = 0.0240 molecules/\AA^3$ (full line) and $\rho = 0.0308 molecules/\AA^3$.

properties over a wide range of temperatures, densities and pressures, such as are often required in a variety of applications in the chemical and geological sciences, existing potentials for water will need to be modified.

Acknowledgments

For the work presented in this article, the author acknowledges the collaboration of numerous colleagues, especially A.K.Soper, F.Bruni and P.Postorino.

References

1. P.Postorino, R.H.Tromp, M.A.Ricci, A.K.Soper, and G.W.Neilson, *Nature*, 366, 668 (1993).
2. P.Postorino, M.A.Ricci, and A.K.Soper, *J.Chem.Phys.*,101, 4123, (1994).
3. R.H.Tromp, P.Postorino, G.W.Neilson, M.A.Ricci, and A.K.Soper, *J.Chem.Phys.*, 101, 6210, (1994).
4. F.Bruni, M.A.Ricci, and A.K.Soper, *Phys.Rev.B*, 54, 11584 (1996)
5. A.K.Soper, F.Bruni, and M.A.Ricci, *J.Chem.Phys.*, 106, 247 (1997)
6. The calculations are reviewed in: F.H.Stillinger, *Adv.Chem.Phys.*, 31, 1 (1975).
7. H.J.C.Berendsen, J.R.Grigera, and T.P.Straatsma, *J.Phys.Chem.*, 91, 6269 (1987).
8. W.L. Jorgensen, J.Chandrasekhar, J.D.Madura, R.W.Impey, and M.L.Klein, *J.Chem.Phys.* 79, 926 (1983) for the TIP4P model.
9. Foor a brief review, see: R.Frattini, M.A.Ricci, G.Ruocco, and M.Sampoli, *J.Chem.Phys.* 92, 2540 (1990).
10. J.Brodholt and B.Wood, *J.Geophys.Res.*, 98, 519 (1993).
11. A.G.Kalinichev and J.D.Bass, *Chem.Phys.Lett.*, 231, 301 (1994)
12. D.Bratko, L.Blum, and A.Luzar, *J.Chem.Phys.* 83, 6367 (1985); L.Blum, F.Vericat, and D.Bratko, *J.Chem.Phys.* 102, 1461 (1995).
13. P.A.Egelstaff, in *Methods of Experimental Physics. Neutron Scattering*, ed. D.L.Price and K.Sköld, 23 Part B, 405 (Academic Press, London 1987).
14. H.H.Paalman, C.J.Pings, *J.Appl.Phys.* 33, 2635 (1962); A.K.Soper, P.A.Egelstaff, *Nucl.Instrum.Methods* 178, 415 (1980).
15. V.F.Sears,*Adv.Phys.*
24, 1 (1975); J.R.D.Copley, *Comput.Phys.Commun.* 7, 289 (1974); 9, 59 (1975); 20, 459 (1980).
16. A.K.Soper, *Advanced Neutron Sources 1988*, IOP Conferences Series, ed. D.K.Hyer (IOP, Bristol, 1989) vol.97.

17. P.Postorino, M.Nardone, M.A.Ricci, and M.Rovere, *J. of Molecular Liquids* 64, 221 (1995).
18. P.Zetterström, A.K.Soper, and P.P.Schofield, *Molec.Phys.* (in press).
19. A.K.Soper and A.Luzar, *J.Chem.Phys.*, 97, 1320, (1992)
20. G.Löffler, H.Schreiber, and O.Steinhauser, *Ber.Bunsenges.Phys.Chem.*, 98, 1575, (1994)
21. P.Jedlovszky and R.Vallauri, *J.Chem.Phys.* (in press).
22. A.K.Soper, M.Nardone, and C.Andreani, *Phys.Rev.E.* ,47, 2598 (1993).
23. H.E.Stanley, and J.Teixeira, *J.Chem.Phys.* 73, 3404 (1980).
24. Yu.E.Gorbaty, and A.G.Kalinichev, *J.Phys.Chem.* 99, 5336 (1995).
25. P.A.Giguere, *J.Chem.Phys.* 87, 4835 (1987).
26. A.K.Soper, *Chem.Phys.* 202, 295 (1996).
27. A.A.Chialvo, and P.T.Cummings, *J.Chem.Phys.* 101, 4426 (1994).
28. R.D.Mountain, *J.Chem.Phys.* 90, 1866 (1989).
29. P.T.Cummings, H.D.Cochran, J.M.Simonson, R.E.Mesmer, and S.Karaborni, *J.Chem.Phys.* 94, 5606 (1991).
30. J.Brodholt, M.Sampoli, and R.Vallauri, *Molec.Phys.* 86, 149 (1995).
31. J.Brodholt, R.Vallauri, F.Bruni, M.A.Ricci, and A.K.Soper, to be published.

Static and Dynamic Nonlocal Dielectric Response of Water

A.A.Kornyshev and G.Sutmann
Institut für Energieverfahrenstechnik (IEV)
Forschungszentrum Jülich GmbH (KFA)
D-52425 Jülich, Germany

Information about the static and dynamic linear nonlocal dielectric response of liquid water, based on computer simulations, diffraction experiments and phenomenological models, is summarized together with the implications of this response in the nonlocal electrostatics of solutions and the theory of charge transfer processes. Aspects of nonlinear response are discussed.

1 Introduction. Basic definitions

Nonlocal electrostatics, based on the constitutive relation between the electrostatic induction, $\boldsymbol{D}$, and the medium polarization, $\boldsymbol{P}$,

$$P_\alpha(\boldsymbol{r},t) = \int d\boldsymbol{r}'dt'\chi_{\alpha\beta}(\boldsymbol{r},\boldsymbol{r}',t-t')D_\beta(\boldsymbol{r}',t') \tag{1}$$

is an efficient tool to take into account the structure of the medium in electrostatic calculations [1,2]. Indeed, the nonlocal kernel in the integral means that polarization at a point $\boldsymbol{r}$ is determined not only by the value of the external field at the same point, but also by values at other spatial coordinates, $\boldsymbol{r}'$, in the vicinity of $\boldsymbol{r}$. The range of nonlocality is determined by the correlation range of the medium bound charge density fluctuations, i.e. by the structure of the medium. The nonlocality in time has a causality constraint, which is due to the retardation of the response caused by the inertia of the atomic and molecular degrees of freedom. Generally the *spatial and temporal dispersion* cannot be decoupled, and the information about the static and dynamic linear response of the medium is given by a complicated *dielectric response* tensor-function, $\chi_{\alpha\beta}(\boldsymbol{r},\boldsymbol{r}',t-t')$.

If the medium is homogeneous, as liquids in the bulk are,

$$\chi_{\alpha\beta}(\boldsymbol{r},\boldsymbol{r}',t-t') = \chi_{\alpha\beta}(\boldsymbol{r}-\boldsymbol{r}',t-t'), \tag{2}$$

and it is convenient to apply a Fourier transform in space and time

$$f(\boldsymbol{k},\omega) = \int d\boldsymbol{r}dtf(\boldsymbol{r},t)e^{i(\boldsymbol{kr}-\omega t)} \tag{3}$$

because the constitutive relation

$$P_\alpha(\boldsymbol{k},\omega) = \chi_{\alpha\beta}(\boldsymbol{k},\omega)D_\beta(\boldsymbol{k},\omega) \tag{4}$$

becomes local in Fourier space. For an isotropic medium the response tensor can be written in terms of the longitudinal and transversal components in $\boldsymbol{k}$ -space,

$$\chi_{\alpha\beta}(\boldsymbol{k}) = \frac{k_\alpha k_\beta}{\boldsymbol{k}^2}\chi_{\|}(\boldsymbol{k}) + (\delta_{\alpha\beta} - \frac{k_\alpha k_\beta}{\boldsymbol{k}^2})\chi_{\perp}(\boldsymbol{k}). \tag{5}$$

In electrostatics or in the dynamic phenomena where the effects due to retardation of light can be neglected ($\omega/k << c$), only the longitudinal component of the tensor is of practical interest since it is only $\chi_{\|}$, which determines the result for the electric potential. Indeed, by Fourier transformation of Maxwell's equation one finds: $\nabla\boldsymbol{D}(\boldsymbol{r},t) = 4\pi\rho_e(\boldsymbol{r},t) \rightsquigarrow \hat{\boldsymbol{k}}\boldsymbol{D}(\boldsymbol{k},\omega) = 4\pi\rho_e(\boldsymbol{k},\omega)/ik$ (here $\rho_e(\boldsymbol{k},\omega)$ is the space-time Fourier transform of the external charge distribution in the medium). From this relationship the electric potential follows as

$$\Phi(\boldsymbol{k},\omega) = \frac{4\pi}{k^2}\frac{\rho_e(\boldsymbol{k},\omega)}{\epsilon_{\|}(k,\omega)} \tag{6}$$

where

$$\epsilon_{\|}(k,\omega) = \frac{1}{1-\chi_{\|}(k,\omega)} \tag{7}$$

is the ***longitudinal dielectric function with spatial dispersion.*** In the particular case of a static and spherically symmetric charge distribution centered in the coordinate origin, $\Phi(k) = 4\pi\rho_e(k)/k^2\epsilon_{\|}(k)$, where $\epsilon_{\|}(k) = \epsilon_{\|}(k,\omega=0)$ is the so called ***static dielectric function.*** The external charge distribution $\rho_e(k) = \int_0^\infty dr 4\pi r^2 \rho_e(r)\sin(kr)/kr$ and the potential distribution in space (r is the radius from the origin of the coordinate system) is given by

$$\Phi(r) = \frac{2}{\pi}\int_0^\infty dk \frac{\rho_e(k)}{\epsilon_{\|}(k)}\frac{\sin(kr)}{kr}. \tag{8}$$

Dealing with the parallel components of the response tensor, the subscript "$_{\|}$" is often omitted, which will be done hereafter. The static dielectric function $\epsilon(k)$, or the static response function $\chi(k)$, are the quantities characterizing the response of the surrounding medium to external charge sources imbedded into the medium, and their k-dependence is the representative of the structure of the medium. Of course, in reality the charge sources have a finite size and may perturb the structure of the medium, so that the full tensor $\chi_{\alpha\beta}(\boldsymbol{r},\boldsymbol{r}',t-t')$ or $\chi_{\alpha\beta}(\boldsymbol{k},\boldsymbol{k}',t-t') = \delta_{\alpha\beta} - \epsilon^{-1}_{\alpha\beta}(\boldsymbol{k},\boldsymbol{k}',t-t')$ must be used. The inhomogeneity and anisotropy of the medium may come not only from imbedded charge sources (in some cases the corresponding effect was found to be small), but due to boundaries and interfaces. The nonlocal electrostatics has developed approximate or exact methods to handle this problem [3]. In this article we concentrate on the nonlocal dielectric properties of a homogeneous medium, and, when applying it to the calculation of the screened potentials of ions in solution, will adopt an approximation of an *immersed charge distribution*. The latter envisages an ion as a charge distribution imbedded into a liquid, not perturbing its structure. This is a zero order approximation which, however, works in many cases very well [2,3].

The first-principle or model calculations of the nonlocal dielectric functions of solids have a long history [4,5,6]. In polar liquids the use of dielectric functions has started with phenomenological models. The simplest one, suggested for $\epsilon(k)$, has a Lorentzian form [2]

$$\epsilon(k) = \epsilon_* - \frac{\epsilon - \epsilon_*}{1 + \frac{\epsilon}{\epsilon_*}\lambda^2 k^2} \tag{9}$$

which follows from an exponential decrease of the correlation function of bound charge density fluctuations in real space. Here, λ is the correlation length of the low frequency medium polarization fluctuations; ϵ and ϵ_* are the macroscopic and short range dielectric constants ($\epsilon > \epsilon_*$; ϵ_* corresponds to the distortion polarization of the molecules, i.e. vibrational motions).

The potential produced by a point charge located in the coordinate origin ($\rho_e(k) = 1$), gives, according to Eq. 8, a result which can be rewritten in terms of a Coulomb law, Eq. 10, with an *effective* dielectric constant dependent on the distance r from the charge [2]

$$\Phi(r) = \frac{q}{\epsilon_{eff}(r) r}. \tag{10}$$

In a medium with a dielectric function given by Eq. 9, it reads

$$\epsilon_{eff}(r) = \frac{\epsilon}{1 + (\frac{\epsilon}{\epsilon_*} - 1)\exp\left(-\frac{r}{\lambda}\right)} \tag{11}$$

i.e. at large distances $\epsilon_{eff}(r)$ approaches the macroscopic dielectric constant ϵ, while at short distances it decreases to the value of ϵ_*. This was in agreement with the wisdom of solution chemists who often assumed the effective dielectric constant of the solvent to be much smaller near the ion than in the bulk [7]. Usually this was interpreted as the result of dielectric saturation. At the same time, Eq. 1 follows within the linear response approximations, as a consequence of the *structure* of the solvent. It, thus, rationalizes the reduction of the dielectric constant as a result of structural constraints. The Lorentzian model for $\epsilon(k)$ was therefore widely used in nonlocal electrostatics of solutions [2,3]. It appeared to give basically non-controversial results for different electric field-determined quantities in water for approximately the same sets of parameters for the Lorentzian model: $\lambda \approx 3 - 4\mathring{A}, \epsilon_* \approx 5, \epsilon \approx 80$. Note also the *encouraging parallelism* with the band structure calculations of $\epsilon(k)$ for solid dielectrics which fit well the Lorentzian shape of the so called Inkson model[8].

The development of statistical-mechanical theories of polar liquids allowed to perform microscopic calculations of $\epsilon(k)$ and $\epsilon(k,\omega)$, which did not confirm this simple picture[9,10,11]. Independently, the progress in computer simulations and the development of model potentials for polar molecules made it possible to compute these functions for polar liquids[12,13]. The first results were discouraging for the Lorentzian model[12,13,14]. In all cases $\epsilon(k)$ did not decrease, but went up to divergence in the small k range, and then became negative. Only in one work[11] the correct limiting behavior was obtained where $\epsilon(k)$ approached unity from above with $k \to \infty$, as it should be. Indeed, $\chi(k \to \infty)$ must vanish, because there cannot be any response to Fourier components with a wave length shorter than all the medium characteristic coherence lengths. The wrong limiting behavior was understood as an artifact of point-like dipolar models of molecules constituting the liquid[12,14,15] (for a detailed analysis see Ref. [16]). It appeared, however, also from more involved molecular models[12,13]. Later in this article, we explain the nature of this artifact.

According to Kirzhnitz et al. [17], the negative values of $\epsilon(k)$ do not contradict existing physical principles. Indeed, in the macroscopic limit ($k = 0$) one can imagine plane capacitor experiments, in which either the macroscopic electric field $\boldsymbol{E}$ or the induction $\boldsymbol{D}$ are controlled and either $(\epsilon - 1)$ or $(1 - 1/\epsilon)$ are response functions. Since the latter must be positively defined, this leads to the constraint $\epsilon \geq 1$. However, for any finite k only $\chi(k) = 1 - 1/\epsilon(k)$ is the response function, which is positively defined in thermodynamic equilibrium. No *Gedanken experiment* has yet been formulated to show that $\epsilon(k) - 1$ is also a response function and consequently, the only forbidden band of $\epsilon(k)$-values lies between 0 and 1.

First results with negative dielectric constants where obtained for computer simulations of molten salts[18]. The phenomenon was called *overscreening*, since negative $\epsilon(k)$ simply means that $\chi(k) > 1$. In the k-region, where overscreening occurs, the Fourier amplitude of the medium polarization field, $4\pi\boldsymbol{P}_{\boldsymbol{k}}$, is larger than the corresponding amplitude of the external field, $\boldsymbol{D}_{\boldsymbol{k}}$. This is possible at the cost of other k-harmonics, typical of *underscreening* ($\chi(k) < 1$), so that the total balance is preserved: the net bound charge, q_b, of the medium never exceeds the charge, q_e, causing the external field ($q_b = -\chi(k=0)q_e$). The overscreening can be rationalized, physically, as a *spatial resonance*, associated with certain quasi-periodical arrangements in the medium, *preferred* by the short range structure. Of course, separate harmonics of the electrostatic field (standing waves) cannot be realized in nature, but are a mathematical fiction. Thus, the overscreening only tells us how large could be the response to a given Fourier component of the external field. The total external field (induction) varies in space subject to the solution of the Poisson equation for the given distribution of charge sources. Different harmonics have different *weight* in the external field, i.e. different Fourier amplitudes. If the field is such that the resonance k-components dominate in it, the manifestations of the overscreening will be huge. If, on the contrary, such components have negligible amplitudes, their effect would be smeared out. We shall illustrate in this paper how this happens.

Besides the fact, that $\epsilon(k)$ exhibits a negative region, the shape of it is not yet fully understood. Indeed, the fact that $\epsilon(k)$ at small k first increased and not decreased (as in the Lorentzian model), seems to be counter-intuitive. The stability of this result to a weak variation of the model parameters and the accuracy of calculation was questioned by a set of estimates [19]. Together with Ph. Bopp and S. Leikin, we therefore performed systematic investigations of $\epsilon(k)$ and $\epsilon(k,\omega)$ of water [16,20,21,22], as the most important solvent, and of their implications [21,23,24,25] (based essentially on molecular dynamics computer simulations) [16,20]. The results of these investigations, are summarized in this review article.

2 Calculation of response functions. Fluctuation dissipation theorems

There are two ways to calculate the response function with spatial dispersion. The first one is to use the fluctuation dissipation theorem (FDT) [26], which gives a relationship between the response function and the correlation function of bound charge density fluctuations. The latter quantity may be evaluated either within statistical theories of polar liquids or in computer simulations. The second one is a *direct way*: one may calculate the medium polarization field as a function of an external standing wave field $\boldsymbol{D_k}\exp(-i\boldsymbol{kR})$. The proportionality coefficients between the polarization field $4\pi\boldsymbol{P_k}\exp(-i\boldsymbol{kR})$ and the standing wave field will give then the response function for a given $\boldsymbol{k}$. Both approaches are equivalent within the linear response approximation but the direct approach may also help to detect nonlinear effects and establish the limits of validity of the linear response approximation. We shall focus on this point later (see Sec. 12).

The general form of FDT reads [26,27]

$$Im\{\chi(k,\omega)\} = \frac{2\pi}{\hbar}\left(1 - e^{-\beta\hbar\omega}\right) S(k,\omega) \quad (12)$$

where $\hbar$ is the Planck constant, $\beta = 1/k_B T$, with k_B the Boltzmann constant and T the temperature, and

$$S(k,\omega) = \frac{1}{k^2}\langle \rho_b(\boldsymbol{k},\omega)\rho_b^*(\boldsymbol{k},\omega)\rangle \quad (13)$$

is the dynamic bound charge density structure factor. The static dielectric response function is obtained from the Kramers-Kronig relationship [27]

$$1 - \frac{1}{\epsilon(k)} = \frac{2}{\pi}\int_0^\infty \frac{d\omega}{\omega} Im\{\chi(k,\omega)\}. \quad (14)$$

If there are no frequencies in the spectrum of $S(k,\omega)$ which lie in the *quantum range*, i.e. $\omega >> 1/\beta\hbar$, one can expand the exponential function in Eq. 12 and, according to Eq. 14, come to the linearized, classical form of FDT

$$1 - \frac{1}{\epsilon(k)} = 4\pi\beta S(k) \quad (15)$$

This form is much simpler than Eq. 12, since it requires only information about the static correlations, $\langle \rho_b(\boldsymbol{k})\rho_b^*(\boldsymbol{k})\rangle$. However, the spectrum of water is typical of strong absorption in the infra-red region, so that the results based on the classical FDT may appear to be inaccurate. It is therefore important to compare the results which follow from the classical and general forms of FDT.

3 Molecular Dynamics Computer Simulation

We focus on the Bopp-Janćso-Heinzinger variant of the central force model of water [28]. In the BJH model the interaction potential is devided into two parts. The intramolecular part allows the molecule to vibrate and is chosen to reproduce experimentally measured vibrational H-H and O-H frequencies; it includes three body interactions. The intermolecular part consists of Coulomb interactions of effective charges on oxygen and hydrogen atoms and the short range non-Coulombic interactions chosen to reproduce the partial structure factors and thermodynamic parameters of water.

The molecular dynamics simulation results [20] which we discuss below were obtained with 200 molecules in a cubic box of length $L = 18.1\mathring{A}$ at $T = 290K$.

Electrostatic interactions were handled with the help of the Ewald summation method in tabulated form while a spherical cut-off with a *shifted-force-potential* was applied to all short range interactions. The time step in the MD simulation run was $0.4fs$ and the total trajectory length was $800ps$, i.e. $2 \cdot 10^6$ time steps.

The bound charge density fluctuations, ρ_b, were obtained in a *direct* way via the Fourier sum

$$\rho_b(\boldsymbol{k}) = \frac{1}{\sqrt{N}} \sum_{j,\alpha} ez_\alpha e^{i\boldsymbol{k}(\boldsymbol{r}_j + \boldsymbol{\delta r}_{j\alpha})} \tag{16}$$

which extends over all sites j with partial charges ez_α ($z_H = -z_O/2 = 0.3298$). In Eq. 16, $\boldsymbol{r}_j$ is position vector of the center of mass and $\boldsymbol{\delta r}_{j\alpha}$ is the vector, pointing to site α of molecule j. If one chooses the wavevector according to the periodicity constraint, $\boldsymbol{k} = 2\pi\boldsymbol{n}/L$ ($\boldsymbol{n} = (n_x, n_y, n_z)$; $n_i = (1, 2, \ldots)$) one avoids artefacts in the calculation of correlation functions (in the small wavenumber region) due to finite upper limits in the Fourier integral. According to the system parameters, the smallest wavenumber accessible in these simulations is $0.346 Å^{-1}$.

4 Static partial structure factors, bound charge density correlation functions, and the static dielectric response

The static bound charge density correlation function may be splitted into the molecular and distinct parts

$$S(k) = S^{(m)}(k) + S^{(d)}(k) \tag{17}$$

where

$$S^{(m)}(k) = \frac{2n_O z^2 e^2}{k^2} \left(3 + f_{HH}(k) \frac{\sin(kd_{HH})}{kd_{HH}} - 4 f_{OH}(k) \frac{\sin(kd_{OH})}{kd_{OH}} \right) \tag{18}$$

$$S^{(d)}(k) = \frac{4n_O z^2 e^2}{k^2} (h_{OO}(k) + h_{HH}(k) - 2h_{OH}(k)) \tag{19}$$

$$= \frac{4z^2e^2}{Vk^2} \left\langle \sum_{l \neq l'} e^{i\boldsymbol{k}(\boldsymbol{r}^l_H - \boldsymbol{r}^{l'}_H)} + \sum_{l \neq l'} e^{i\boldsymbol{k}(\boldsymbol{r}^l_O - \boldsymbol{r}^{l'}_O)} - 2 \sum_{l \neq l'} e^{i\boldsymbol{k}(\boldsymbol{r}^l_O - \boldsymbol{r}^{l'}_H)} - 2 \sum_{l \neq l'} e^{i\boldsymbol{k}(\boldsymbol{r}^l_H - \boldsymbol{r}^{l'}_O)} \right\rangle \tag{20}$$

Here, the product $\boldsymbol{k}\boldsymbol{r}^l_i$ denotes the projection of the position vector $\boldsymbol{r}^l_i$ of the particle l of species i on the direction of the wave vector $\boldsymbol{k}$. n_O is the molec-

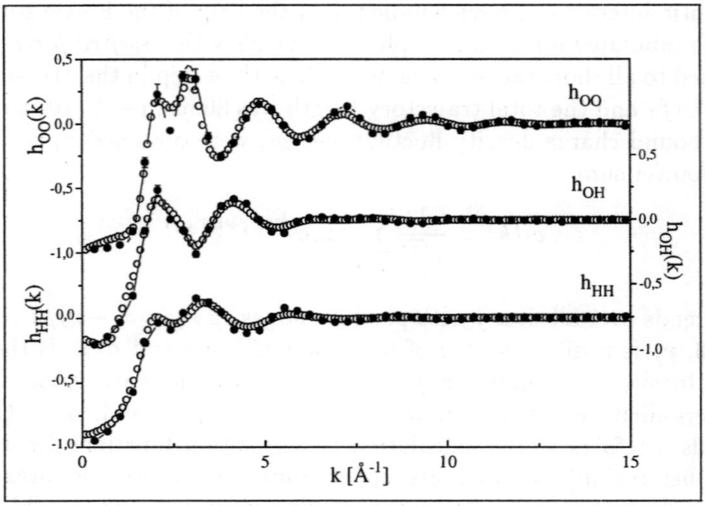

Figure 1: Simulated partial structure factors $h_{ij}(k)$, compared to neutron diffraction data of Ref.[20]

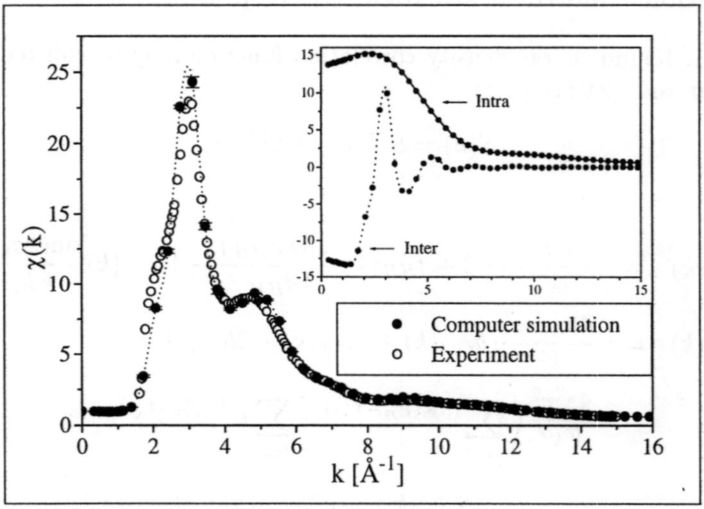

Figure 2: Comparison between the dielectric response function $\chi(k)$ from computer simulation and *experiment* [20].

ular number density in the fluid; z, the partial charge quantity of the hydrogens; $h_{ij}(r) = g_{ij}(r) - 1$, the site-site pair correlation function and d_{ij}, the mean intramolecular distances. V is the volume of the simulation cell and the f_{ij} are distribution functions, characterizing the deviation of the site-site distances from their mean values, thereby including atomic form factors. For rigid molecules with point like charges on the sites, $f_{ij} = 1$.

The simulated partial structure factors [20] are shown in Fig. 1 together with experimental neutron diffraction data of Soper et al [29]. The agreement is excellent for not too small k.

Using these results and the model of the charge distribution of the BJH water molecule, one may plot the simulated and *experimental* $S(k)$, or the corresponding values of the response function calculated via the classical form of FDT (Eq. 15). The results are displayed in Fig. 2 where the inset shows separately the infrastructure of the plot: the simulated intramolecular and intermolecular contributions. We see the overscreening ($\chi(k) > 1$) in a wide range of $1.3Å^{-1} \leq k \leq 12.5Å^{-1}$ with a sharp main maximum at $k = 3Å^{-1}$ and a satellite maximum centered around $k = 5Å^{-1}$. According to the inset, the both maxima are due to intermolecular correlations, but the overall overscreening effect is due to a subtle compensation between the intra and intermolecular contributions. As it is seen from Fig. 1, the peak positions ($3Å^{-1}$, $5Å^{-1}$) correspond to maxima in $h_{OO}(k)$ and to minima in $h_{OH}(k)$. Since the negative values of h_{OH} enters into $S(k)$ (Eq. 19) the maxima in $\chi(k)$ appear to be mainly the result of the interplay between h_{OO} and h_{OH} contributions. In real space these wave-numbers correspond to the periodicity unit $2\pi/k \approx 2Å$ and $1.2Å$. While, for h_{OH}, the distance of $2Å$ could be related to the hydrogen bond length, it is not clear what this length could mean for h_{OO}.

5 Dielectric function. The problem of small k

The corresponding plot for the static nonlocal dielectric function, $\epsilon(k)$, (calculated via Eq. 15 at $\omega = 0$) shows a number of remarkable features [20]:

(i) In the small k-range $\epsilon(k)$ is positive and *goes down* with the increase of k (but not up!), then passes a minimum and rises to divergence (at the point where $1 - \chi(k) = 0$).

(ii) In the overscreening range $\epsilon(k)$ is negative, then at the second point where $1-\chi(k)$ vanishes it changes sign through divergence and then approaches the correct physical limit of 1 from above, as it should be ($\epsilon(k \to \infty) = 1$ since there are no electronic degrees of freedom included in the BJH water model).

Note, that there are no reliable neutron diffraction data in the very small k-range (the scatter is too large), so that we may judge about the $\epsilon(k)$ behavior in

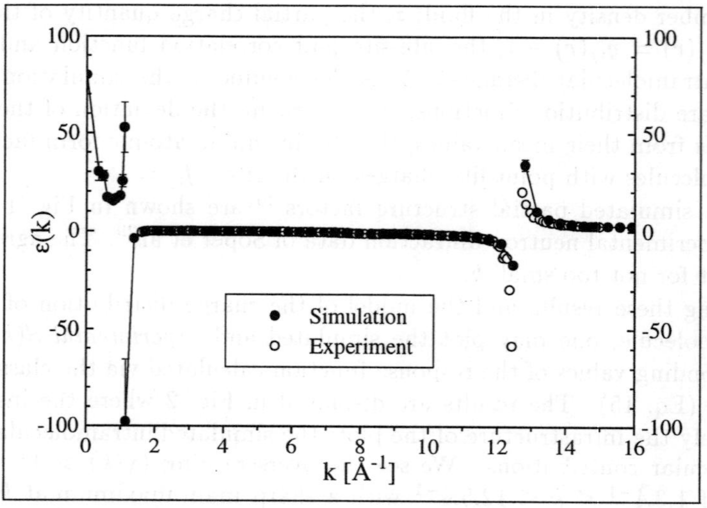

Figure 3: The simulated nonlocal dielectric function $\epsilon(k)$ [20], compared to *experimental* values, based on neutron diffraction data of Ref. [29].

this region only on the basis of computer simulations. If we fit the decreasing values in the small k-range to a Lorentzian function, we obtain an effective value for the correlation length $\lambda \approx 1\AA$. A larger value of λ would correspond to a steeper decrease of $\epsilon(k)$. As we definetely have not enough points in the small k-range we cannot give a conclusive statement about the value of λ. This problem could only be solved if the system size would be considerably increased. Note however, that a doubling of the simulation box, i.e. increasing the number of molecules by a factor of 8, would increase the computer time, roughly, by a factor of ≈ 60. Taking into account that the results of Ref. [20], obtained at ZAM KFA Jülich, took ≈ 200 hours on a CRAY Y-MP M94/4256, an extension to smaller k-values would hardly be possible without new computational schemes.

6 The artifacts of the dipole-dipole correlation function approximation

In order to understand the qualitative differences between the results of Ref.[20] and the previous works, the expression for the bound charge density fluctua-

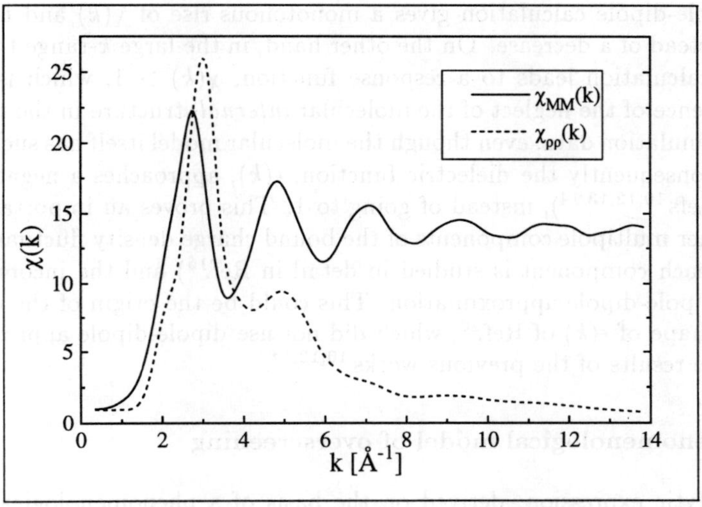

Figure 4: The simulated dielectric response function: calculated in the dipole-dipole-approximation (dashed line) and in the full bound-charge density correlation function approach (solid line).

tions (Eq. 16) was expanded into a multipolar series [16]

$$\frac{\rho_b(\boldsymbol{k})}{ik} = \frac{1}{\sqrt{N}}\sum_{j,\alpha} e z_\alpha e^{i\boldsymbol{k}\boldsymbol{r}_j}\Big\{ \hat{\boldsymbol{k}}\boldsymbol{\delta r}_{j\alpha} + \frac{ik}{2}\hat{\boldsymbol{k}}\hat{\boldsymbol{k}} : \boldsymbol{\delta r}_{j\alpha}\boldsymbol{\delta r}_{j\alpha} - \frac{k^2}{6}\hat{\boldsymbol{k}}\hat{\boldsymbol{k}}\hat{\boldsymbol{k}} \vdots \boldsymbol{\delta r}_{j\alpha}\boldsymbol{\delta r}_{j\alpha}\boldsymbol{\delta r}_{j\alpha} \pm \dots \Big\} \quad (21)$$

to give for the correlation function

$$\begin{aligned}\frac{\langle \rho_b(\boldsymbol{k})\rho_b^*(\boldsymbol{k})\rangle}{k^2} &= \langle M^{\|}(\boldsymbol{k})M^{\|*}(\boldsymbol{k})\rangle + \langle \mathcal{Q}^{\|}(\boldsymbol{k})\mathcal{Q}^{\|*}(\boldsymbol{k})\rangle + \langle \mathcal{O}^{\|}(\boldsymbol{k})\mathcal{O}^{\|*}(\boldsymbol{k})\rangle + \quad (22)\\ &+ \langle M^{\|}(\boldsymbol{k})\mathcal{Q}^{\|*}(\boldsymbol{k})\rangle + \langle M^{\|}(\boldsymbol{k})\mathcal{O}^{\|*}(\boldsymbol{k})\rangle + \langle \mathcal{Q}^{\|}(\boldsymbol{k})\mathcal{O}^{\|*}(\boldsymbol{k})\rangle + \dots\end{aligned}$$

where the terms in sequence correspond to the dipole-dipole, quadrupole-quadrupole, octupole-octupole, etc., contributions + the corresponding cross-terms. Fig. 4 displays some of the results obtained in Ref.[16]: the full $\langle\rho_b\rho_b\rangle$-calculation of the response function is compared to the calculation limited to the first term on the *rhs* of Eq. 22, i.e. the dipole-dipole contribution. Drastic differences are found both in the small and high k-ranges. In the small k-range the dipole-dipole calculation gives a monotonous rise of $\chi(k)$ and thereby of $\epsilon(k)$, instead of a decrease. On the other hand, in the large k-range the dipole-dipole calculation leads to a response function, $\chi(k) > 1$, which is a direct consequence of the neglect of the molecular *internal* structure in the treatment of the simulation data, even though the molecular model itself has such a structure. Consequently the dielectric function, $\epsilon(k)$, approaches a negative value (as in Refs.[10,12,13,14]), instead of going to 1. This proves an important role of the higher multipole components of the bound charge density fluctuations (the role of each component is studied in detail in Ref.[16]) and the incompleteness of the dipole-dipole approximation. This could be the origin of the difference in the shape of $\epsilon(k)$ of Ref.[20], which did not use dipole-dipole approximation, from the results of the previous works [10,12,14].

7 Phenomenological model of overscreening

An analytic expression, derived on the basis of a phenomenological Landau Hamiltonian of the liquid parameterizes approximately the simulated behavior of $\epsilon(k)$ and $\chi(k)$ [21]. The Hamiltonian considers two independent polarization modes, $\boldsymbol{P}_1$ and $\boldsymbol{P}_2$, of which the higher frequency mode is coupled to a *hidden* scalar order parameter (local density fluctuations in the liquid, for example). From symmetry constraints it is clear, that the only way to couple a scalar field to a vectorial field is via the product of the gradient of the density fluctuations and the polarization fluctuations. All other terms in the Hamiltonian are standard for a medium far from any critical points: they are quadratic in vectorial and scalar fields and their first gradient corrections. Surprisingly, this simple model describes the main effects. It leads to an expression for the response function [21],

$$\chi(k) = \left(\frac{1}{\epsilon_\infty} - \frac{1}{\epsilon}\right)\left\{\frac{1-c}{1+\lambda_1^2 k^2} + \frac{c(1+l^2k^2)}{(1+\lambda_2^2k^2)(1+l^2k^2) - L^2k^2}\right\} \tag{23}$$

where the coefficient c determines the *weight* of the low and high frequency modes:

$$(1-c)\left(\frac{1}{\epsilon_\infty} - \frac{1}{\epsilon}\right) = \frac{1}{\epsilon_*} - \frac{1}{\epsilon} \quad \text{and} \quad c\left(\frac{1}{\epsilon_\infty} - \frac{1}{\epsilon}\right) = \frac{1}{\epsilon_\infty} - \frac{1}{\epsilon_*} \tag{24}$$

Here, ϵ_∞ is the dielectric constant of the electronic degrees of freedom and ϵ_* is the intermediate dielectric constant at the frequencies separating the low frequency modes, not coupled to the hidden order parameter, and the high frequency modes, coupled to it; λ_1 is the correlation length of the non-coupled polarization, λ_2 the length which would be equal to the correlation length in the high frequency polarization if it were uncoupled from the density fluctuations; l, would be equal to the correlation length of the scalar order parameter fluctuations if they would be uncoupled to the high frequency polarization; L, the dimensional coupling constant which characterizes the coupling of the high frequency polarization and the hidden scalar order parameter fluctuations. When $L = 0$, Eq. 23 reduces to a two-mode Lorentzian approximation [2]. The formal stability of the quadratic Hamiltonian requires that $L < l + \lambda_2$ and it is easy to find [21] that at $\lambda_2 < L < l + \lambda_2$ the second term of the *rhs* of Eq. 23 has a maximum at $k_* = \sqrt{L/l - 1}/l$ of a height proportional to $l^2/(l^2 - (L - \lambda_2)^2)$ which may, thus, take large values, or even diverge, depending on how close is L to $l + \lambda_2$. Thus, depending on the parameters, Eq. 23 may describe the overscreening effect with corresponding negative values of $\epsilon(k)$.

Due to the simplicity of the model, it cannot be regarded as a full description of a *real* polar system. However, the approach has its own validity, since it

(i) gives a compact and elegant analytical approximation for the response functions which can cover, depending on parameters, various cases of $\epsilon(k)$ behavior - from the Lorentzian shape to the overscreening;

(ii) gives a fairly good fit of the simulation data (c.f. Fig.4 of Ref.[21]);

(iii) could be of a rather general nature, reflecting the essential symmetry constraints.

It thus can be used as a simple analytical framework for describing dielectric screening in polar liquids and its account in various nonlocal electrostatic calculations [2] (c.f. the calculations of screening functions and reorganization energies in Ref.[21]).

8 Screening functions and reorganization energies. How the overscreening effects vanish

In the framework of nonlocal electrostatics the expressions for the reorganization energy of outer sphere charge transfer reactions between symmetric ions [30]

$$E_r = \frac{(\Delta q)^2}{\pi} \int_0^\infty dk \left(\frac{1}{\epsilon_\infty} - \frac{1}{\epsilon} \right) \left\{ \rho_{a_1}(k)^2 + \rho_{a_1}(k)^2 - 2\rho_{a_1}(k)\rho_{a_2}(k) \frac{\sin(kR)}{kR} \right\} \tag{25}$$

and the interaction energy between these ions

$$W_{ab}(R) = \frac{q^2}{\pi} \int_0^\infty \frac{dk}{\epsilon(k)} \frac{\sin(kR)}{kR} \rho_{a_1}(k)\rho_{a_2}(k) \tag{26}$$

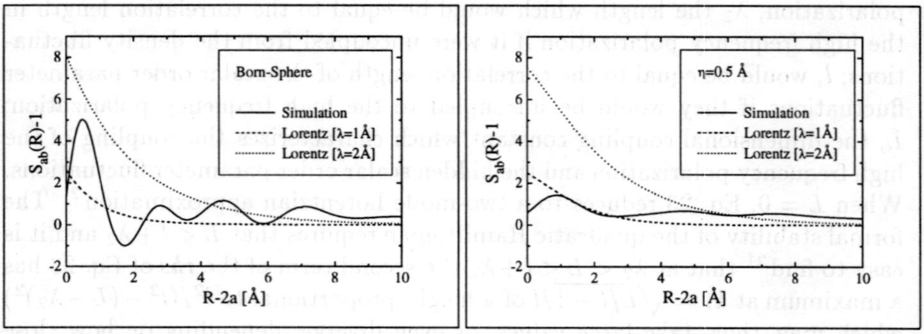

Figure 5: Interaction screening functions $S_{ab}(R)$ for Born spheres (a) and smeared Born spheres (b) with *radii* $a_i = 2\AA$. In both figures we compare results from calculations, which use the simulated $\chi(k)$ and Lorentzian $\chi(k)$ respectively.

have a simple form in the immersed charge distribution model $\rho(k)$. Here, $\rho_{a_i}(k)$ is the Fourier transform of the charge distribution of ion i ($i = 1,2$) with radius a_i and R, the distance, separating the centers of the two charge distributions. From Eq. 26 one may define the screening function

$$S_{ab}(R) = W_{ab}(R)\epsilon R \tag{27}$$

which is a measure for the deviation from Coulomb's law, i.e. it shows the effect of nonlocal effects on the ionic interaction. For very large distances R, S_{ab} will approach unity since on large spatial scales, nonlocal effects become negligible. Eqs. 25,26,27 give us the possibility to study the interplay between the $\epsilon(k)$-spectrum and the form-factors of the charge distributions representing the ions. We consider two types of form-factors:

Born spheres (BS) where the charge is located on an infinitely thin sphere of radius a_i [2]

$$\rho_{BS}(k) = \frac{\sin(ka_i)}{ka_i} \tag{28}$$

Smeared Born spheres (SBS) where the charge density has a peak at the *radius* distance a_i [23] with an *exponential width* η_i

$$\rho_{SBS}(k) = \frac{1}{1 + (\eta_i/a_i)^2(2 - e^{-a_i/\eta_i})} \times \left(\frac{\sin(ka_i)}{ka_i} \frac{1}{1 + \eta_i^2 k^2} + \frac{\eta_i^2}{a_i^2} \frac{2\cos(ka_i) - e^{-a_i/\eta_i}}{(1 + \eta_i^2 k^2)^2} \right). \tag{29}$$

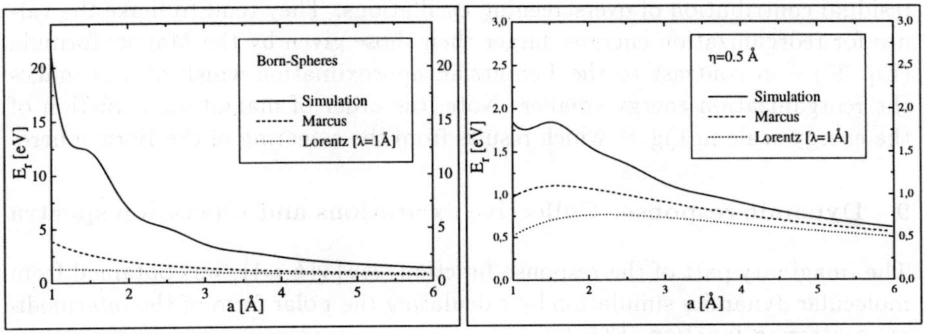

Figure 6: Reorganization energy E_r for two ions in contact as a function of their radii. (a) Bornspheres; (b) smeared Bornspheres. In both figures we compare results from calculations, using the simulated $\chi(k)$ and the Lorentzian $\chi(k)$ respectively as input as well as the limiting case without spatial dispersion (leading to the Marcus formulae).

Note that the Marcus formula for reorganization energy [31]

$$E_r^{Marcus} = \frac{(\Delta q)^2}{2}\left(\frac{1}{\epsilon_\infty} - \frac{1}{\epsilon}\right)\left(\frac{1}{a_1} + \frac{1}{a_2} - \frac{2}{R}\right) \tag{30}$$

follows from Eq. 25 for the Born sphere model of ions and with the neglect of spatial dispersion ($\epsilon(k) = \epsilon$).

Figs. 5 and 6 show, respectively, the screening functions and reorganization energies [24] for the two types of formfactors and for the simulated response function. For comparison, we also show calculations on the basis of a Lorentzian $\chi(k)$.

Overscreening may be understood as a resonance effect and since every resonance is broadened if the external perturbation is smeared, the overscreening effect should be reduced. This is well seen from the plot for the screening function, based on the simulated $\chi(k)$ (Fig. 5). For Born spheres ($\eta = 0$) we observe large oscillations around 0, i.e. strong overscreening, which is due to the resonance in k-space. This effect is strongly reduced, when we introduce a continuous charge distribution with $\eta = 0.5\AA$. There are still oscillations but all values lie in the positive range and are closer to the monotonously decreasing S_{ab}, given by the Lorentzian $\chi(k)$. The tendency is to smear out any resonance effects when we move from Born spheres to smeared Born spheres. A similar smearing of the resonance effect is seen for reorganization energies. The natural question, *whether we can simply apply the Lorentzian approximation since the resonance phenomena for smeared ions diminishes* cannot be answered generally. For realistic smearing parameters there is always some residual contribution of overscreening oscillations. They tend to make the values for reorganization energies larger then those given by the Marcus formula (Eq. 30) – in contrast to the Lorentzian approximation which always makes the reorganization energy smaller. Note, the order of magnitude reduction of the energy scale in Fig. 6, which results from the smearing of the Born sphere.

9 Dynamic response. Collective excitations and relaxation spectra

The imaginary part of the response function, $Im\{\chi(k,\omega)\}$, was obtained from molecular dynamics simulation by calculating the polar form of the intermediate scattering function, $F(k,t)$,

$$F(k,t) = \frac{1}{N_t}\sum_{i=1}^{N_t} \rho_b(\boldsymbol{k}, t+\delta t_i)\rho_b^*(\boldsymbol{k}, \delta t_i). \tag{31}$$

Here, N_t is the number of time steps δt, over which averaging was performed. $F(k,t)$ may then be Fourier transformed to give the frequency dependent structure factor, $S(k,\omega)$, from where the response function is then obtained with the help of the classical form of the FDT, Eq. 15 (for a discussion of the general form, Eq. 12, see Sec. 10). In Fig. 7 we show some of the spectra, representative for the small k-range. Characteristic features are a broad band in the small frequency range, corresponding to Debye-like reorientational motions of the molecules, and a sharp peak in the high frequency range. The latter one corresponds to the librational band of the molecules. Increasing k (c.f. Figs. 2a,b of Ref.[16]) the low frequency part is strongly increased – the maximum value of it, at $k = 3.15\mathring{A}^{-1}$, is about ≈ 50 times larger than the corresponding value at small k. This effect is a manifestation of the resonance peak of the static $\chi(k)$. However, it is interesting to note, that the spatial resonance affects mainly the low frequency motions. With a further increase of k, $> 6\mathring{A}^{-1}$, the two distinct frequency bands merge and the maximum of $Im\{\chi(k,\omega)\}$ shows a continuous shift to higher frequencies, indicative of single particle behavior.

A more detailed analysis of the small k-range shows, that the spectrum may be fitted approximately by two Lorentzian functions in the small frequency range (with an order of magnitude difference in the positions of the maxima) and a resonance function in the librational region. Concerning the two low frequency bands, one may speculate that the lower frequency component is due to Debye rotations of clusters (domains) of molecules while the higher frequency component is due to rotations of non-associate molecules, which resembles a, somewhat outdated, two-state picture of water.

An interesting phenomenon is observed for small values of k in the high

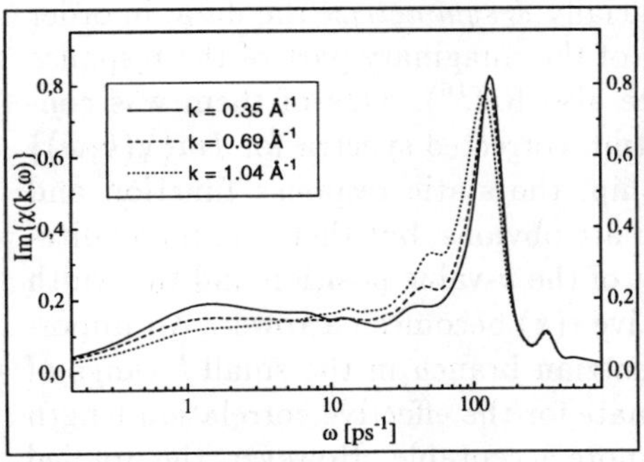

Figure 7: Imaginary part of the response function for the three smallest wavenumbers, accessible in the simulation.

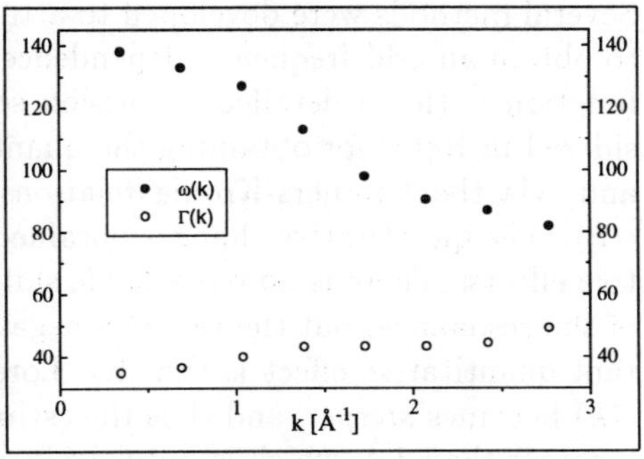

Figure 8: Dispersion relation for the high frequency collective excitation: the *libron*.

frequency, librational band of $Im\{\chi(k,\omega)\}$. With an increase of k, the maximum of this band is shifted gradually from $\approx 140ps^{-1}$ to $\approx 80ps^{-1}$. Since the resonance is relatively sharp, one may speak about the k-dependence of the frequency of a *collective* excitation (see Fig. 8), the *libron*. As is seen from Fig. 8, the dispersion for this *orientational phonon* has the shape of an optical mode. In the same figure we show the decrement $\Gamma(k)$ of the libron, which shows, that it is strongly damped, but not overdamped. This mode, thus, may propagate through water at certain wave-vectors. Overdamping will occur in a higher k-range, where the collective mode decays into single particle motions. A detailed analysis of the libron shows [16], that its main contribution comes from the R_y-librations [16,32], which are the fastest angular oscillations of the water molecules.

10 The problem of quantum corrections

The use of the classical FDT is not appropriate in the libration part of the spectrum, since $\omega > 1/\hbar\beta$ and consequently the exponential in Eq. 12 cannot be expanded. One way to handle this problem is, of course, to apply directly Eq. 12 to the simulated $S(k,\omega)$. This naive approach, however, requires some *quantum nature* from a classical simulation. The problem becomes obvious, when considering the symmetry requirements on $Im\{\chi(k,\omega)\}$. Indeed, to preserve an odd ω-dependence, $Im\{\chi(k,\omega)\} = -Im\{\chi(k,-\omega)\}$, the polar structure factor $S(k,\omega)$ must obey the detailed ballance principle [27], which *it does not* in a classical simulation. This is an old problem emerged in the times of interpretation of the first data on inelastic neutron scattering in liquids [33,34].

Several methods were developed to artificially *desymmetrize* the data, in order to obtain an odd frequency dependence of the imaginary part of the response function [35] (for a detailed discussion see also Ref.[16]). One of them was considered in Ref.[16] for obtaining the quantum-corrected spectra for $Im\{\chi(k,\omega)\}$ and, via the Kramers-Kronig relationship, the static response function and $\epsilon(k)$. The quantitative changes obtained are obvious, but there are no qualitative effects. There is no remarkable shift of the k-value position and the width of the resonance, but the range of negative $\epsilon(k)$ becomes narrower. An important quantitative effect is that the Lorentzian branch in the small k-range of $\epsilon(k)$ becomes *steeper*, and thus the estimate for the effective correlation length is larger than 1Å, which is, intuitively, more acceptable. However, the applied method of quantum corrections may show only a tendency of what might be the effect of a correct treatment of the high frequency branch. A more conclusive answer may only be given from a full quantum-dynamics simulation, which is far away from computational facilities at the moment.

11 Femtosecond solvation dynamics

The data obtained for $Im\{\chi(k,\omega)\}$ were used to calculate the femtosecond hydration dynamics [25], which may be characterized through the normalized time-dependent correlation function $C(t)$

$$C(t) = \frac{E(t) - E(\infty)}{E(0) - E(\infty)} \tag{32}$$

Here, $E(t) = \hbar\omega(t)$, where $\omega(t = 0)$ is the emitted frequency at time $t = 0$ (moment of excitation), when the polarization of the solvent is still the same as it was when the solute was in the ground state, and $\omega(t = \infty)$ is the frequency emitted when the solvent polarization fully relaxes in the new equilibrium position for the excited solute (Fig. 9 shows schematically the free energy profile of the system as a function of solvent polarization). The time-dependent solvation free energy may be expressed through the imaginary part of the solvent response function and the change of induction $\boldsymbol{D}$ of the solute [36]

$$E(t) = \frac{1}{2\pi^2} \sum_k |\Delta \boldsymbol{D}(\boldsymbol{k})|^2 \int_0^\infty \frac{d\omega}{\omega} Im\{\chi(k,\omega)\} \cos(\omega t) \tag{33}$$

We note, that Eq. 33 may be applied only for solvation by classical polarization. If there are contributions of quantum modes the upper limit of integration must be replaced by $4\hbar\beta$; the modes with $\omega > 4\hbar\beta$ contribute to the tunneling factor in the transition probability [37] , which, of course, cannot be described

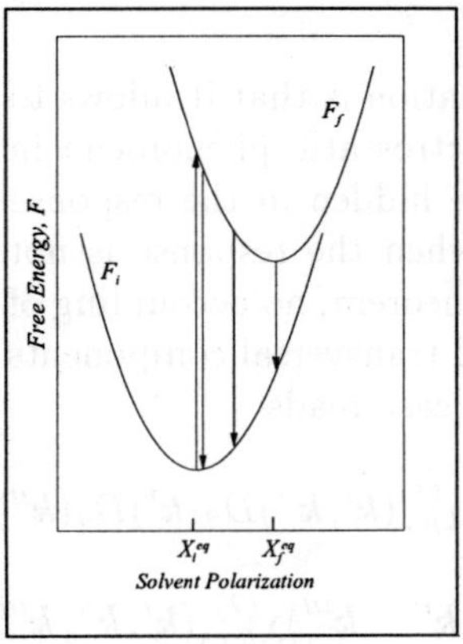

Figure 9: Fluorescence experiment.

within the classical MD-simulation. The simplest way to model the effect of an electron excitation on $\Delta \boldsymbol{D}$ is to assume that the electron cloud swells from a Born sphere with radius a_i in the initial state to a sphere of a_f in the excitated state (a more detailed consideration of charge distribution models, including smearing, will be given in Ref.[25]). This model of two Born spheres could be a first order approximation for the dynamcis of the solvated electron, undergoing the transition between two localized states. For an illustration of the effect, we assume $a_i = 3\mathring{A}$ and vary a_f in the range between $3.6\mathring{A}$ and $6\mathring{A}$. The results for the $C(t)$-function (see Fig. 10), show an effect discussed already in Ref.[36]: a *strong dependence of the character of relaxation on the size of the charge distribution in the initial and final states of the probe molecule.* Since the relaxation pattern is usually attributed only to the dynamic properties of the liquid, effects of these kind are referred to the nonlinearity of the solvent response to the electric field of the solute. However this effect *emerges already within the linear response theory* and is due to spatial dispersion. Indeed, the shape of the charge distribution affects the inductions in Eq. 33, and since k and ω are coupled through the response function $Im\{\chi(k,\omega)\}$, variations in induction affect $C(t)$. The strongest effect is on the short time relaxation.

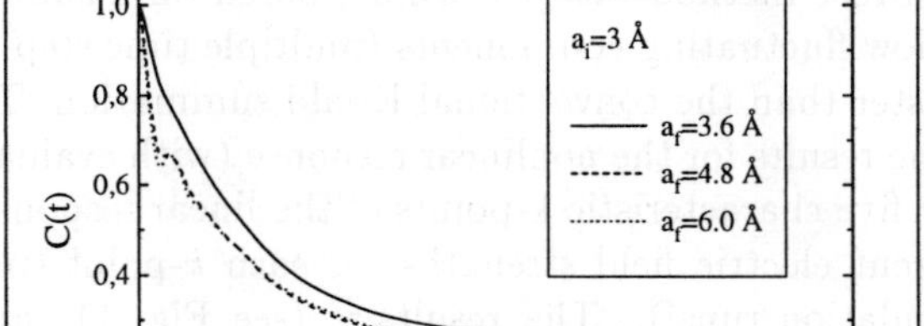

Figure 10: Energy correlation function $C(t)$, calculated on the basis of the simulated dynamic response function.

12 Nonlinear effects

The great advantage of the linear response approximation is that it allows to develop a general closed form theory for various electrostatic phenomena in solutions, where all the properties of the medium are hidden in the response function. The situation is much more complicated when the response is not linear. There is no close form fluctuation dissipation theorem, no decoupling of different $\boldsymbol{k}$-modes, no decoupling into longitudinal and transversal components of the response, etc. The constitutive relation in this case reads

$$\begin{aligned} P_\alpha(\boldsymbol{k}) &= \chi^{(0)}_{\alpha\beta}(\boldsymbol{k})D_\beta(\boldsymbol{k}) + \int d\boldsymbol{k}'d\boldsymbol{k}''\delta(\boldsymbol{k}-\boldsymbol{k}'-\boldsymbol{k}'')\chi^{(1)}_{\beta\gamma}(\boldsymbol{k}',\boldsymbol{k}'')D_\beta(\boldsymbol{k}')D_\gamma(\boldsymbol{k}'') \\ &+ \int d\boldsymbol{k}'d\boldsymbol{k}''d\boldsymbol{k}'''\delta(\boldsymbol{k}-\boldsymbol{k}'-\boldsymbol{k}''-\boldsymbol{k}''')\chi^{(2)}_{\beta\gamma\lambda}(\boldsymbol{k}',\boldsymbol{k}'',\boldsymbol{k}''') \\ &\times D_\beta(\boldsymbol{k}')D_\gamma(\boldsymbol{k}'')D_\lambda(\boldsymbol{k}''') + \ldots \end{aligned} \tag{34}$$

As it was discussed in Sec.2, we may use a direct way to evaluate the system response: create an external standing wave $\boldsymbol{D_k}\exp(-i\boldsymbol{kR})$ and calculate the corresponding polarization field. Though it, of course, will not give all off-diagonal nonlinear response tensors, we shall be able to evaluate deviations from the linear response.

From equilibrium MD simulations one may evaluate the bound charge density fluctuations in real- or $\boldsymbol{k}$-space. Applying FDT in the linear response approximation one may calculate the whole k-spectrum of $\chi(k)$ in the form of correlation functions from only one simulation run. In the direct calculation of the response, however, every new $\boldsymbol{k}$-point is to be obtained within a new simulation. The more points we want to have the longer the computational time. In Ref.[22] a different MD routine (with respect to that, described in Sec. 3) appropriate for the direct method was performed, based on a subdivision of forces into fast and slow fluctuating components (multiple time step method), which is ≈ 8 times faster than the conventional Ewald summation. This made possible to obtain the results for the nonlinear response (with evaluation of diagonal terms only) for five characteristic k-points of the linear response spectrum and for $7-8$ different electric field strengths for each k-point (more than 35 independent simulation runs!). The results [22] (see Fig. 11) are shown in the form of a dependence of the ratio of bound charge and external charge density ($\rho_b(k)/\rho_e(k)$, i.e. the response function in the linear case) on the Fourier amplitude of the external field. We find that the nonlinearity appears to be the stronger, the closer the k-values to the main resonance of the linear response function. A phenomenological theory, based on the same ideas, as discussed in Sec.8, but with introduction of anharmonic, fourth order terms in the Hamilto-

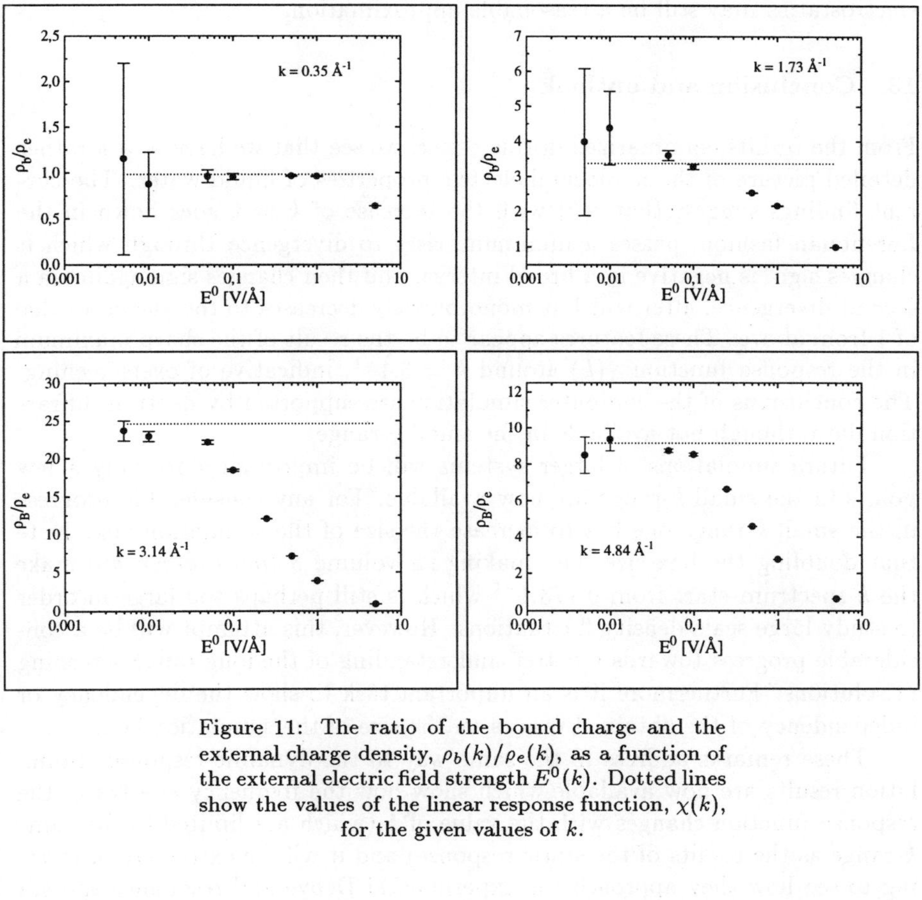

Figure 11: The ratio of the bound charge and the external charge density, $\rho_b(k)/\rho_e(k)$, as a function of the external electric field strength $E^0(k)$. Dotted lines show the values of the linear response function, $\chi(k)$, for the given values of k.

nian of the polarization coupled to the hidden order parameter was developed [22]. Surprisingly, it appears to be enough to rationalize the simulation data!

From Fig. 11 we see that the deviation from linearity for k close to the main resonance starts to be significant already at fields of the order of $0.2V/\AA$. These are discouraging results for the linear response theory, if it is applied to the screening of ions in solutions. Indeed, the Coulomb field of a $2\AA$ singly charged ion at the center of a first shell water molecule is $\approx 1V/\AA$. The estimate may be reduced twice if we take into account the screening by electronic polarizability, which was not explicitly considered in the simulation. However, it would still be a large field. At small k the nonlinear effects start only at the fields of the order of $3-5V/\AA$. As the smearing of the ionic charge distribution (see Sec. 8) diminishes the contribution of the resonance wave numbers, linear electrostatics may still be a reasonable approximation.

13 Conclusion and outlook

From the results, summarized in this paper, we see that we have now a rather detailed picture of the nonlocal dielectric properties of liquid water. The current findings suggest that $\epsilon(k)$ with the increase of k first goes down in the Lorentzian fashion, passes a minimum, rises to divergence through which it changes sign, is negative in a broad interval and then changes sign again via a second divergence, after which it monotonously decreases to the vacuum value of 1 from above. These features appear to be the result of the sharp maximum in the response function $\chi(k)$ around $k = 3\mathring{A}^{-1}$, indicative of overscreening. The conclusions of the computer simulation are supported by neutron diffraction data, though not available in the small k-range.

Future simulations of larger systems will be important since only a few points in the small k-range are now available. For any considerable progress in the small k-range one has to increase the size of the simulation box. Note that doubling the box size, i.e. making its volume *8 times larger* will make the k-spectrum start from $0.173\mathring{A}^{-1}$ which is still perhaps too large in order to study large scale density fluctuations. However, this attempt will be a considerable progress towards a better understanding of the long range screening in solutions. Furthermore it is an important task to show the dependency or independency of the obtained results on the size of the simulation box.

These remarks address in the same way to the dynamic response. Simulation results are now available which show how the frequency spectra of the response function changes with the value of k (which are limited to the same k-range as the results of the static response) and it will be extremely interesting to see how they approach the experimental Debye and resonance spectra with $k \to 0$.

Other directions for further research are:

- Variation of temperature and pressure.
- Study of inhomogeneous and anisotropic systems (water near an ion or near the electrode: How large are the off-diagonal elements of the response tensor? How good is the immersed charge approximation?).
- Study of nonlinear effects near an ion or an electrode.
- The explicit account for electronic polarizability (it will require a strong modification of the water model for the MD-simulation).

- Full quantum or semi-quantum dynamics simulations which are able to treat the high frequency modes with account for quantum tunneling.

The latter two points are, strictly speaking, necessary for the correct treatment of the librational and vibrational modes which belong to the quantum part of the water frequency spectrum. Though there are efficient tricks to treat the results of a purely classical simulation in the quantum frequency range, the treatment cannot be considered rigorous.

Acknowledgements

Special thanks are to Ph. Bopp and S. Leikin with whom many of the original results covered in this review have been obtained, including those which are in Press. The article benefited from stimulating discussions with H. Capellmann, D. Kossakowski, A. Kuznetsov and A. Nitzan. The support and encouragement of U. Stimming is gratefully appreciated. The review summarizes the results of the project supported by the German Science Foundation (DFG, the program "Fast Molecular Processes in Liquids", grant *Sti 74/3-2*). The reviewed computer simulations were performed on a CRAY YM-P M94/4256 and IBM RS6000 workstations at the Central Institute for Applied Mathematics (ZAM) of Research Center Jülich and at the Laboratory of Molecular Physical Chemistry at the University of Bordeaux.

References

1. A.A.Kornyshev, Electrochim.Acta **26**, 1 (1981).
2. A.A.Kornyshev, in *The chemical physics of solvation*, edited by R.R.Dogonadze, E.Kalman, A.A.Kornyshev, and L.Ulstrup (Elsevier Science Publishers B.V., Amsterdam, 1985), Vol. Part A,ch.3, p. 77.
3. M.A.Vorotyntsev and A.A.Kornyshev, *Electrostatics of Media with Spatial Dispersion* (Nauka, Moscow, 1993).
4. J.Lindhard, Kgl.Danske Vidensk.Selskab Mat.-Fys.Medd., **28**, 8 (1954); J.Hubbard, Proc.Roy.Soc., **A243**, 336 (1957); L.I.Sham, Proc.Roy.Soc., **283**,33 (1965); A.O.E.Animalu, Phil.Mag., **11**, 379 (1965); H.Ehrenreich and M.H.Cohen, Phys.Rev., **115**, 786 (1959); P.K.W.Vinsome and M.Jaros, J.Phys., **C3**, 2140 (1970); P.K.W.Vinsome and D.Richardson, J.Phys., **C4**, 2650 (1971); S.P.Singhal, Phys.Rev:, **B12**, 564 (1975).
5. V.D.Gorobchenko and E.G.Maksimov, Usp.Fiz.Nauk **130**, 65 (1980).
6. W.Harrison, *Solid State Theory* (McGraw Hill, New York, 1970).
7. B.Case, in *Reaction of Molecules of Electrodes*, edited by N.S.Hush (Wiley Interscience, London, 1971); D.G.Knox and J.J.Kozak, Mol.Phys. **33**,

811 (1977); J.Liszi and I.Ruff, in *The chemical physics of solvation* edited by R.R.Dogonadze, E.Kalman, A.A.Kornyshev and J.Ulstrup (Elsevier Science Publishers B.V. Amsterdam, 1985), Part A, p.119.
8. J.C.Inkson, J.Phys. **C5**, 2599 (1972).
9. R.F.Loring and S.Mukamel, J.Chem.Phys. **87**, 1272 (1987).
10. A.Chandra and B.Bagchi, J.Chem.Phys. **90**, 1832 (1989); A.Chandra and B.Bagchi, J.Chem.Phys. **91**, 3056 (1989).
11. F.O.Raineri, Y.Zhou, and H.L.Friedman, Chem.Phys. **152**, 201 (1991); F.O.Raineri, H.Resat, and H.L.Friedman, J.Chem.Phys. **96**, 3068 (1992).
12. T.Fonseca and B.M.Ladanyi, J.Chem.Phys. **93**, 8148 (1990).
13. D.Bertolini and A.Tani, Mol.Phys. **75**, 1065 (1992).
14. M.Neumann, J.Chem.Phys. **85**, 1567 (1986).
15. B.Bagchi, J.Chem.Phys. **95**, 467 (1991).
16. Ph.A.Bopp, A.A.Kornyshev and G.Sutmann, J.Chem.Phys. (submitted).
17. O.V.Dolgov, D.A.Kirzhnits, and E.G.Maksimov, Rev.Mod.Phys. **53**, 81 (1981).
18. A.Fasolino, M.Parrinello, and M.P.Tosi, Phys.Lett. **66A**, 119 (1978); M.P.Tosi, in *Condensed Matter Physics Aspects of Electrochemistry*, edited by M.P.Tosi and A.A.Kornyshev (World Scientific, Singapore, 1991), p. 68.
19. A.A.Kornyshev, D.A.Kossakowski, and M.A.Vorotyntsev, in *Condensed Matter Physics Aspects of Electrochemistry*, edited by M.P.Tosi and A.A.Kornyshev (World Scientific, Singapore, 1991), p. 92.
20. Ph.A.Bopp, A.A.Kornyshev, and G.Sutmann, Phys.Rev.Lett. **76**, 1280 (1996).
21. A.A.Kornyshev, S.Leikin, and G.Sutmann, Electrochim.Acta **42**, 849 (1997).
22. A.A.Kornyshev and G.Sutmann (in preparation).
23. A.A.Kornyshev and G.Sutmann, J.Chem.Phys. **104**, 1524 (1996).
24. A.A.Kornyshev and G.Sutmann, Electrochim.Acta, 1997 (in press).
25. Ph.A.Bopp, A.A.Kornyshev and G.Sutmann (in preparation).
26. D.N.Zubarev, *Nonequilibrium Statistical Thermodynamics* (Plenum Press, New York, 1974).
27. L.E.Reichl, *A Modern Course in Statistical Physics* (Edward Arnold Ltd, Texas, 1980).
28. Ph.A.Bopp, G.Janćso, and K.Heinzinger, Chem.Phys.Lett. **98**, 129 (1983); G.Janćso, Ph.A.Bopp, and K.Heinzinger, Chem.Phys. **85**, 377 (1984).

29. A.K.Soper and J.Turner, Int.J.Mod.Phys. **B7**, 3049 (1993); A.K.Soper, J.Chem.Phys. **101**, 6888 (1994).
30. R.R.Dogonadze, A.A.Kornyshev, and A.M.Kuznetsov, Theor.Math. Phys. **15**, 407 (1973).
31. R.A.Marcus, J.Chem.Phys. **24**, 966 (1956).
32. Ph.A.Bopp, Chem.Phys. **106**, 205 (1986); G.Pálinkás, I.Bakó, K.Heinzinger, and Ph.A.Bopp, Mol.Phys. **73**, 897 (1991).
33. P.Schofield, Phys.Lett. **4**, 39 (1960).
34. P.A.Egelstaff, Adv.Phys. **11**, 203 (1962).
35. J.Borysow, M.Moraldi, and L.Frommhold, Mol.Phys. **56**, 913 (1985).
36. A.A.Kornyshev, A.M.Kuznetsov, D.K.Phelps, and M.J.Weaver, J.Chem.Phys. **91**, 7159 (1989).
37. A.M.Kuznetsov, *Charge Transfer in Physics, Chemistry and Biology* (Gordon and Breach Publishers, Amsterdam, 1995)

PAIR CORRELATIONS AND THE NONLOCAL STATIC DIELECTRIC PROPERTIES OF MOLECULAR LIQUIDS

MUNIR S. SKAF
Instituto de Química, UNICAMP,
Campinas, SP 13083-970, Brazil

The $h^{110}(r)$ functions and the site-site pair correlations $h_{ij}(r)$ of water, methanol, dimethyl sulfoxide and acetonitrile, obtained from molecular dynamics simulations, are used to calculate these liquids' dielectric constants and static longitudinal dielectric responses over wide ranges of wave vectors k. Our calculation of the dielectric constants are in execellent agreement with other simulation results for the same models, but require much shorter simulations. Regarding the k-dependent dielectric response, we show that the sharp peaks exhibited by these liquids arise from charge-charge correlations at intermediate to large ranges (typically from 5 to 15 Å). For water and methanol it is shown that the secondary peaks located at $k \approx 5.0$Å^{-1} are due to H-bonding. In addition, we have calculated the dielectric response for CO_2 and compare it to that exhibited by the the highly polar liquids.

1 Introduction

Understanding the dielectric behavior of polar liquids is a problem of long-standing interest to chemists[1,2] because these properties are fundamental to characterize the reactional environment of chemical processes taking place in the bulk of solutions. From the theoretical point of view, there is interest in determining not only the dielectric constants and orientational Kirkwood g-factors of realistic models of polar liquids, but also their dielectric response at finite wave vectors k, which are particularly important to theories of nonlocal electrostatics[3] devoted to the study of thermodynamics of ionic solutions,[3-5] solvent effects on charge transfer reactions,[6,7] and a variety of related phenomena.[3] In this context, the longitudinal dielectric permittivity $\varepsilon_L(k)$ is the most relevant quantity. Experimentally, $\varepsilon_L(k)$ can be determined through the charge-charge structure factor, indirectly measured by small angle neutron scattering.

While reliable calculation of dielectric constants, $\varepsilon(0)$, and Kirkwood g-factors, g_K, of polar liquids usually require lengthy computer simulations,[8,9] the calculation of $\varepsilon_L(k)$ is expected to be more easily performed because the underlying polarization response at finite wavelengths is mainly determined from correlations of shorter range. In spite of this, there are only a few studies reported in the literature on the properties of $\varepsilon_L(k)$ from a molecular point of view.[10-14] Although it is now well established that $\varepsilon_L(k) \rightarrow 1$ as $k \rightarrow \infty$ when the molecular spatial charge distribution is taken into account,[10] there are other questions regarding the behavior of $\varepsilon_L(k)$ for molecular liquids, for which our understanding is less complete. One of them concerns the origin of the sharp peak in the dielectric response function, $\chi(k) = 1 - 1/\varepsilon_L(k)$, which has been recently discussed by Bopp *et al.*[14] for liquid water, but appear in other liquids as well.[10,12]

In the light of this scenario, we show first, that accurate estimates for $\varepsilon(0)$ and g_K for different polar liquids can be obtained through the $\ell = 0$ Fourier-Hankel transform of the dipolar symmetry projection[1] $h^{110}(r)$ using only a fraction on the simulation runs commonly used in the calculation of those quantities. The method is applied to acetonitrile (MeCN), dimethylsulfoxide (DMSO), and the H-bonding liquids, methanol (MeOH) and water, whose dielectric constants have been obtained previously through different methods,[15-17] with the exception of DMSO. Secondly, we investigate some of the general features of $\varepsilon_L(k)$ over wide ranges of k for these liquids. In particular, we present a very simple rationale explaining the appearance of a sharp peak in the k-dependent longitudinal dielectric response we find for the simulated models and argue that this should be common to liquids. For the H-bonding liquids we also identify the origin of the secondary, satellite[14] peaks of $1/\varepsilon_L(k)$. In addition to these highly polar liquids, we have also simulated liquid CO_2 and compare the behavior of its k-dependent longitudinal dielectric response to that presented by the polar liquids.

2 Theoretical Framework

The approach to the dielectric properties we adopt here is based entirely on the calculation of the liquid structure through computer simulations. At finite k values, the static longitudinal dielectric response is evaluated directly through the charge-charge structure factor $S(k)$, which according to Raineri *et al.*[6,10] are related by

$$1 - \varepsilon_L^{-1}(k) = \frac{1}{k_B T \varepsilon_0 k^2} S(k) \tag{1}$$

where

$$S(k) = \sum_{ij} q_i q_j \left[\rho\, w_{ij}(k) + \rho^2 h_{ij}(k)\right] \tag{2}$$

Here $\rho = N/V$ is the number density and q_i the site charge. According to this expression $S(k)$ has an intramolecular contribution — given in terms of $w_{ij}(k) = \delta_{ij} + (1-\delta_{ij})\, j_0(k\,\ell_{ij})$, with ℓ_{ij} the *intramolecular* distance between sites *i* and *j*, and $j_0(x)$ the spherical Bessel function of order 0 — and an intermolecular contribution, given by the Fourier transform of the site-site pair correlations $h_{ij}(r)$, which are easily calculated from computer simulations. For large enough values of *k*, typically larger than 0.8 Å^{-1} for the systems we have simulated here, very accurate $\varepsilon_L(k)$ functions can be obtained from Eqs.1 and 2, as we shall see below.

In the $k \to 0$ limit, Eqs.1 and 2 reduce to the well-known formula for the dielectric constant $\varepsilon(0)$ of polar liquids derived by Hoye and Stell,[18] and by Chandler.[19] In principle, one could obtain estimates for $\varepsilon(0)$ from the $k \to 0$ limit of Eqs.1 and 2, taking advantage of the fact that the $h_{ij}(r)$ functions are short-ranged[18,19] (i.e., they decay like r^{-6} at large *r*). It turns out, however, that as $k \to 0$, Eqs.1 and 2 become increasingly sensitive to numerical errors associated with the Fourier transforms $h_{ij}(k)$ of the site-site correlation funtions, which thus turn this route somewhat impractical as means of estimating $\varepsilon(0)$. Alternatively, there are several other *bona fide* expressions for $\varepsilon(0)$ in terms of the $k \to 0$ limit of the transforms of pair distribution functions.[1,18,19] Among them, the infinite system Kirkwood relation (also valid for simulations using Ewald sums):

$$\frac{\left[\varepsilon(0)-1\right]\left[2\,\varepsilon(0)+1\right]}{9\,y\,\varepsilon(0)} = g_K = \lim_{k\to 0}\, 1 + \frac{\rho}{3} h^{110}(k) \tag{3}$$

where $y = \rho\mu^2/(9k_BT\varepsilon_0)$, with μ the molecular dipole, and $h^{110}(k)$ the $\ell = 0$ Fourier-Hankel transform of $h^{110}(r)$, proves to be particularly useful as means of obtaining $\varepsilon(0)$ from computer simulations. Its success can be partially attributed to the functional form of Eq.3, but also to the fact that the projections $h^{110}(r)$ are short-ranged[18] as well, so that $h^{110}(k)$ can be easily calculated numerically even at small *k*, provided the system is not too small.

3 Molecular Dynamics Simulations

The five liquids we simulate here are described by rigid, non-polarizable interaction site models in which the methyl groups, when present, are treated as united atoms. For MeCN we use the 3-site model of Edwards *et al.*,[15] for DMSO we use a model developed by Liu *et al.*,[20] for MeOH the H1 model of Haughney *et al.*,[21] for water we use SPC/E,[17] and for liquid CO_2 we use a 5-site model.[22] In each case the intermolecular interactions are given by the site-site sums of 6-12 Lennard-Jones plus Coulombic potentials. The molecular geometries and model parameters may be found in the references cited above. For DMSO, we have simulated 864 molecules in a cubic box with periodic boundary conditions, whereas for the other liquids 500 molecules were used. The dimensions of the simulation cells were set to match the experimental liquid densities at atmospheric pressure and 298 K, except for CO_2 where the temperature is 220 K. The simulations were run in the NVE ensemble with Ewald sums for the long range interactions. About 15,000 equilibrium timesteps were used to calculate the $h_{ij}(r)$ and $h^{110}(r)$ functions for each system.

4 Results and Discussion

4.1 Kirkwood g-factors and Dielectric Constants

The projection $h^{110}(r)$ gives the average relative orientation between the dipole vectors of a pair of molecules separated by a distance r and is the most relevant function for the calculation of g_K and $\varepsilon(0)$ in this work. Our simulation results for $h^{110}(r)$ for the four polar liquids are shown in Figure 1. There one notices that the $h^{110}(r)$ functions for both water and MeOH are mostly positive, while for DMSO and MeCN one sees antiparallel alignment of dipoles at smallest intermolecular separations due to the predominance of dipolar forces in these liquids‡ as opposed to the H-bonding structures[9] which characterize water and MeOH. Figure 1 also reveals that at large intermolecular separations, the $h^{110}(r)$ functions for DMSO and MeCN present quite apparent oscillations. Such feature is less prominent in water and MeOH because these liquids are less packed (i.e., they have smaller

‡The molecular shape may strongly affect the short-range behavior of $h^{110}(r)$ for MeCN.[15]

reduced density, $\rho^* = \rho\,\sigma^3$, compared to DMSO and MeCN). These oscillations at large r have in general a negative impact on the numerical accuracy of the $h^{110}(k)$ transform at small k. Therefore, Eq.3 is expected to work less well for liquids consisting of large molecules (large ρ^*). For the system sizes we simulate here, the correcting factor[1] for the simulated $h^{110}(r)$ function, which scales as $1/N$, yields negligible contributions within the numerical accuracy of the transforms.

The $\ell = 0$ Fourier-Hankel transform of $h^{110}(r)$ can be readily calculated numerically using standard techniques. The result is more commonly presented as the k-dependent short-range portion of the generalized Kirkwood g-factor $g_K^S(k) = 1 + \frac{1}{3}\,\rho\, h^{110}(k)$, whose $k \to 0$ limit yields the usual Kirkwood g-factor g_K, as expressed in Eq.3. Our $g_K^S(k)$ results for the four polar liquids are depicted in Figure 2. The resulting Kirkwood g-factors can be read off the y-axis, where we find $g_K = 1.6 \pm 0.15$ for DMSO, 2.0 ± 0.1 for MeOH, 2.5 ± 0.1 for water, and 1.1 ± 0.2 for MeCN. Using these values in Eq.3, we find for the dielectric constants: $\varepsilon(0) = 44 \pm 4$, 24 ± 2, 70 ± 3, and 32 ± 5, for DMSO, MeOH, SPC/E water, and MeCN, respectively. The error bars have been estimated according to [12]. These results are in excellent agreement with other computer simulation[15-17] estimates for the same quantities within similar accuracies. One advantage of this approach is that we need just a fraction of the lengthy runs usually required for the convergence of the mean-squared dipole densities.

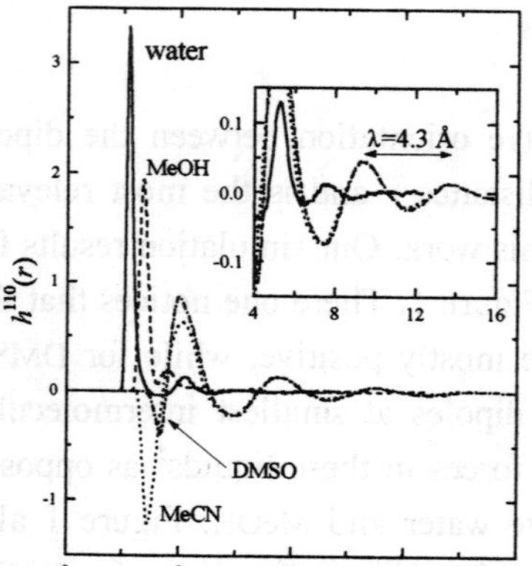

Figure 1. Projections h^{110}. Insert shows large r details for water, DMSO and MeCN. A typical wavelength λ (see text) for DMSO indicated.

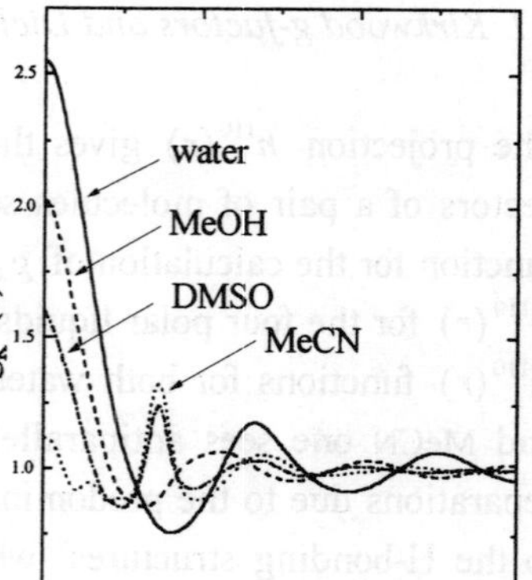

Figure 2. Generalized Kirkwood g-factor vs. k. First local maxima are located at $k^* = 2\pi/\lambda$.

Before proceeding to the discussion of the dielectric response at finite wave vectors in terms of $\varepsilon_L(k)$, it is interesting to notice in Figure 2 that the $g_K^S(k)$ functions for these four polar liquids present well defined peaks at $k^* > 0$, whose locations depend on the liquid. The presence of a peak in $g_K^S(k)$ at a finite but small wave vector k^*, indicates an underlying fluctuation of "wavelength" $\lambda = 2\pi / k^*$ in the structure of medium. Since k^* is a small wave vector, such feature of the liquid structure should be found at intermediate to large intermolecular separations, and therefore, is not expected to be *directly* associated to the molecular size (i.e., the nearest neighbor distance) or to other structural features at short range such as H-bonding distances. It should be remembered that oscillations in the pair distribution functions reflect the *local particle number density fluctuations seen by a molecule located at the origin.* In fact, this is characteristic of moderate to highly dense liquids and amorphous materials: λ corresponds to the spatial extent of a density fluctuation while the range over which such fluctuation persists defines essentially a correlation length in the medium. In Figure 1 this appears as oscillations in the $h^{110}(r)$ functions at intermediate to large values of r. [Of course, here the fluctuations are also modulated by the relative orientation between the dipoles]. Both DMSO and MeCN, exhibit density fluctuations with a typical wavelength of λ =4.3 Å in the range between roughly 8 and 20 Å, leading to k^*=1.46 Å^{-1}, which is precisely to the location of the peaks in the $g_K^S(k)$ for these two fluids depicted in Figure 2. Notice that the "wavelength" of such fluctuations may be different at different ranges. For instance, the distance between the first two minima of DMSO's $h^{110}(r)$ function is only 3.4 Å. Furthermore, notice that DMSO and MeCN present different short range structures (Figure 1) leading to different behavior in the corresponding $g_K^S(k)$'s at larger k. Similar considerations regarding the position of the local maxima in $g_K^S(k)$ for water and MeOH also apply, but the oscillations in their $h^{110}(r)$ are less well defined, indicating a larger spread of λ values for their dipole-dipole fluctuations. Consequently, the first peaks in their $g_K^S(k)$ are broader than in DMSO and MeCN.

4.2 Longitudinal Dielectric Response of Polar Liquids

We now turn attention to the k-dependent longitudinal dielectric response, calculated directly through Eqs.1 and 2. Our results for the four polar liquids are displayed in Figure 3 as $1/\varepsilon_L(k)$. In all four cases, one may see spurious oscillations in $1/\varepsilon_L(k)$ at small k values due to large numerical errors associated with the Fourier transforms. At wave vectors larger than 1.0, 0.7, 1.2, and 1.0 Å^{-1} for water, MeOH, DMSO, and MeCN, respectively, the results are very accurate (the error bars would barely show in this scale). These curves show some general features which are now well established in the literature.[10] Namely, $\varepsilon_L(k)$ for each liquid goes to 1 as $k \to \infty$ after crossing a singularity at large k whose location depends on the liquid. A singularity at small k is also common to polar liquids. However, these cannot be precisely detected from these calculations because the method is not accurate at small k. Similarly $\varepsilon(0)$ cannot be obtained from the $k \to 0$ limit of our $\varepsilon_L(k)$.

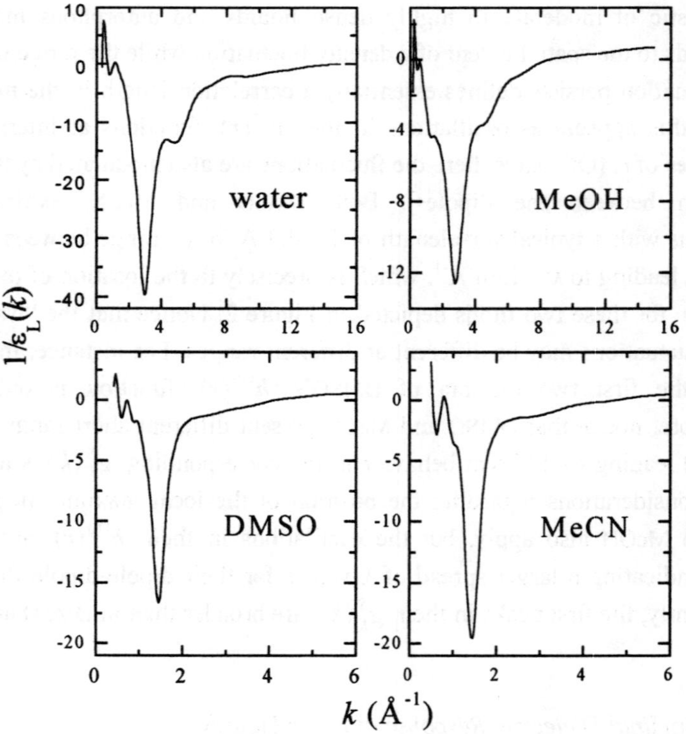

Figure 3. Inverse the *k*-dependent longitudinal dielectric permittivity vs. *k*.

The question we would like to address here concerns the origin of the sharp peak in the dielectric response function, $\chi(k) = 1 - 1/\varepsilon_L(k)$, implicit in the data shown in Figure 3. The main peaks in $\chi(k)$ are approximately located at 3.0, 2.8, 1.46, and 1.46Å^{-1}, for water, MeOH, DMSO, and MeCN, respectively, and are essentially due to the intermolecular contributions. For DMSO and MeCN the location of the peaks in $\chi(k)$ coincide with those found in $g_K^S(k)$ (see Figure 2). This indicates that the corresponding charge-charge local fluctuations in these two liquids are essentially those we found for the dipole-dipole fluctuations through the $h^{110}(r)$ functions in the range from 8 to 20 Å, that is, their typical wavelength should lie around 4.3Å. On the other hand, we also see from Figure 3 that the main peak in the dielectric response function for the H-bonding liquids present contributions from the charge-charge local fluctuations beyond those shown by the purely dipolar type, since $\chi(k)$ and $g_K^S(k)$ peak at different wave vectors. To further investigate this, we have displayed in Figure 4 the charge-charge weighted sums of site-site pair correlation functions. The line style follows that of Figure 1. The curves for DMSO and MeCN have been shifted down by 1.5 units, from which one may see a fluctuation of "wavelength" 4.3Å at intermediate to large values of r.

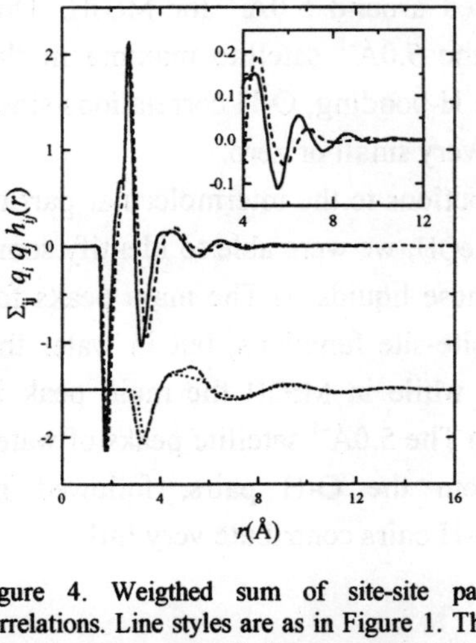

Figure 4. Weigthed sum of site-site pair correlations. Line styles are as in Figure 1. The insert shows large r details for water and MeOH.

Table 1. Positions of the consecutive minima and maxima of the weighted sum of site-site functions for water and MeOH, in Å.

Water		MeOH	
Valeys	Peaks	Valeys	Peaks
1.75	2.73	1.86	2.78
3.23	4.50	3.48	4.61
5.53	6.43	5.87	6.86
7.54	8.62	8.09	(9.6)
9.60	10.67	—	—

Details of the weighted sums for water and MeOH are shown in the insert, while the precise location of their minima and maxima are shown in Table 1. From there, one sees that the separation between consecutive valeys or consecutive peaks changes in going from small to larger values of r for both water and MeOH. In water, the dominating "wavelength" of the charge-charge fluctuations in the range between 4.0 and 11.0Å is roughly 2.08Å, which would give a peak around $k^*=3.0Å^{-1}$ in the intermolecular part of $\chi(k)$. In MeOH the average "wavelength" is about 2.23Å in the range between 4.5 and 9.0Å, indicating that the intermolecular part of $\chi(k)$ should peak roughly at $2.8Å^{-1}$. This is very much in agreement with the $\chi(k)$ data shown in Figure 3.

It should be noted, however, that the intermolecular part of $\chi(k)$ arises from an integration over all values of r and, therefore, the rationale we have provided above is only semi-quantitative. To further substantiate our argument, we have calculated the intermolecular structure factor for water and MeOH truncating the correlations beyond a certain cutoff value r_c. From an auxiliary simulation with 2048 MeOH molecules, we find that using $r_c > 9$ Å has a negligible effect[22] on the main peak of $\chi(k)$ and that $r_c < 5$ Å leads to noticeable changes in its position and shape. For water, we find that $r_c < 4.6$ Å strongly affect the main peak. However, the secondary peak in $\chi(k)$ for water, located at $5.0Å^{-1}$, remains practically unaffected unless $r_c < 2.3$ Å. That is, the satellite[14] peak is essentially recovered as long as r_c includes most of the first peak of the $h_{OH}(r)$ site-site function. The same holds true for the satellite shoulder also located around $5.0Å^{-1}$ for MeOH. This clearly indicates that, unlike the main peak, the $5.0Å^{-1}$ satellite maxima in the $\chi(k)$ of water and MeOH arise mainly from the H-bonding, O-H correlations since below 2.3Å the other site-site functions are still very small or zero.

By looking at the individual site-site contributions to the intermolecular part of the dielectric response functions of water and MeOH, we were able to identify some additional features of the $\chi(k)$ functions for these liquids: i) The main peaks for water and MeOH have contributions from all site-site functions, but in water the H-H correlations contribute to a lesser extent, while in MeOH the main peak is mostly determined from the O-O correlations. ii) The $5.0Å^{-1}$ satellite peaks of water and MeOH have dominant contributions from the O-H pairs, followed in importance by the O-O correlations, while the H-H pairs contribute very little.

4.3 Dielectric Response of CO_2

We have seen that the longitudinal dielectric response functions for the simulated polar liquids share many common features. In order to illustrate what would be the general shape of this response for a non-dipolar molecular liquid, we have calculted $1/\varepsilon_L(k)$ for liquid CO_2 at 220K. The result is shown Figure 5, where the dashed line is the intermolecular contribution, the dotted line the intramolecular part, and the solid line is the total $1/\varepsilon_L(k)$. The general shape of the dielectric response for this non-dipolar fluid is very similar to that exhibited by the polar liquids of Figure 3. The difference lies essentially in the fact that while for all four polar liquids we have studied here, the main peak comes from the intramolecular correlations at intermediate ranges, for CO_2 the shape and intensity of the main peak is almost entirely determined from the intramolecular charge distribution. The only visible contributions from intermolecular correlations are the small glitches on the main peak between 1.5 and 4.0Å. Notice that the intensity of the main peak in the longitudinal dielectric response of liquid CO_2 at 220 K is nearly twice as that found for MeOH at room temperature but smaller than that of water. This feature is evidently incidental. The location of the peak, however, is likely to be related to the distance between the two oxygen sites within the CO_2 molecule which is 2.32Å for the model used in our simulations.

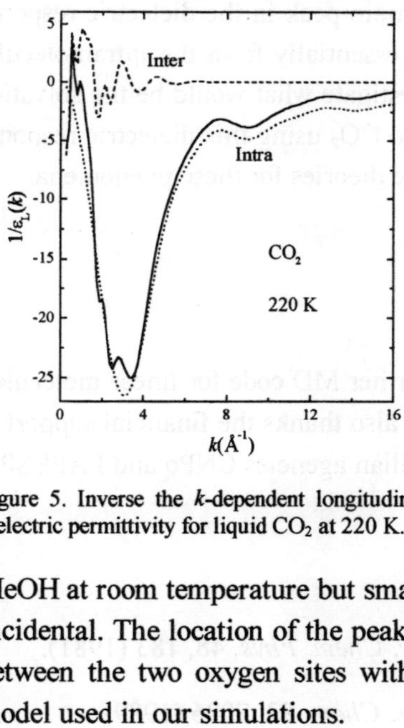

Figure 5. Inverse the *k*-dependent longitudinal dielectric permittivity for liquid CO_2 at 220 K.

5 Summary and Concluding Remarks

We described a method for calculating dielectric constants of polar liquids through computer simulations based on the transform of the $h^{110}(r)$ projection which require much shorter simulation runs than the conventional approach via the mean squared dipole densities. This has been illustrated by determining the dielectric constants of highly polar liquids whose $\varepsilon(0)$ were previously determined through other methods.

The *k*-dependence of the static longitudinal dielectric response for highly polar liquids and for liquid CO_2 have also been investigated in detail. We have been able to identify the origin of the main, overresponse peak of the polar liquids as charge-charge fluctuations of certain dominating "wavelengths" which take place at intermediate ranges (typically from 5 to 15Å) in the liquid structure. The satellite peaks at approximatelly 5.0Å^{-1} exhibited by water and MeOH are shown to be a direct consequence of H-bonding.

In contrast to the highly polar liquids, the main peak in the dielectric response of the non-dipolar liquid, CO_2 at 220K, arises essentially from the intramolecular part. It would be an interesting question to investigate what would be the solvation dynamics and other solvation properties in liquid CO_2 using this dielectric response function within the context of nonlocal dielectric theories for these phenomena.

Acknowledgements

The author thanks Professor Branka Ladanyi for her MD code for linear molecules, and José Pedro Rino for fruitful discussions. He also thanks the financial support of ICTP, Universidade de São Paulo, and the Brazilian agencies CNPq and FAPESP.

References

1. G. Stell, G. N. Patey and J. S. Høye, *Adv. Chem. Phys.* **48**, 183 (1981).
2. D. Kivelson and H. L. Friedman, *J. Phys. Chem.* **93**, 7026 (1989).
3. *The Chemical Physics of Solvation*, Part A, edited by R. Dogonadze, E. Kálman, A. A. Kornyshev and J. Ulstrup (Elsevier, 1985).
4. T. Fonseca and B. M. Ladanyi, *in Condensed Matter Aspects of Electrochemistry*, edited by M. Tosi and A. A. Kornyshev (World Scientific, 1991).
5. A. A. Kornyshev and G. Sutmann, *J. Chem. Phys.* **104**, 1524 (1996).

6. F. O. Raineri, Y. Zhou, H. L. Friedman and G. Stell, *Chem. Phys.* **152**, 201 (1991).

7. B. Bagchi and A . Chandra, *Adv. Chem. Phys.* **80**, 1 (1991).

8. S. W. De Leeuw, J. M. Perram and E. R. Smith, *Ann. Rev. Phys. Chem.* **37**, 245 (1986).

9. B. M. Ladanyi and M. S. Skaf, , *Ann. Rev. Phys. Chem.* **44**, 335 (1993).

10. F. O. Raineri and H. L. Friedman, *J. Chem. Phys.* **96**, 3068 (1992).

11. M. S. Skaf and B. M. Ladanyi, *J. Chem. Phys.* **102**, 6542 (1995).

12. M. S. Skaf, *Mol. Phys.* (Vol. 90, Number 1, 1997, to appear).

13. A. D. Trokhymchuk, M. F. Holovko and K. Heizinger, *J. Chem. Phys.* **99**, 2964 (1993).

14. P. Bopp, A. Kornyshev and G. Sutmann, *Phys. Rev. Lett.* **76**, 1280 (1996).

15. D. Edwards, P.A. Madden and I. McDonald, *Mol. Phys.* **51**, 1141 (1984).

16. T. Fonseca and B. M. Ladanyi, *J. Chem. Phys.* **93**, 8148 (1990).

17. M. R. Reddy and M. Berkowitz, *Chem. Phys. Lett.* **155**, 173 (1989).

18. J. S. Høye and G. Stell, *J. Chem. Phys.* **65**, 18 (1976).

19. D. Chandler, *J. Chem. Phys.* **67**, 1113 (1977).

20. H. Liu, F. Müller-Plathe and W. van Gunsteren, *J. Am. Chem. Soc.* **117**, 4363 (1993).

21. M. Haughney, M. Ferrario and I. McDonald, *J. Phys. Chem.* **91**, 4934 (1987).

22. For intermolecular separations larger than approximately 9.0Å the site-site functions of MeOH seem to simply oscillate with statistical noise around unity, showing no evidence of a definite "wavelength" in the local density fluctuations. In water this takes place for $r > 11$ Å.

MOLECULAR MECHANISMS OF SOLVATION DYNAMICS IN POLAR AND NONPOLAR LIQUIDS

B.M. LADANYI
Department of Chemistry, Colorado State University, Fort Collins, CO 80523, USA

We analyze the molecular origin of solvation dynamics arising from a perturbation in solute-solvent electrostatic interactions for dipolar and quadrupolar diatomic solutes in acetonitrile and carbon dioxide liquids. Our focus is on $C(t)$, the time correlation function of the fluctuations in the solute-solvent electrostatic potential, which describes solvation dynamics within the linear response approximation. $C(t)$ is calculated from molecular dynamics trajectory data and expressed in terms of static and dynamical correlations, the latter contained in the solvation velocity autocorrelation, $G(t)$. We demonstrate how these functions can be analyzed in order to determine several aspects of the molecular origin of solvation response: the relative contributions of rotation and translation, single-molecule vs. collective velocity correlations, single-solvent vs. solvent-pair dynamics and contributions from solvent molecules in different solvation shells. Our results confirm the importance of rotational relaxation in electrostatic solvation dynamics in polar and nonpolar solvents. We find that $C(t)$ is much more collective in character than $G(t)$, showing that the collective aspects of the short-time solvation dynamics are due largely to static correlations. We also find that contributions from outer solvation shells mainly impact long-time solvation dynamics and are unimportant in nonpolar solvents.

1. Introduction

Solvation dynamics refers to the rate of solvent reorganization in response to an abrupt change in solute properties. In most of the experimental manifestations of this phenomenon, the solute is a chromophore that undergoes an electronic transition resulting primarily in a change in its charge distribution. The solvent reorganization dynamics is then monitored by measuring the time-evolution of the Stokes shift in the fluorescence spectrum of a dissolved chromophore. Stimulated in part by the advances in laser spectroscopy, which have made it possible to measure the contributions of ultrafast processes to solvation dynamics, there has been a great deal of reasearch activity and progress in this area, much of it summarized in recent reviews.[1-6] Although solvation dynamics is interesting in its own right, its understanding is also necessary in order to account for the effects of solvent dynamics on chemical reactions. Solvation dynamics occurring in response to a change in solute-solvent electrostatic interactions, is closely related to the influence of solvent dynamics on the rates of electron and ion transfer reactions.[7-10]

Recent advances in our understanding of the solvation process have come through an interplay of ultrafast laser spectroscopy, theory and computer simulation. It is through computer simulation studies that the inertial component, which accounts for the major portion of the solvation response in common polar liquids such as water,[11-17] acetonitrile[18-20] and methyl chloride,[21] was first discovered and later confirmed through experimental studies.[22-26] An important goal for theoretical modelling of solvation dynamics, in addition to the realistic representation of this process, is to determine its molecular origin. In order to understand the molecular mechanisms of solvation dynamics, we need answers to many questions, including: Does the solvent relax mainly through rotation or through translation? Is the solvation process predominantly collective, involving correlated motion of solvent molecules? Which molecules dominate the early solvation response? How does solvation dynamics depend on the range and symmetry of the perturbation in solute-solvent interactions? Is solvation dynamics fundamentally different in polar and nonpolar solvents? Answers to such mechanistic questions are accessible from the analysis of molecular dynamics (MD) simulation data.

In this article, I will describe some of the methods that my coworkers and I are using to obtain mechanistic information from MD simulation. These methods will be illustrated by analyzing and comparing solvation dynamics in acetonitrile and carbon dioxide liquids, two solvents of similar shape, size, mass and moments of inertia, but different charge distributions. This comparative study is motivated by recent experiments[27] which show that solvation in nonpolar solvents such as benzene and dioxane, is quite similar in both the size of the fluorescence Stokes shift and in the scale of its time-evolution to what has been observed in polar solvents. This suggests that solvation dynamics in nonpolar polyatomic solvents might proceed via a similar mechanism. We also investigate here the solute dependence by comparing solvation dynamics for dipolar and quadrupolar diatomic solutes in CO_2.

Solvation dynamics is usually measured by monitoring the time-evolution in the Stokes shift of the fluorescence spectrum of a dissolved chromophore. For chromophores used in these experiments, the change in the molecular geometry on electronic excitation is small, so the Stokes shift arises primarily from the change in the solute-solvent interaction potential broughr about by solute electronic transition. The dynamical variable monitored in the experiment is therefore

$$\Delta E = U_{Ss}^{e} - U_{Ss}^{g}, \tag{1}$$

where U_{Ss}^{e} and U_{Ss}^{g} are solute-solvent potentials for the solute in the electronically excited (e) and ground (g) states. The largest steady-state Stokes shifts are observed for solute-solvent pairs for which ΔE corresponds to a large change in electrostatic interactions, so ΔE is expected to be primarily electrostatic in nature for solvation in polyatomic solvents. In this work we will focus on electrostatic solvation, but will consider solute-solvent pairs with and without permanent dipoles.

In terms of ΔE, the Stokes shift response function is given by

$$S(t)=\frac{\overline{\Delta E(t)}-\overline{\Delta E(\infty)}}{\overline{\Delta E(0)}-\overline{\Delta E(\infty)}}, \tag{2}$$

where overbars denote averages over nonequilibrium trajectories in which the system is perturbed by solute electronic excitation occurring at $t = 0$. $t = \infty$ corresponds to steady-state fluorescence.

For small enough perturbations in solute-solvent interactions on solute electronic excitation, $S(t)$ can be approximated by its linear-response counterpart, the solvation time correlation function (TCF)

$$C(t)=\langle\delta\Delta E(0)\delta\Delta E(t)\rangle/\langle(\delta\Delta E)^2\rangle \tag{3}$$

where $\delta\Delta E=\Delta E-\langle\Delta E\rangle$ is a fluctuation in ΔE in an equilibrated solute-solvent system and where $\langle\ldots\rangle$ denotes an equilibrium ensemble average. The equilibrated system can correspond to the solute in its ground or excited electronic state. MD simulation studies indicate that the approximation $S(t)\cong C(t)$ holds for fairly large perturbations for some solvents,[3-6] including acetonitrile,[18,20] one of the liquids considered here.

The approach that I will discuss focuses on the analysis of $C(t)$ and of the velocity autocorrelation function (VACF) of ΔE,

$$G(t)=\langle\Delta\dot{E}(0)\Delta\dot{E}(t)\rangle, \tag{4}$$

which is related to the solvation TCF by

$$C(t)=1-\langle(\delta\Delta E)^2\rangle^{-1}\int_0^t(t-\tau)G(\tau)\,d\tau. \tag{5}$$

The above equation shows that $C(t)$ is governed by static correlations contained in $\langle(\delta\Delta E)^2\rangle$ and dynamical correlations contained in $G(t)$. As we shall see, both of them play important roles in the behavior of $C(t)$, with the *static* correlations being chiefly responsible for the short-time aspects of its collective character.

Analysis of $G(t)$ is the focus of the instantaneous normal mode (INM) theory of solvation dynamics.[19,29-32] In this approach one constructs the INM solvation spectrum $\rho_{solv}(\omega)$ which is approximately related to $G(t)$ by Fourier transformation

$$G(t)\cong k_B T\int d\omega\,\rho_{solv}(\omega)\cos\omega t. \tag{6}$$

The above relation is exact at short times, making INM analysis an especially valuable tool for the investigation of early solvation dynamics. Using the connection between INM eigenvectors and molecular coordinates and projection operator techniques, $\rho_{solv}(\omega)$ can be decomposed into portions resulting from a variety of molecular processes.[19,29,30] It is through this decomposition that contributions of rotational and translational dynamics to the early solvation response were first determined.[19,29] Specifically, for solvation dynamics in acetonitrile[19] and CO_2,[29] Ladanyi and Stratt found that rotation is the major contributor to electrostatic solvation dynamics in polar and nonpolar liquids. They also showed[29] that translation plays a more important role than rotation when the leading term in the rotational-invariant expansion of the perturbation ΔE in solute-solvent interaction is independent of orientational coordinates, as is the case, for example, for dispersion interactions. INM analysis has also revealed that much of the early solvation response comes from molecules in the first solvation shell, even when ΔE is a perturbation in the dipole-dipole interaction of a polar solute-solvent system.

Although I will refer here to some of these INM results for solvation dynamics, my focus will be on applying a different method, proposed by W.A. Steele,[33] for analyzing collective TCFs in terms of static and dynamical correlations. It is a time-domain method, not restricted to short-time dynamics. Ladanyi and Klein[31] have used it recently to determine the contributions of rotation and translation to polarizability anisotropy and solvation dynamics in acetonitrile. The analysis presented here is extended to include the contributions of velocity self- and cross-correlations to solvation dynamics. In addition, we examine $C(t)$ directly to determine its single-solvent and solvent-pair contributions and to find out how molecules in different solvation shells contribute to solvation dynamics.

The remainder of the article is organized as follows: In Sec. 2, we outline our approach to the study of solvation mechanisms and define the quantities of interest. The potential models for the solutes and solvents are specified in Sec. 3, which also includes information about the MD simulation methods used in our study. Our results are presented in Sec. 4 and our main findings summarized in Sec. 5.

2. Theoretical Background and Definitions of Quantities of Interest

As noted above, our approach to the study of solvation mechanisms is based on the analysis of both $G(t)$ and $C(t)$. The questions that we aim to address concern the relative contributions of of molecular rotational and translational motions to solvation dynamics, the collective vs. single-solvent molecule nature of the solvation response and the relative contributions from solvent molecules in the first and outer solvation shells.

The answer to the first question is available through the analysis of $G(t)$.[19,29-33] As Steele pointed out,[33] for a VACF corresponding to a collective variable such as ΔE, it is possible to identify contributions from velocities corresponding to different degrees of freedom. Specifically, if the solute and solvent molecules are rigid, we can write[31]

$$\Delta\dot{E} = \Delta\dot{E}^{trans} + \Delta\dot{E}^{rot} \tag{6}$$

where

$$\Delta\dot{E}^{trans} = \sum_{j=0}^{N} \sum_{\mu=x,y,z} \frac{\partial \Delta E}{\partial r_{j\mu}} \dot{r}_{j\mu} \quad \text{and} \quad \Delta\dot{E}^{rot} = \sum_{j=0}^{N} \sum_{\mu=\theta,\varphi} \frac{\partial \Delta E}{\partial r_{j\mu}} \dot{r}_{j\mu}. \tag{7}$$

The sums include the solute, $j = 0$, and N solvent molecules. For linear and quasi-linear, such as a 3-site model for acetonitrile, molecules, $r_{j\mu} = \{x_j, y_j, z_j, \theta_j, \varphi_j\}$. For nonlinear molecules, a third angular variable would need to be included into the expression for $\Delta\dot{E}^{rot}$. Eq. (6) gives

$$G(t) = G^{trans}(t) + G^{cross}(t) + G^{rot}(t). \tag{8}$$

Note that $C(t)$ cannot be separated in this way into its rotational, translational and cross-correlation contributions. The problem arises due to the fact that static correlations contribute to the time-evolution of this function. Specifically, substituting Eq. 8 into Eq. 5

$$C(t) = 1 - \left\langle(\delta\Delta E)^2\right\rangle^{-1} \int_0^t (t-\tau)\left[G^{trans}(\tau) + G^{cross}(\tau) + G^{rot}(\tau)\right] d\tau, \tag{9}$$

which shows that the decay rate of $C(t)$ depends on $\left\langle(\delta\Delta E)^2\right\rangle$. Thus $C(t)$ - 1 is separable while the TCF itself is not.

Eq. (7) can also be used to identify the contributions of single-particle dynamics to $G(t)$. They are obtained by constructing autocorrelations of

$$\Delta\dot{E}_j^{trans} = \sum_{\mu=x,y,z} \frac{\partial \Delta E}{\partial r_{j\mu}} \dot{r}_{j\mu} \quad \text{and} \quad \Delta\dot{E}_j^{rot} = \sum_{\mu=\theta,\varphi} \frac{\partial \Delta E}{\partial r_{j\mu}} \dot{r}_{j\mu}, \tag{10}$$

giving, for example,

$$G_s^{trans}(t) = \sum_{j=0}^{N} \left\langle \Delta\dot{E}_j^{trans}(0) \Delta\dot{E}_j^{trans}(t) \right\rangle. \tag{11}$$

$G(t)$ is a sum of single-molecule and pair velocity correlations,

$$G(t)=G_s(t)+G_p(t) \tag{12}$$

Since velocities of different molecules and for different degrees of freedom are uncorrelated at $t=0$,

$$G(0)=G_s(0). \tag{13}$$

Their components are also equal and the rotational-translational cross-correlation vanishes

$$G^{rot}(0)=G_s^{rot}(0), \quad G^{trans}(0)=G_s^{trans}(0), \quad G^{cross}(0)=G_s^{cross}(0)=0. \tag{14}$$

In our model, ΔE corresponds to a sum of pairwise-additive solute-solvent interactions,

$$\Delta E=\sum_{j=1}^{N}\Delta w_{0j} \tag{15}$$

This makes it possible to separate $C(t)$ into contributions from interactions of the solute with a single solvent molecule and with a pair of solvent molecules,

$$C(t)=C_{ss}(t)+C_{sp}(t), \tag{16}$$

where

$$C_{ss}(t)=\left\langle(\delta\Delta E)^2\right\rangle^{-1}\sum_{j=1}^{N}\left\langle\delta\Delta w_{0j}(0)\delta\Delta w_{0j}(t)\right\rangle \tag{17}$$

and $\delta\Delta w_{0j}=\Delta w_{0j}-\left\langle\Delta w_{0j}\right\rangle=\Delta w_{0j}-\langle\Delta E\rangle/N$.

Note that, because of the contribution of solute motion to $G_s(t)$, there is no simple correspondence between it and $C_{ss}(t)$. More explicitly,

$$\Delta\dot{E}_0=\sum_{j=1}^{N}\sum_{\mu}\frac{\partial\Delta w_{0j}}{\partial r_{0\mu}}\dot{r}_{0\mu}, \tag{18}$$

which shows that $\langle\Delta\dot{E}_0(0)\Delta\dot{E}_0(t)\rangle$ will include terms dependent on pairs of solvent molecules. If the solute is stationary, $\Delta\dot{E}_0$ vanishes and $G_s(t)$ contains only single-solvent terms. We will consider nonstationary solutes, which, however have considerably larger masses and moments of inertia than the solvent molecules, so that $\langle\Delta\dot{E}_0(0)\Delta\dot{E}_0(t)\rangle \ll G_s(t)$ and the correspondence between $G_s(t)$ and $C_{ss}(t)$ should hold approximately:

$$G_s(t) \cong -\langle(\delta\Delta E)^2\rangle \frac{d^2 C_{ss}(t)}{dt^2} \tag{19}$$

In addition to examining the rotational vs. translational dynamical origin of contributions to $G(t)$ and the single-particle vs collective aspects of solvation dynamics, we can find out how $C(t)$ is affected by the interaction range. We do this by investigating how much of it comes from the solvent molecules in the first vs. more distant solvation shells. We separate ΔE into first-shell and outer-shell contributions

$$\Delta E = \Delta E^{1st} + \Delta E^{outer}, \tag{20}$$

which leads to

$$\begin{aligned} C(t) = \langle(\delta\Delta E)^2\rangle^{-1}[&\langle\delta\Delta E^{1st}(0)\delta\Delta E^{1st}(t)\rangle + \langle\delta\Delta E^{1st}(0)\delta\Delta E^{outer}(t)\rangle \\ &+\langle\delta\Delta E^{outer}(0)\delta\Delta E^{1st}(t)\rangle + \langle\delta\Delta E^{outer}(0)\delta\Delta E^{outer}(t)\rangle] \\ &= C^{1st}(t) + C^X(t) + C^{outer}(t), \end{aligned} \tag{21}$$

where $C^X(t)$ represents the first-shell - outer-shell cross correlation.

3. Models and Computational Details

We present the results for solvation dynamics of a dipolar diatom in acetonitrile at 293 K and the density of 0.7882 g/cm^3, which corresponds to 1 atm pressure, and for dipolar and quadrupolar diatoms in liquid carbon dioxide at 220 K and the density of 1.128 g/cm^3, corresponding to conditions in the vicinity of the triple point.[34] All the systems contain one solute and 255 solvent molecules and the results presented represent averages over trajectories of 0.8 ns length. Time steps of 0.8 fs were used for the acetonitrile-solute system and 0.6 fs for the CO_2-solute systems. Other details of the simulation methods are as described in Refs. 19 and 29. The solvents and solutes are modelled as rigid molecules and the intermolecular potentials are represented as site-site Lennard-Jones (LJ) + Coulomb interactions, with Lorentz-Berthelot combining rules[35]

$$u_{\alpha\beta}(r) = 4(\varepsilon_\alpha\varepsilon_\beta)^{1/2}\left[\left(\frac{\sigma_\alpha+\sigma_\beta}{2r}\right)^{12} + \left(\frac{\sigma_\alpha+\sigma_\beta}{2r}\right)^{6}\right] + \frac{q_\alpha q_\beta}{4\varepsilon_0\pi r}, \tag{22}$$

where σ_α and ε_α are the LJ diameter and energy, while q_α is the partial charge of site α. For CH_3CN we use a 3-site potential model, developed by Edwards et al.,[36] in which the methyl group, CH_3, is treated as a single site. CO_2 is represented by a

5-site model, with the three atomic sites carrying Coulomb and LJ interactions and 2 outlying sites carrying only partial charges. The model parameters are given in Ref. 29 The dipolar solute has the bond length, mass and LJ potential parameters of Br_2 and site partial charges $e/2$ and $-e/2$. The quadrupolar solute is a model for I_2. The model parameters, which are the same as those used, respectively, in Refs. 19 and 29, are given in Table 1. The table also includes the values of site masses, m_α, site distances from the center-of-mass, d_α.

Table 1: Solute parameters

A. Dipolar solute

site	σ_α / Å	$(\varepsilon_\alpha / k_B)$ / K	q_α / e	m_α / amu	d_α / Å
1	3.800	146	0.5	79.9	-1.14
2	3.800	146	-0.5	79.9	1.14

B. Quadrupolar solute: I_2

site	σ_α / Å	$(\varepsilon_\alpha / k_B)$ / K	q_α / e	m_α / amu	d_α / Å
1(I)	4.009	170.6	0.3277	126.9	-1.335
2	0.0	0.0	-0.6554	0.0	0.0
3(I)	4.009	170.6	0.3277	126.9	1.335

To calculate $C(t)$ and related quantities, we take the energy gap ΔE to be equal to the electrostatic portion of the solute-solvent potential. This gives

$$\Delta w_{0j} = \frac{1}{4\pi\varepsilon_0} \sum_{\alpha=1} \sum_{\beta=1} \frac{q_\alpha q_\beta}{\left|\mathbf{r}_{0\alpha} - \mathbf{r}_{j\beta}\right|}, \tag{23}$$

where $\mathbf{r}_{0\alpha}$ is the position of solute site α and $\mathbf{r}_{j\beta}$ is the position of site β on solvent molecule j. In order to calculate the rotational and translational contributions to $G(t)$ and $G_s(t)$, we obtain analytic expressions for $\partial\Delta E/\partial r_{j\mu}$ and evaluate $\Delta\dot{E}^{trans}$ and $\Delta\dot{E}^{rot}$ by using $r_{j\mu}$ and $\dot{r}_{j\mu}$ obtained in the course of MD simulation.

4. Results

4.1. Solvation Velocity Autocorrelation Function

We consider first the question of the dynamical origin of the solvation VACF $G(t)$ by separating it into components due to rotation, translation and rotation-translation cross correlations, according to Eqs. 6 - 8. The same type of analysis is applied to $G_s(t)$, the contribution to $G(t)$ from molecular velocity autocorrelations.

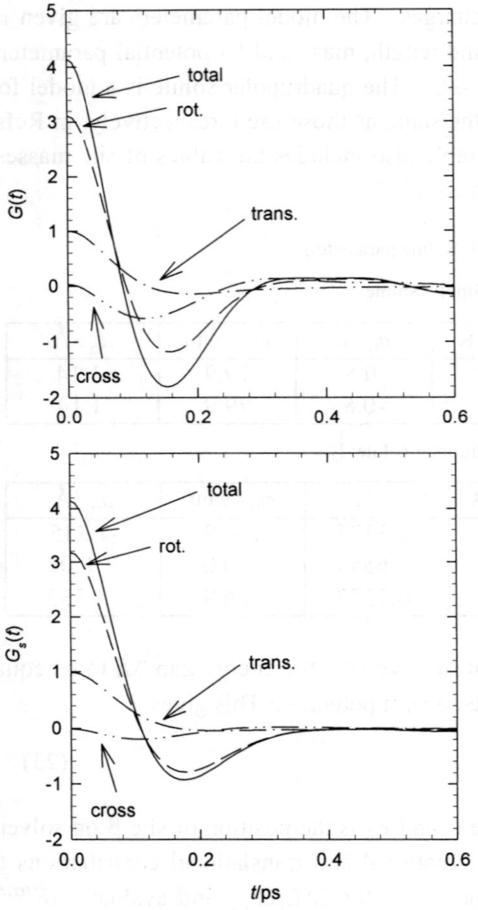

Figure 1: $G(t)$ (top panel) and $G_s(t)$ (bottom panel) and their rotational (dashed line), translational (dash-dotted line) and crosscorrelation (dash-double dotted line) components for the dipolar solute in acetonitrile

The decomposition of $G(t)$ and $G_s(t)$ into rotational, translational and cross-correlation components for the dipolar solute in acetonitrile is shown in Figure 1. The results for $G(t)$ had been reported earlier by Ladanyi and Klein.[31] Because of the lack of velocity cross-correlations for different degrees of freedom at $t = 0$, $G(0) = G_s(0)$ and the same equality holds for their components, as pointed out in Eqs. 13 and 14. The rotational contribution to $G(0)$ is seen be to be the largest, amounting to 75% of $G(0)$, well in excess of the 40% contribution expected on the basis of equipartition. The rotational component is the major contributor to both functions at longer times as well. The translational motion contributing to $G(t)$ is less hindered than rotational motion, exhibiting essentially no recoil, as evidenced by the near absence of a negative region in $G^{trans}(t)$. $G^{cross}(t)$ is relatively small. Comparison of the top and bottom panels of the figure indicates that it is due primarily to intermolecular velocity correlations, given that $G_s^{cross}(t)$ is nearly zero. This comparison also shows that the negative portions of $G(t)$ and $G^{rot}(t)$ are more pronounced than the corresponding portions of $G_s(t)$ and $G_s^{rot}(t)$. Thus a significant fraction of this feature in the solvation VACF is intermolecular in nature. This issue will be discussed further for all three solute-solvent systems.

In Figure 2 is depicted $G(t)$ and its components, $G^{rot}(t)$ $G^{trans}(t)$, and $G^{cross}(t)$ for dipolar (top panel) and quadrupolar (bottom panel) solutes in CO_2. Comparison of these results with Figure 1 indicates that, for the dipolar solute, $G(t)$ and its components look quite similar to the corresponding functions for dipole solvation in acetonitrile. This similarity extends to the relative conribution of rotational dynamics. In this case, we find that $G^{rot}(0)$ is 74% of $G(0)$, very close to 75% found for this solute in acetonitrile.

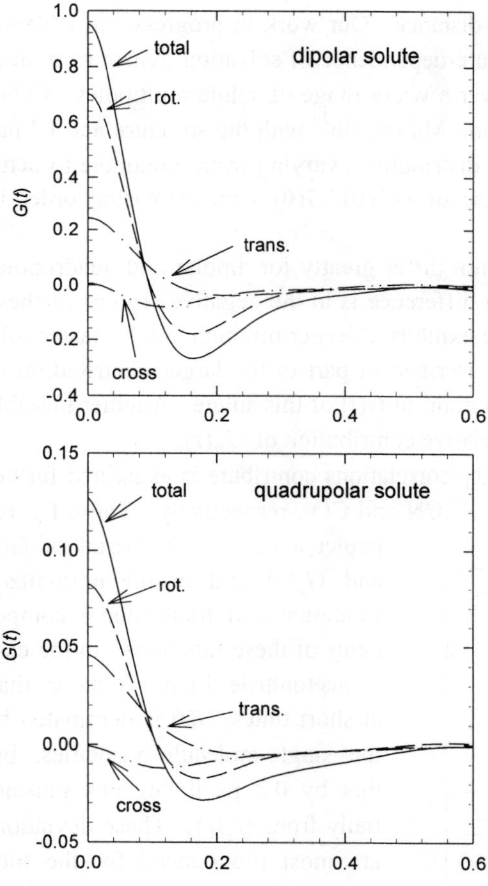

Figure 2: Dipolar (top panel) and quadrupolar (bottom panel) solutes in CO_2: $G(t)$ (full line) is decomposed into its translational (dash-dotted line), rotational (dashed line) and their cross-correlation (dash-double dotted line) components.

For the quadrupolar solute, rotational motion is still the largest contributor to solvation dynamics, but not as large as for the diatomic solute, with $G^{rot}(0)$ now 65% of $G(0)$. This is consistent with the predictions of Ladanyi and Stratt[29] on how the interaction multipolar character and range would affect the relative contributions of rotation and translation to solvation dynamics. Their argument is based on the multipole expansion of Δw_{0j}. This is an expansion in d/r, where d is an intermolecular distance (here the C-O bond length or the solute half bond length) and r the center-of-mass intermolecular separation. They showed that rotation would contribute more than translation as long as the leading term in the expansion was not independent of d/r, but that the relative contribution of rotation would diminish for shorter-ranged interactions, because the effective intermolecular distance would then not be much larger than *d*. In the present case of solvation in CO_2, the range of the leading multipolar contribution to ΔE decreases from r^{-4} to r^{-5} in going from a dipolar to a quadrupolar solute, thus decreasing the effective intermolecular distance. Our work in progress[37] in collaboration with M. Maroncelli on the solute-dependence of solvation dynamics in acetonitrile shows an analogous trend over a wider range of solute multipoles. Using model solutes developed by Kumar and Maroncelli[20] with the structure and LJ parameters of benzene and with charge distributions varying from monopole to octupole, we find a nearly linear increase of $G^{tr}(0)/G(0)$ with increasing order in leading solute multipole moment.

The behavior of $G(t)$ vs. t does not differ greatly for dipolar and quadrupolar solutes in CO_2. The most prominent difference is in the negative regions of these functions. $G(t)$ for the dipolar solute exhibits a deeper minimum and a faster subsequent decay to zero. This can be ascribed in part to the larger contribution of $G^{rot}(t)$, the main source of this minimum, to $G(t)$ of this solute. Another possible source of these differences is in the relative contribution of $G_p(t)$.

How much intermolecular velocity correlations contribute is examined further in Figures 3 and 4, for solvation in CH_3CN and CO_2, respectively. These figures depict a direct comparison of $G(t)$ and $G_s(t)$ and of the normalized rotational and translational components of these functions. In the case of acetonitrile, Figure 3 shows that, at short times, $G(t)$ is dominated by the single-molecule velocities, but that by 0.2 ps it deviates substantially from $G_s(t)$. These deviations are most pronounced for the total $G(t)$, reflecting the fact that $G^{cross}(t)$ is largely an intermolecular velocity correlation, i.e., $G^{cross}(t) \cong G_p^{cross}(t)$.

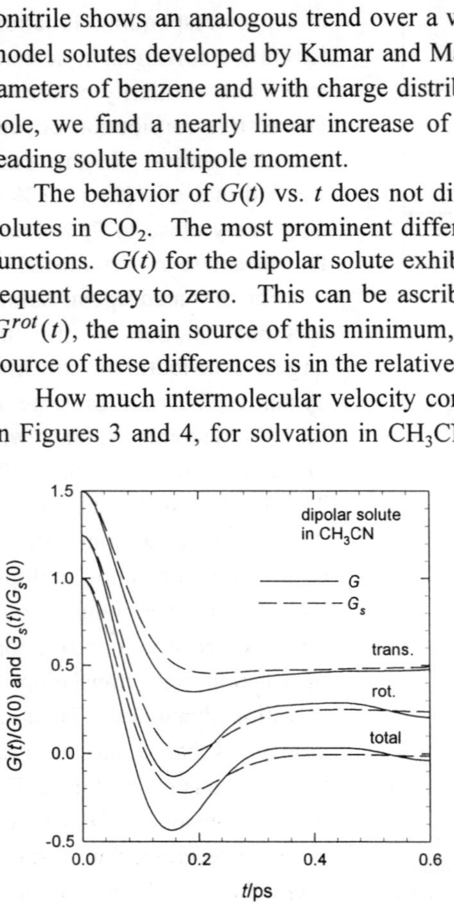

Figure 3: A comparison of $G(t)/G(0)$/ (full line) and $G_s(t)/G_s(0)$ (dashed line) for the dipolar diatom in acetonitrile. The normalized rotational and translational components of these functiona are also compared. They are shifted vertically by 0.25 and 0.50, respectively.

Figure 4 shows an analogous comparison for dipolar and quadrupolar solutes in CO_2. We see that in CO_2 the extent of deviation of

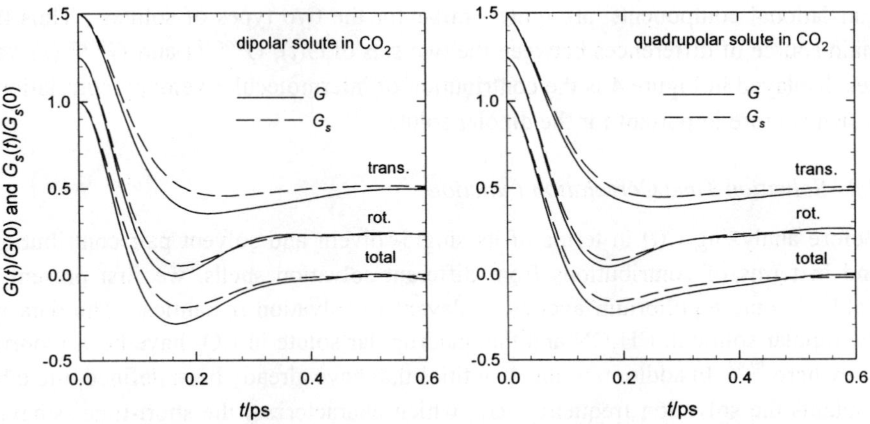

Figure 4: Same as Figure 3, but for dipolar (top panel) and the quadrupolar (bottom panel) diatoms in CO_2.

$G(t)$ from $G_s(t)$ is smaller than in acetonitrile, even for the dipolar solute. The type of deviation that we observe is, however, quite similar: The presence of intermolecular velocity correlations leads to a more pronounced negative region in $G(t)$, $G^{rot}(t)$ and $G^{trans}(t)$ than in their single-molecule VACF counterparts. For the two solutes in CO_2, the intermolecular contribution to the solvation VACF is larger for the dipolar than for the quadrupolar solute.

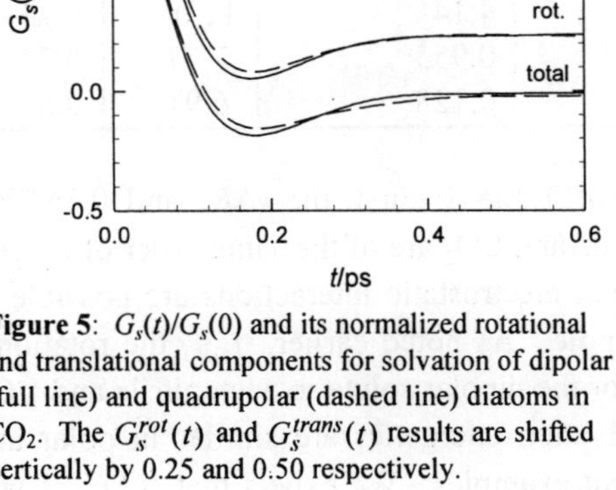

Figure 5: $G_s(t)/G_s(0)$ and its normalized rotational and translational components for solvation of dipolar (full line) and quadrupolar (dashed line) diatoms in CO_2. The $G_s^{rot}(t)$ and $G_s^{trans}(t)$ results are shifted vertically by 0.25 and 0.50 respectively.

In order to find out to what extent intermolecular correlations influence the observed differences between solvation of dipolar and quadrupolar solutes in CO_2, we compare in Figure 5 $G_s(t)$, $G_s^{rot}(t)$, and $G_s^{trans}(t)$ for these solutes. This comparison shows that all three functions when normalized to their $t = 0$ values, even $G_s(t)/G_s(0)$, which contains a different mix of rotational and translational components, are quite similar for the two types of solutes. Thus the main source of differences between the two sets of *G*(*t*), $G^{rot}(t)$ and $G^{trans}(t)$ values displayed in Figure 4 is the contribution of intermolecular velocity correlations, which is more important for the dipolar solute.

4.2. Solvation Time Correlation Function

Before analyzing *C*(*t*) in terms of its single-solvent and solvent pair contributions and in terms of contributions from different solvation shells, we first present in Table 2 some equilibrium averages relevant to solvation dynamics. The data for the dipolar solute in CH_3CN and the quadrupolar solute in CO_2 have been reported elsewhere.[19,29] In addition to the quantities that have already been defined, the table contains the solvation frequency, ω_s, which characterizes the short-time behavior of *C*(*t*):

$$\lim_{t\to 0} C(t) = 1 - \tfrac{1}{2}\omega_s^2 t^2 + O(t^4), \qquad (24)$$

and is given by

$$\omega_s^2 = \langle \Delta\dot{E}^2 \rangle / \langle (\delta\Delta E)^2 \rangle = G(0) / \langle (\delta\Delta E)^2 \rangle, \qquad (25)$$

and $f_{rot} = G^{rot}(0)/G(0)$, the fraction of ω_s^2 due to rotational dynamics.

Table 2: Equilibrium averages relevant to solvation dynamics

solvent-solute	$\langle \Delta E \rangle$ (10^{-20} J)	$\langle (\delta\Delta E)^2 \rangle$ (10^{-40} J^2)	$\langle (\Delta\dot{E})^2 \rangle$ ($10^{-14}\,J^2 s^{-2}$)	ω_s (ps^{-1})	f_{rot}
CH_3CN-dipole	-8.51	3.42	4.14	11.0	0.75
CO_2-dipole	-5.69	1.58	0.953	7.77	0.74
CO_2-quadrupole	-0.767	0.268	0.128	6.91	0.65

There are several notable features of Table 2: First, the $\langle \Delta E \rangle$ and $\langle (\delta\Delta E)^2 \rangle$ values for the dipolar solute in acetonitrile and CO_2 are of the same order of magnitude, indicating that strong solute-solvent electrostatic interactions are possible in CO_2 despite its lack of a permanent dipole. As noted earlier, f_{rot}, the rotational contribution to ω_s^2, is almost identical for the dipolar solute in acetonitrile and CO_2. Thus many aspects of electrostatic solvation energetics are similar in polar and nonpolar solvents, at least for the present examples. We expect that $\langle (\Delta\dot{E})^2 \rangle$ will be roughly proportional to the temperature,[19,29] which means that some of the dif-

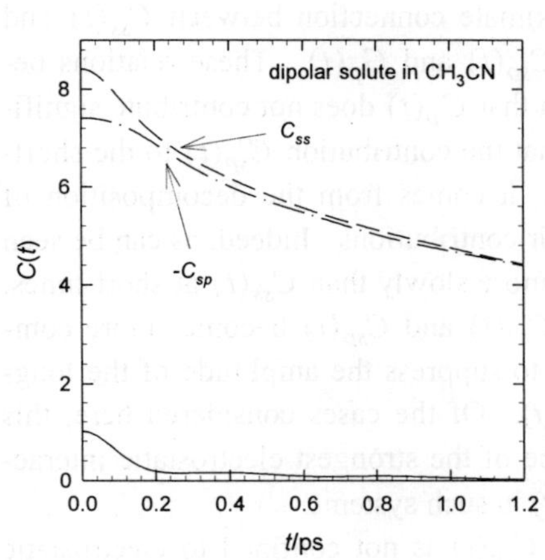

Figure 6: The solvation TCF, $C(t)$ (full line), for the dipolar solute in acetonitrile. Also shown is its single-solvent component $C_{ss}(t)$ (dashed line) and $-C_{sp}(t)$, the negative of its solvent-pair component (dash-dotted line).

ference in its value and in the value of ω_s for the two solvents is due to the fact that CH_3CN is at 293 K while CO_2 is at 220 K.

We turn now to the analysis of $C(t)$ in terms of single-solvent and solvent pair contributions. These results are displayed in Figures 6 and 7. Both figures show that $C_{ss}(t)$ is positive and $C_{sp}(t)$ negative and that $C(t)$ arises from their partial cancellation. In the figures, we show the negative of $C_{sp}(t)$ in order to make the cancellation effects more explicit. The three solute-solvent systems differ quite dramatically in how close to complete this cancellation is. It comes closest to being complete in the case of the dipolar solute in acetonitrile and the least close for the quadrupolar solute in CO_2. The relative size of $C_{sp}(t)$ reflects the contribution of solvent-solvent correlations to $C(t)$. We see that this contribution is appreciable in all three cases and that it is most important for the polar solute-solvent system.

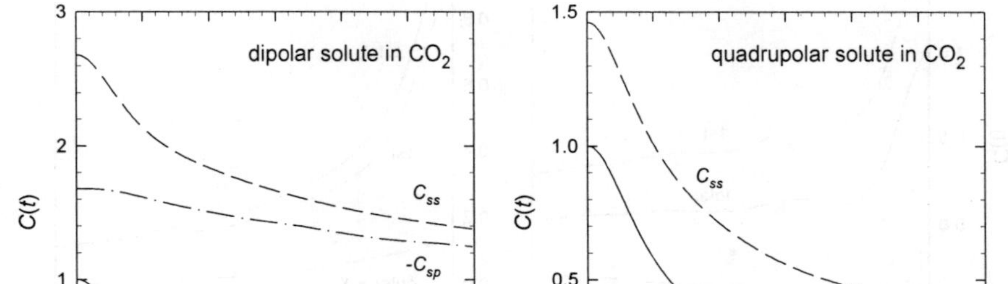

Figure 7: Same as figure 6, but for dipolar (left panel) and quadrupolar (right panel) solutes in CO_2.

As Eq. 19 indicates, there is an approximate connection between $C_{ss}(t)$ and $G_s(t)$. An analogous one exists between $C_{sp}(t)$ and $G_p(t)$. These relations become exact if the solute is stationary. Given that $G_p(t)$ does not contribute significantly to $G(t)$ at short times, we conclude that the contribution $C_{sp}(t)$ to the short-time portion of $C(t)$ is static in nature, i.e., it comes from the decomposition of $\langle(\delta\Delta E)^2\rangle$ into single-solvent and solvent pair contributions. Indeed, as can be seen from Figures 6 and 7, $C_{sp}(t)$ varies much more slowly than $C_{ss}(t)$ at short times. At longer times, the cancellation between $C_{ss}(t)$ and $C_{sp}(t)$ becomes more complete. Thus an important role of $C_{sp}(t)$ is to suppress the amplitude of the long-time, exponentially decaying portion of $C(t)$. Of the cases considered here, this suppression is most effective in the presence of the strongest electrostatic interactions and accounts for the rapid decay of $C(t)$ in such systems.

The partial cancellation of $C_{ss}(t)$ and $C_{sp}(t)$ is not confined to electrostatic solvation. It has been observed by Saven and Skinner[38] for ΔE corresponding to a perturbation in solute-solvent LJ potential in an apolar system of spherically symmetric LJ molecules. Furthermore, similar cancellation effects occur for other TCFs for which the dynamical variable is a sum of intermolecular interactions. Prominent examples are interaction-induced components of collective dipoles and polarizabilities,[39] which contribute to the experimentally-observable relaxation processes in pure liquids.

Finally, we turn to the contributions to solvation dynamics from solvent mole-

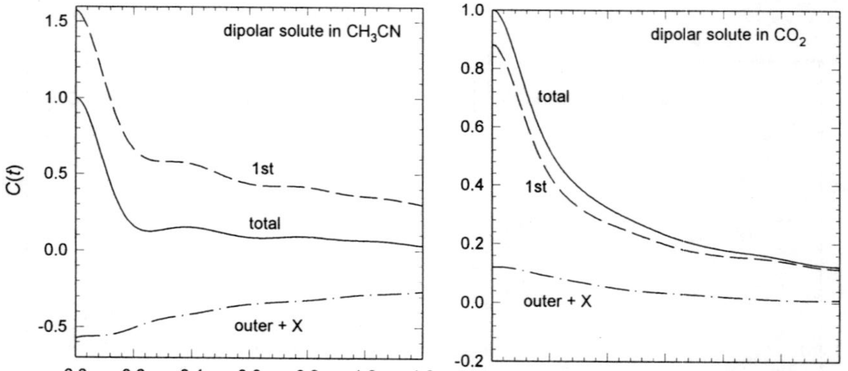

Figure 8: Decomposition of $C(t)$ (full line) for the dipolar solute in acetonitrile (left panel) and CO_2 (right panel) into contributions from the first solvation shell (dashed line) and from $C^{outer}(t)+C^X(t)$ (dash-dotted line).

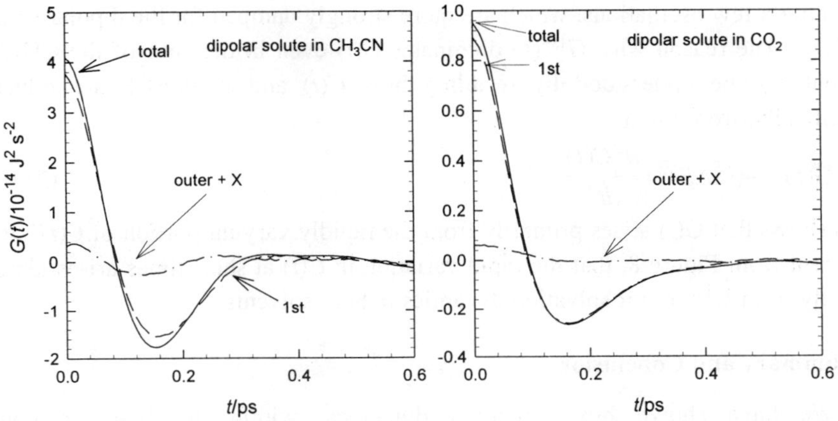

Figure 9: Same as Figure 8, but for the solvation VACF, $G(t)$.

cules in different solvation shells. The molecules were identified as belonging to the first solvation shell using a distance criterion based on the location of the first minimum of the pair distribution functions between the solvent center-of-mass and solute atomic sites. We found that the first solvation shell contained about 15 solvent molecules in the presence of the dipolar solute. More precisely, we found that it contained, on average, 14.8 solvent molecules in the case of this solute in CH_3CN and 15.2 solvent molecules for the same solute in CO_2.

Figure 8 shows the resulting contributions to $C(t)$ for this solute in both solvents. We find that the situation is quite different for CH_3CN than for CO_2. In the former case, the contributions from outer shells and from the first shell-outer shell cross correlations are appreciable, especially at long times, where the positive $C^{1st}(t)$ is nearly cancelled by the negative $C^{outer}(t) + C^X(t)$. As was shown in Ref. 19, $C^{outer}(t)$ for this system is positive throughout this time interval, which means that $C^X(t)$ is large and negative. In the case of the CO_2 solvent, $C(t)$ is always dominated by $C^{1st}(t)$ and $C^{outer}(t) + C^X(t)$ makes a small, positive contribution. The results for the quadrupolar solute in CO_2 are not displayed, but they show that the first shell dominates $C(t)$ even more strongly than for the dipolar solute in this solvent.

While different shell contributions to $C(t)$'s for the two systems are quite different, their contributions to $G(t)$'s are strikingly similar, as shown in Figure 9. In both solvents, $G(t)$ is strongly dominated by the first shell contribution, while $G^{outer}(t)+G^X(t)$ is always small. It is positive at short times and then decays to zero after a few oscillations, which are more strongly damped for the dipolar solute in CO_2. The reason why $G^{1st}(t)$ dominates $G(t)$ even in the case of the CH_3CN solvent can be understood by recalling how $C(t)$ and its VACF are related. Specifically, from Eq. 5

$$G(t) = -\langle(\delta\Delta E)^2\rangle \frac{d^2C(t)}{dt^2}. \tag{25}$$

This shows that $G(t)$ arises primarily from the rapidly varying portion of $C(t)$. It is apparent from Figure 8, that the rapid variation in $C(t)$ at short times arises almost entirely from $C^{1st}(t)$ for solvation dynamics in both solvents.

5. Summary and Conclusion

We have shown how solvation dynamics, within the linear response approximation, can be analyzed by focusing not only on the solvation TCF, $C(t)$, but also on the solvation VACF, $G(t)$, which is related to it, equivalently, by Eqs. 5 and 25. The study of $G(t)$ makes it possible to determine the relative contributions of molecular rotation and translation to solvation dynamics as well as to find out how much intermolecular velocity correlations contribute. Generally, $G(t)$ is more sensitive than $C(t)$ to the rapidly-varying portions of the solvation response and can therefore provide us with additional insights into the molecular origin of its inertial and oscillatory portions. Our main finding is that these mechanisms are strikingly similar in the present examples of electrostatic solvation in polar and nonpolar solvents. Differences in the behavior of $C(t)$ are more prominent, but they are due primarily to slowly-varying processes contributing to solvation dynamics.

More specifically, our study of $G(t)$ indicates that for all solute-solvent systems we considered, solvation dynamics occurs predominantly through rotation. This is consistent with the conclusions reached earlier on the basis of INM analysis of electrostatic solvation in polar and nonpolar liquids.[19,29] The relative contribution of rotations diminishes in going from a dipolar to a quadrupolar solute, but remains significantly larger than 40%, which would correspond to equal participation of all degrees of freedom. We note that this conclusion is valid for molecular chromophores which do not appreciably change their size and shape on electronic excitation. Thus it definitely does not apply to the hydrated electron, which becomes substantially larger and more elongated in going from the ground *s*-state to the excited *p*-state.[40]

Eqs. 13 and 14 indicate that at very short times, solvation dynamics is not collective, i.e., it does not include intermolecular velocity correlations. At longer times, the contribution of $G_p(t)$ is no longer zero and its relative importance varies with the strength of multipolar interactions in the system. Thus for the three systems studied, intermolecular velocity correlations contribute the most for the dipolar solute in acetonitrile and the least for the quadrupolar solute in CO_2. For solvation in CO_2, we find that single molecule processes contributing to solvation dynamics are very similar for dipolar and quadrupolar solutes as evidenced by the fact that the normalized single-molecule VACFs $G_s^{rot}(t)/G_s^{rot}(0)$ $G_s^{trans}(t)/G_s^{trans}(0)$, and $G_s(t)/G_s(0)$ are nearly identical for the two solutes.

In analyzing $C(t)$, we first separated it into $C_{ss}(t)$ and $C_{sp}(t)$, the contributions from single solvent molecules and pairs of solvent molecules and then into contributions from the first and outer solvation shells. The first separation shows that $C_{sp}(t)$ is negative and, at short times, significantly more slowly-varying than $C_{ss}(t)$. At long times both functions decay exponentially at similar rates. $C(t)$ results from their partial cancellation, which comes closer to being complete at longer times. The extent of cancellation also depends quite dramatically on the strength of electrostatic interactions, being nearly complete for the dipolar solute in acetonitrile and much smaller for the quadrupolar solute in CO_2.

Separation of $C(t)$ into contributions from the first solvation shell, outer shells and first shell-outer shell cross correlations also reveals significant differences between solvation in CO_2 and CH_3CN. In the former case, $C^{1st}(t)$ dominates the solvation response, even for the dipolar solute. In the latter case, $C^{1st}(t)$ is the major contributor to the short-time dynamics, but the slowly varying contribution of $C^{outer}(t)+C^X(t)$ plays an important role at longer times, where it cancels almost entirely the exponentially decaying portion of $C^{1st}(t)$. A similar analysis of $G(t)$ reveals a very different picture. Here we find striking similarities in the *dynamical* correlations contributing dipolar diatom solvation in the two solvents. In both cases, $G^{1st}(t)$ strongly dominates the solvent response, indicating that the rapidly varying portions of solvation dynamics can be ascribed to nearby solvent molecules in polar and nonpolar solvents.

Acknowledgments

I thank Professors Mark Maroncelli and Richard M. Stratt for helpful discussions. This work was supported in part by the grant CHE-9520619 from the National Science Foundation. Acknowledgment is also made to the Donors of the Petroleum Research Fund, administered by the American Chemical Society, for partial support of this research.

References

1. J.D. Simon, *Acc. Chem. Res.* **21**, 128 (1988); B. Bagchi, *Annu. Rev. Phys. Chem.* **40**, 115 (1989).
2. W. Jarzeba, G.C. Walker, A.E. Johnson, and P.F. Barbara, *Chem. Phys.* **152**, 57 (1991); P.F. Barbara and W. Jarzeba, *Adv. Photochem.* **15**, 1 (1990).
3. M. Maroncelli, *J. Mol. Liq.* **57,** 1 (1993).
4. M. Maroncelli, P.V. Kumar, A. Papazyan, M.L. Horng, S.J. Rosenthal, and G.R. Fleming, in *Ultrafast Reaction Dynamics and Solvent Effects*, eds. P.J. Rossky and Y. Gauduel (AIP Conference Proceedings 298, AIP, New York, 1994).
5. B.M. Ladanyi and M.S. Skaf, *Annu. Rev. Phys. Chem.* **44**, 335 (1993).
6. R.M. Stratt and M. Maroncelli, *J. Phys. Chem.* **100**, 12981 (1996).
7. J.T. Hynes, in *Ultrafast Dynamics of Chemical Systems*, ed. J.D. Simon (Kluwer, Dordrecht, 1994); B.B. Smith, H.J. Kim, D. Borgis, and J.T. Hynes, in *Dynamics and Mechanism of Photoinduced Electron Transfer*, eds. N. Mataga, T. Okada, and H. Masuhara (Elsevier, Amstredam, 1992).
8. P.J. Rossky and J.D. Simon, *Nature* (*London*) **370**, 263 (1994).
9. H. Heitele, *Angew. Chem. Int. Ed. Eng.* **32**, 359 (1993).
10. P.F. Barbara, T.J. Meyer, and M. Ratner, *J. Phys. Chem.* **100**, 13148 (1996).
11. M. Maroncelli and G.R. Fleming, *J. Chem. Phys.* **89**, 5044 (1988).
12. J.S. Bader and D. Chandler, *Chem. Phys. Lett.* **157**, 501 (1989).
13. R.M. Levy, D.B. Kitchen, J.T. Blair, and K. Krogh-Jespersen, *J. Phys. Chem.* **94**, 4470 (1990); M. Belhadj, D.B. Kitchen, J.T. Blair, K. Krogh-Jespersen, and R.M. Levy, *J. Phys. Chem.* **95**, 1082, (1991).
14. K. Ando and S. Kato, *J. Chem. Phys*. **95**, 5966 (1991).
15. P. Muiño and P. R. Callis, *J. Chem. Phys.* **100**, 4093 (1994).
16. D.B. Bursulaya, D.A. Zichi, and H.J. Kim, *J. Phys. Chem*. **99**, 10069 (1995); *J. Phys. Chem*. **100**, 1392 (1996).
17. M.S. Skaf and B.M. Ladanyi, *J. Phys. Chem*. **100**, 18258 (1996).
18. M. Maroncelli, *J. Chem. Phys.* **94**, 2084 (1991).
19. B.M. Ladanyi and R.M. Stratt, *J. Phys. Chem*. **99**, 2502 (1995).
20. P.V. Kumar and M. Maroncelli, *J. Chem Phys*. **103**, 3038 (1995).
21. E.A. Carter and J.T. Hynes, *J. Chem. Phys.*_**94**, 5961 (1991).

22. S.J. Rosenthal, X. Xie, M. Du, and G.R. Fleming, *J. Chem. Phys.*_**95**, 4715 (1991).
23. S.J. Rosenthal, R. Jimenez, G.R. Fleming, P.V. Kumar, and M. Maroncelli, *J. Mol. Liq*. **60**, 25 (1994).
24. R. Jimenez, G.R. Fleming, P.V. Kumar, and M. Maroncelli, *Nature (London)* **369**, 471 (1994).
25. J. Gardecki, M.L. Horng, A. Papazyan, and M. Maroncelli, *J. Mol. Liq.* **65/66**, 49 (1995); M.L. Horng, J. Gardecki, A. Papazyan, and M. Maroncelli, *J. Phys. Chem*. **99**, 17311 (1995).
26. D. Bingemann and N. Ernsting, *J. Chem. Phys.* **102**, 2691 (1995).
27. L. Reynolds, J.A. Gardecki, S.J.V. Frankland, M.L. Horng, and M. Maroncelli, *J. Phys. Chem.* **100**, 10337 (1996).
28. B.M. Ladanyi and R.M. Stratt., *J. Phys. Chem.* **100**, 1266 (1996).
29. R.M. Stratt and M. Cho, *J. Chem. Phys*. **100**, 6700 (1994).
30. B.M. Ladanyi and S. Klein, *J. Chem. Phys.* **105**, 1552 (1996).
31. R.M. Stratt, *Acc. Chem. Res*. **28**, 201 (1995).
32. T.S. Kalbfleisch, L.D. Ziegler, and T. Keyes, *J. Chem. Phys*. **105**, 7034 (1996).
33. W.A. Steele, *Mol. Phys*. **61**, 1031 (1987).
34. J.T. Kennedy and G. Thodos, *J. Chem. Eng. Data.* **5,** 293 (1960).
35. D.M.F. Edwards, P.A. Madden, and I.R. McDonald, *Mol. Phys*. **51,** 1141 (1984).
36. M.P. Allen and D.J. Tildesley, *Computer Simulation of Liquids* (Oxford, New York, 1989).
37. M. Maroncelli, N. Patel, and B.M. Ladanyi, manuscript in preparation.
38. J.G. Saven and J.L. Skinner, *J. Chem. Phys.* **99**, 4391 (1993).
39. G. Birnbaum and B. Guillot, in *Collision- and Interaction-Induced Spectroscopy*, eds. G.C. Tabisz and M.N. Neuman (Kluwer, Dordrecht, 1995).
40. B.J. Schwartz and P.J. Rossky, *J. Chem. Phys.* **101**, 6902 (1994); *J. Chem. Phys.* **101**, 6917 (1994); *J. Chem. Phys.* **105**, 6997 (1996).

MOLECULAR DYNAMICS COMPUTER SIMULATIONS OF A SOLUTION OF I_2 IN PYRIDINE

S.REILING*, M.BESNARD, PH.LALANNE, PH.A.BOPP
Laboratoire de Spectroscopie Moléculaire et Cristalline, Université Bordeaux I, F-33405 Talence CEDEX, France

* *present address: TRIPOS Inc., 1699 South Hanley Road, St. Louis, MO 63144-2917, USA*

We report preliminary results of molecular dynamics (MD) computer simulations of a molecule of I_2 dissolved in Pyridine. Literature potentials are used for the intermolecular Pyridine-Pyridine and intramolecular I-I interactions; the Pyridine molecule itself is modeled as a rigid entity. A newly constructed interaction model is employed for the I_2-Pyridine interactions. It involves polarization of the I_2 as well as the transfer of charge from the Pyridine. We observe large fluctuations in the time dependent polarization of the I_2-molecule, which are reflected in the I-N-radial distribution function. The infrequent occurrence of very large values in the fluctuations of the dipole moment, and consequently of large partial charges on the I_2-molecule, can be interpreted as the initial steps which will eventually lead to the ionization reaction.

1 Introduction

Computer simulations of pure Pyridine and of pyridinic solutions have been reported in several instances[1,2]. Solutions of I_2 in Pyridine as well as ternary systems with an additional inert or polar solvent have also been extensively investigated by means of far infrared and Raman spectroscopy[3–8]. We focus here our interest on the binary system I_2 in Pyridine only, and more particularly on the dynamical behavior of the "I_2-Pyridine-complex".

The experiments in binary systems have been previously interpreted in terms of the existence of such I_2-Pyridine complexes and the subsequent formation of (I-Pyridine)$^+$-cations and I^--ions[9]. More recently, a mechanism involving I_2-Pyridine$_2$ complexes with subsequent formation of (I-Pyridine$_2$)$^+$-cations and I_3^--ions has been put forward[5]. Finally, the formation of I_3^--ions is observed in the experiments. The characteristic times for these steps vary over several orders of magnitude and can be estimated as follows:

$I_2 + Py \rightleftharpoons Py{:}I_2 + Py$	femtoseconds to picoseconds
or	
$I_2 + Py \rightleftharpoons Py_2{:}I_2 + Py$	femtoseconds to picoseconds
$Py_2{:}I_2 + Py \rightleftharpoons (Py_2I)^+ + I^- + Py$	nanoseconds
$I^- + I_2 + Py \rightleftharpoons I_3^- + Py$	seconds to days

The first reactions should be fast, with expected time constants in the domain of picoseconds. The dissociation of the complex and the subsequent separation of the ions is slower by several orders of magnitude. The last process has been inferred from an increase of the intensity of the peak in the far infrared at $\approx 139\ cm^{-1}$. It is observed to occur on time scales extending over days.

Obviously, only the processes involving the formation of the complexes and, possibly, the onset of the ion-formation are open to investigation by means of molecular dynamics (MD) computer simulations.

2 Modeling and Simulations

The Pyridine-Pyridine interaction potential is taken from previous work[2]. It consists of site-site potentials of Coulomb and Buckingham type between united atoms. An additional site represents the additional electron density next to the Nitrogen atom. A new I_2-Pyridine model[12] was developed from very extensive ab-initio quantum mechanical calculations on the I_2 molecule and the I_2-Pyridine[10] and I_2-Pyridine$_2$[11] supermolecules. The details of these calculation will not be repeated here; essentially, a few all-electron calculations were first carried out on I_2 and I_2-Pyridine in order to select and calibrate a basis set with effective core potential (ecp). This set was then used in the "routine" runs to generate the energies for the I_2-Pyridine and I_2-Pyridine$_2$ supermolecules. The calculations had to be carried to MP2-level, and the basis set superposition error (BSSE) was corrected.

These energies were then used to derive an analytical I_2-Pyridine interaction potential. Besides the usual site-site pair potentials, it takes into account the polarization of the Iodine and the transfer of charge from the "lone-pair" site of the Pyridine to the I_2-molecule. This is achieved by actually transferring charge, according to geometric criteria, during the simulation, between the site representing the 'lone pair' electron on the Pyridine and the Iodine. The geometric rules were determined by a fitting procedure to the ab-initio energies obtained for the I_2-Pyridine and I_2-Pyridine$_2$ supermolecules, taking into account the Pyridine-Pyridine model potential mentioned above. While we feel that the characteristic features and orders of magnitude of the interactions between Iodine and Pyridine are well reproduced by this Ansatz, together with the literature Pyridine-Pyridine potential, further work would be required to achieve an exact quantitative agreement. As a further approximation the Pyridine molecules are represented as rigid bodies. The I_2-molecule, however, is flexible with an intramolecular Morse potential from Herzberg[13].

The simulated system consists of one I_2-molecule (thus no I_2-I_2 interaction potential is needed) and 208 Pyridine molecules, representing a dilute solution.

A rectangular box with periodic boundary conditions is used for the simulations. The shifted force potential method with a cut off radius of 11 Å is employed for the computation of the interactions; the length of the required simulation runs prevents the usage of a more accurate scheme for the treatment of the long range interactions. The density of the systems is set to 0.982 g/cm^3 (pure Pyridine at room temperature). The Verlet algorithm with a step size of 2 fs is used for the integration of the equations of motion; the rigid molecules are treated by a method due to Ciccotti et al.[14]. The temperature is held constant (at room temperature) with a stochastic collision algorithm developed by Kast[15]. The system is first set up at a lower density than the desired one, then it is slowly compressed at a constant rate. After the desired density of the system has reached, it is further equilibrated for more than 80 ps prior to starting the first production run and data collection. The simulation has now been extended over a total time of more than 1000 ps. All simulations were performed on IBM RS/6000 workstations.

3 Results

Only preliminary results concerning the polarization of I_2 dissolved in pyridine will be reported here; full details will be given in a subsequent publication[12]. Figure 1 shows the dipole moment μ induced by the environment on the I_2-molecule as a function of time from a part of the simulation run of 500 ps duration. The average value of $|\mu|$ is about 1.5 D, the dipole moment fluctuates around zero with frequent changes of sign, the mean time between sign inversion being about 0.25 ps. In a few instances, times up to $\approx$40 ps between sign inversions have been observed. Large fluctuations with values of $|\mu|$ up to more than 9 D occur at time intervals of about 100 ps. These time scales are essentially in agreement with experimental findings[6,7].

Figure 2 shows the distribution of μ, averaged over the entire simulation run. In order to estimate roughly the statistical uncertainties of the results, the symmetrized distribution and the difference between the measured and symmetrized distributions are also shown. The distribution seems to be monomodal with a half width at half height of about 3 D. A slight shoulder can be suspected at $\mu \approx -3$ D; it would take much longer simulations to ascertain the existence of such a weak feature.

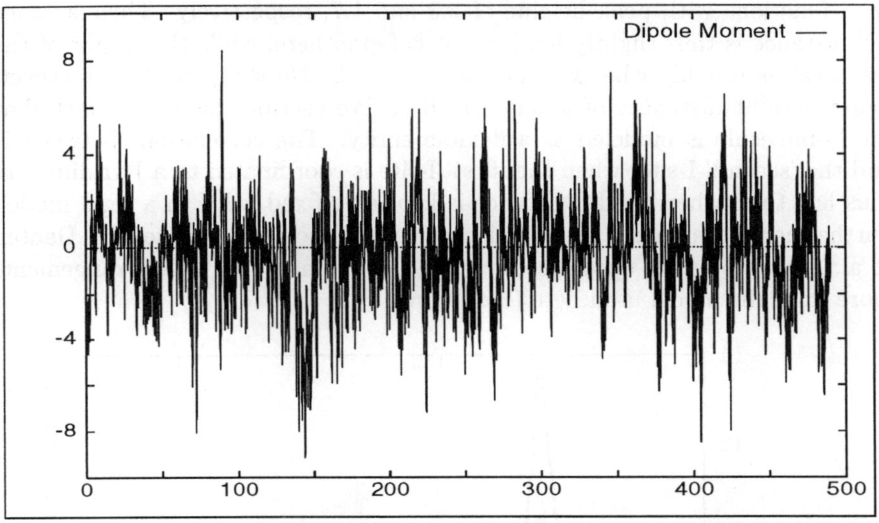

Figure 1: Dipole moment μ of the I_2-molecule, in Debye, as a function of time, in ps. The sign of μ refers to the direction of the dipole moment (say, pointing from I-atom 1 to 2, or vice versa).

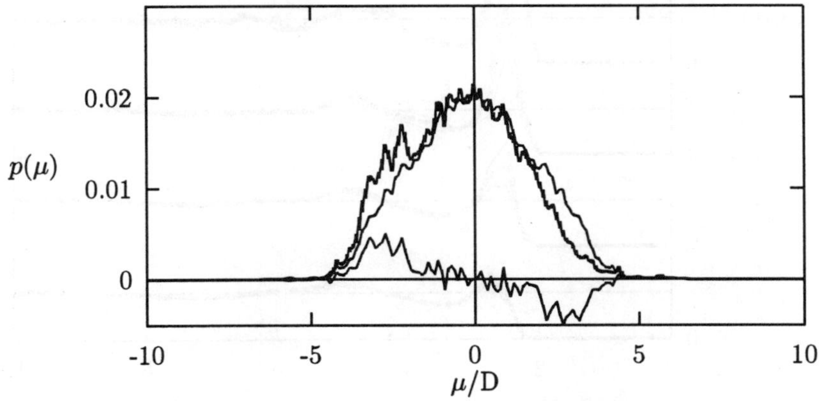

Figure 2: Distribution of the instantaneous molecular dipole moments μ of I_2 (fat), symmetrized distribution, and difference between the two.

Figure 3 shows the radial pair distribution function $g_{\text{I-N}}(r)$, calculated for all I-N pairs for various values of the instantaneous dipole moment μ of the I_2 molecule. A pronounced increase of the height of the first peak at ≈ 2.5 Å is seen with increasing μ. Within the statistical accuracy no shift of the peak position is detectable. Danten et al.[2] report peaks at 2.9 Å and 5.5 Å in their

g_{I-N} functions with peak heights of 2.3 and 1.7, respectively. Their average I-N distance is thus slightly longer that is found here, while the height of the first peak is roughly what we find for $\mu \approx$ 4-5 D. No sharp peak is, however, found here at distances of about 5 to 6 Å. We ascribe this to the fact that the I_2-molecule is modeled as a flexible entity. The correlations between N and the 'second' I-site, when the 'first' I-site is coordinated to a Pyridine, are thus less than when the I-I distance is kept at a fixed value in a rigid model. Furthermore, the charge distribution on the I_2-molecule selected by Danten et al.[2], two dipoles in opposite direction, favors linear N..I-I..N arrangements more than the model used here (*vide infra*).

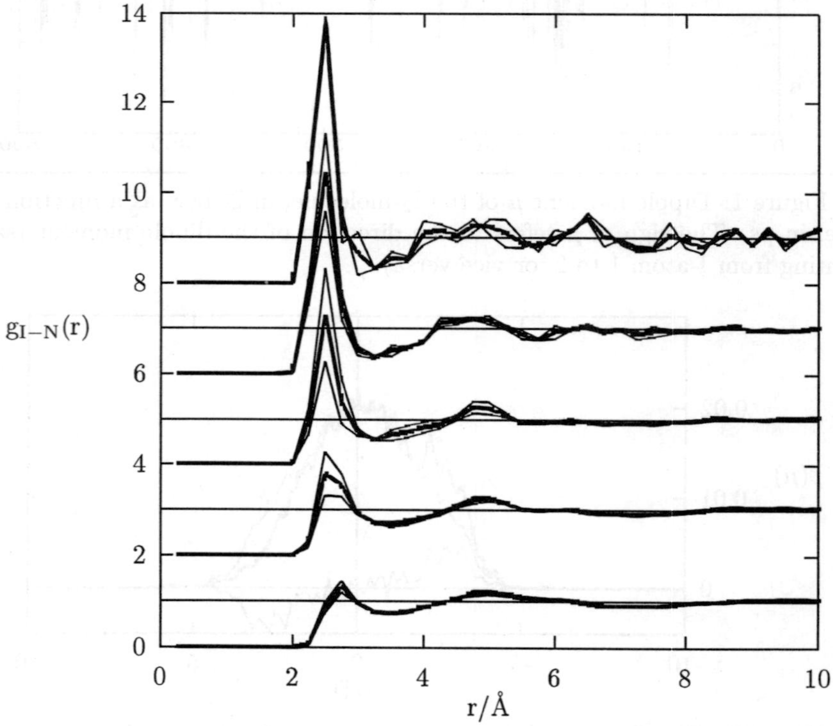

Figure 3: Radial distribution functions $g_{I-N}(r)$, computed separately for different instantaneous values of the molecular dipole moment of I_2. The curves have been shifted by two units for better visibility: from bottom to top: $0\,D < |\mu| < 2\,D$, $2\,D < |\mu| < 4\,D$, $4\,D < |\mu| < 6\,D$, $6\,D < |\mu| < 8\,D$, $8\,D < |\mu| < 10\,D$. The thin lines are the functions computed for each of the two I-atoms separately and are thus a rough indication of the statistical reliability of the $g(r)$s.

Figure 4 shows the normalized distribution of the angles Θ between the I-I vector and the I..N vector, computed for all Pyridine molecules with $r_{I-N} <$ 3.25 Å for the same values of $|\mu|$ as in Figure 3. The vectors are oriented in such a way that Θ=180° corresponds to a linear I-I..N arrangement. Similar to Figure 3, the results obtained for the two moieties of the I_2-molecules have been plotted together with the four lower curves in Figure 4 as an estimate of their statistical reliability.

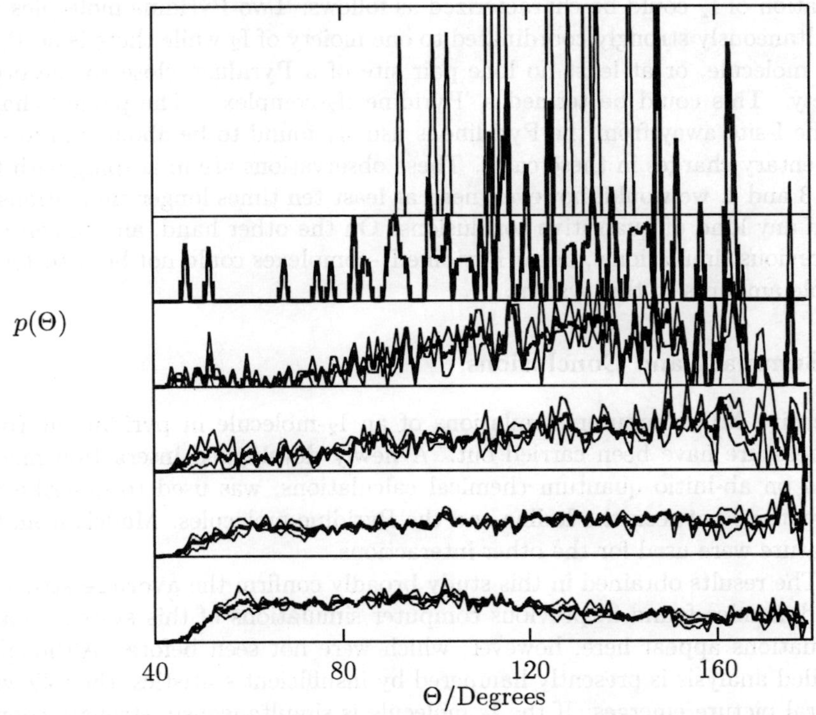

Figure 4: Normalized distribution, in arbitrary units, of the angle Θ, as defined in the text, computed separately for different values of the molecular dipole moment of I_2, as in Figure 3. The topmost curve is for 9 D < $|\mu|$ < 10 D. The thin lines in the lower plots are the results obtained separately for the two moieties of the I_2-molecule.

It is seen that at low values of $|\mu|$ there seems to be no clearcut preferential orientation of the Pyridine with respect to the I-I-direction. (The region Θ <40° is prohibited for steric reasons.) A slight depletion in the region 130°< Θ <160° could, however, be suspected. At intermediate $|\mu|$-values a maximum starts building up at large Θ-values; it shifts to smaller angles with increasing dipole moment. At the largest μ-values, very infrequent events on the timescale of the simulation, only I-I..N angles in the range from about 90° to 150° seem to occur.

Unfortunately, we do not yet have a sufficient number of such events to attempt a statistical treatment. From observations of the builtup of the histograms for figure 4 during the simulation, and from molecular graphics, we suspect that the fluctuations in the solvation shell which lead to a high polarization of I_2 could be characterized as follows: Two Pyridine molecules are simultaneously strongly coordinated to one moiety of I_2 while there is no Pyridine molecule, or at least no lone pair site of a Pyridine, close to the other moiety. This could be termed a 'Pyridine$_2$:I_2-complex'. The partial charge on the I-site away from the Pyridine is usually found to be about -0.5 to -0.6 elementary charges in these cases. These observations are in keeping with figures 3 and 4, we would, however, need at least ten times longer simulations to reach any kind of quantitive conclusions. On the other hand, and in contrast to previous simulations[2], linear Pyridine:I_2-complexes could not be detected in sizable amounts in this system.

4 Summary and Conclusions

Extensive MD computer simulations of an I_2-molecule in pyridine at room temperature have been carried out. A newly developped interaction model, based on ab-initio quantum chemical calculations, was used to describe the interactions between the Iodine and the Pyridine molecules. Models from the literature were used for the other interactions.

The results obtained in this study broadly confirm the **average** structure and dynamics found in previous computer simulations of this system. Large fluctuations appear here, however, which were not seen before. Although a detailed analysis is presently hampered by insufficient statistics, the following general picture emerges: If the I_2 molecule is simultaneously strongly coordinated by two Pyridine molecules on one side, while there is no stong coordination on the other side, an arrangement which could be called a Pyridine$_2$:I_2 complex arises, accompanied by a strong polarization of the I_2. In contrast to previous simulations, linear Pyridine:I_2-complexes were not detected in sizable amounts. The fluctuations of the pyridinic environment which lead to the Pyridine$_2$:I_2-compexes seem to occur on a timescale of roughly 100 ps. They can be suspected to be the ones favorable for a subsequent formation of ionic products.

Acknowledgments

This work is part of the research effort coordinated by the groupement de recherche du CNRS "Acte Chimique Elémentaire en Phase Liquide" (GDR 1017). S.R. thanks the European Union (contract No. ERBCHBGCT930388) and the GDR for a postdoctoral fellowship. We also thank Y. Danten for useful discussions and acknowledge Région Aquitaine, the CNRS (IDRIS computer center), and J. Brickmann (TH-Darmstadt) for providing facilities for our computations.

References

1. I. Bakó, T. Radnai, and G. Pálinkás, *Z.Naturforsch.* **51a**, 859 (1996)
2. Y. Danten, B. Guillot, and Y. Guissani, *J.Chem.Phys.* **96**, 3795 (1992)
3. M. Besnard, N. Del Campo, and J. Yarwood, *J.Molec.Liq.* **48,** 183 (1991)
4. T. Tassaing, M. Besnard, and J.Yarwood, Far IR vibrational spectroscopic studies of the chemical dynamics of the pyridine/iodine system in solution, submitted for publication
5. T. Tassaing and M. Besnard, Ionization reaction in Iodine-Pyridine solutions: What can we learn from conductivity measurements? submitted for publication
6. T. Tassaing, M. Besnard, and J.Yarwood, Vibrational spectroscopic studies of the chemical dynamics in charge-transfer complexes of the type iodine-pyridine: I:experimental results, submitted for publication
7. T. Tassaing, M. Besnard, and J.Yarwood, Vibrational spectroscopic studies of the chemical dynamics in charge-transfer complexes of the type iodine-pyridine: II: Intermolecular dynamics from far-IR bands, submitted for publication
8. T. Tassaing, M. Besnard, and J.Yarwood, Vibrational spectroscopic studies of the chemical dynamics in charge-transfer complexes of the type iodine-pyridine: III: A mid-infrared study of the internal modes of the donor molecule, submitted for publication
9. J.M. Nigretto and M. Jozefowicz, *Electrochim.Acta* **19,** 809 (1974); S. Aronson and S,B. Wilensky, *J.Polym.Sci.* **A 26,** 1259 (1988)
10. S. Reiling, M. Besnard, and Ph.A.Bopp, Theoretical Studies on the Pyridine-I_2 Charge-Transfer Complex I: Ab-Initio Calculations on I_2 and Pyridine–I_2, submitted for publication
11. S. Reiling, M. Besnard, and Ph.A.Bopp, Theoretical Studies on the Pyridine-I_2 Charge-Transfer Complex II: Ab-initio calculations on Pyridine–I_2 and Pyridine–I_2–Pyridine, work in progress

12. S. Reiling, M. Besnard, and Ph.A.Bopp, Theoretical Studies on the Pyridine-I_2 Charge-Transfer Complex III: Development of a Force Field and First Simulations, work in progress
13. G. Herzberg, Molecular Spectra and Molecular Structure I, D. van Nostrand, New York 1945
14. G. Ciccotti, M. Ferrario, and J.-P. Ryckaert, *Mol.Phys.* **47**, 1253 (1982), K. Nicklas, PhD-thesis, Technische Hochschule Darmstadt 1993
15. S. Kast, K. Nicklas, H.-J. Bär, and J.Brickmann, *J.Chem.Phys.* **100,** 566 (1994)

SOME EFFECTS OF VIBRONIC COUPLING IN COVALENTLY LINKED TRANSITION METAL DONOR-ACCEPTOR COMPLEXES

JOHN F. ENDICOTT,* MURIELLE A. WATZKY, ARIEL V. MACATANGAY, SELMA E. MAZZETTO, XIAOQING SONG AND TIONE BURANDA

DEPARTMENT OF CHEMISTRY, WAYNE STATE UNIVERSITY
DETROIT, MI 48202-3489

Introduction

In this report we summarize our observations concerning the effects of strong D/A coupling on the spectroscopic and electrochemical properties of some simple, CN^- bridged transition metal D/A complexes. Consideration of these observations has led us to propose that they can be more readily accommodated by a vibronic approach than by more conventional two state models [1-6].

A number of recent studies [7-13] of strongly coupled donor-acceptor complexes have indicated that standard Hush and Mulliken formalisms [14,15], are not always adequate to account for the observations. Very recently, Reimers and Hush have argued that a vibronic approach may be necessary to account for at least some of these observations [16]. Solomon and co-workers [13] have found spectroscopic evidence implicating metal-ligand skeletal vibrational modes in the coupling of Fe(III) and Fe(II) in an oxy-bridged D/A complex. Several vibronic models have been proposed [17-21], but these have been used almost exclusively to describe details of the donor-acceptor-charge-transfer (DACT) spectrum of the Creutz-Taube ion [21], and they have not been adapted to treat the range of experimental observations described here. We have used a simple, semi-classical vibronic model for this purpose.

The results of our observations on these CN^- bridged D/A complexes can be organized as follows: (i) we have found that the unusual shift of the CN^- stretching frequency to lower energies in CN^- bridged D/A complexes is proportional to β_{DA}^2 (for β_{DA} the electronic coupling matrix element) and is associated mostly with the symmetric combination of stretching frequencies (in centrosymmetric complexes); (ii) the ground state stabilization energy, of CN^- bridged D/A complexes is much greater than expected based on the DACT absorption band; (iii) there is a great deal of mixing between spectroscopic states; (iv) the results of our preliminary studies of back electron transfer (following photoexcitation of the D/A complexes) have not conformed to simple expectations .

A. **Complexes employed.** In our studies of the effects of strong terminal metal–central metal D/A coupling on the properties of CN^- bridged complexes, we have employed series of complexes of the types $(L)M_c(CNRu(NH_3)_5)_2^{n+}$ and $(L)(CN)M_c(CNRu(NH_3)_5)^{n+}$. The most extensive work has employed complexes with L = $(bpy)_2$ or $(phen)_2$ and M_c = Ru(II), Fe(II) (for D/A complexes with $Ru(NH_3)_5^{3+}$) or M_c = Cr(III), Co(III), Rh(III) (for D/A complexes with $Ru(NH_3)_5^{2+}$). To some degree, the complexes with M_c = Rh(III) have served as reference compounds, and we have also employed the complexes $(bpy)_2(CN)Ru(CNRh(NH_3)_5)^{3+}$, $(bpy)_2Ru(CNRh(NH_3)_5)_2^{6+}$ and $(bpy)_2Ru(CNRu(NH_3)_5)(CNRh(NH_3)_5)^{6+}$ as reference compounds. Current research is emphasizing complexes with (L) = a tetraazamacrocyclic (non-bridging) ligand (MCL). This series of complexes has the advantage of fewer, and consequently better separated CT absorbances. We have prepared and characterized the $Ru(NH_3)_5^{2+}$ cyano-ruthenates of: (i) the $M_c(MCL)(CN)_2^+$ complexes with M_c = Cr(III), Co(III), Cr(III) for MCL = [14]aneN_4, *ms* -Me_6[14]aneN_4 (all *trans* -complexes); (ii) the *cis* -M_c(*rac* -Me_6[14]aneN_4)$(CN)_2^+$ complexes with M_c = Rh(III), Rh(III); (iii) and *trans* -Co(MCL)$(CN)_2^+$ complexes with MCL = *ms* -Me_6[14]aneN_4, [15]aneN_4, *rac* -Me_6[14]dieneN_4, Me_4[14]tetraeneN_4 or Me-Me_2[14]dienatoN_4 (see figure 1).

B. **Ground state properties of the D/A complexes.** The ground state properties relevant to this discussion are those which are altered by means of the electronic coupling between the donor and acceptor. The D/A coupling-dependent properties examined are: (1) the CT absorption spectra; (2) the CN^- stretching frequency; and (3) the half-wave potential of the $Ru(NH_3)_5^{3+,2+}$ couple (in a few complexes we have also examined halfwave

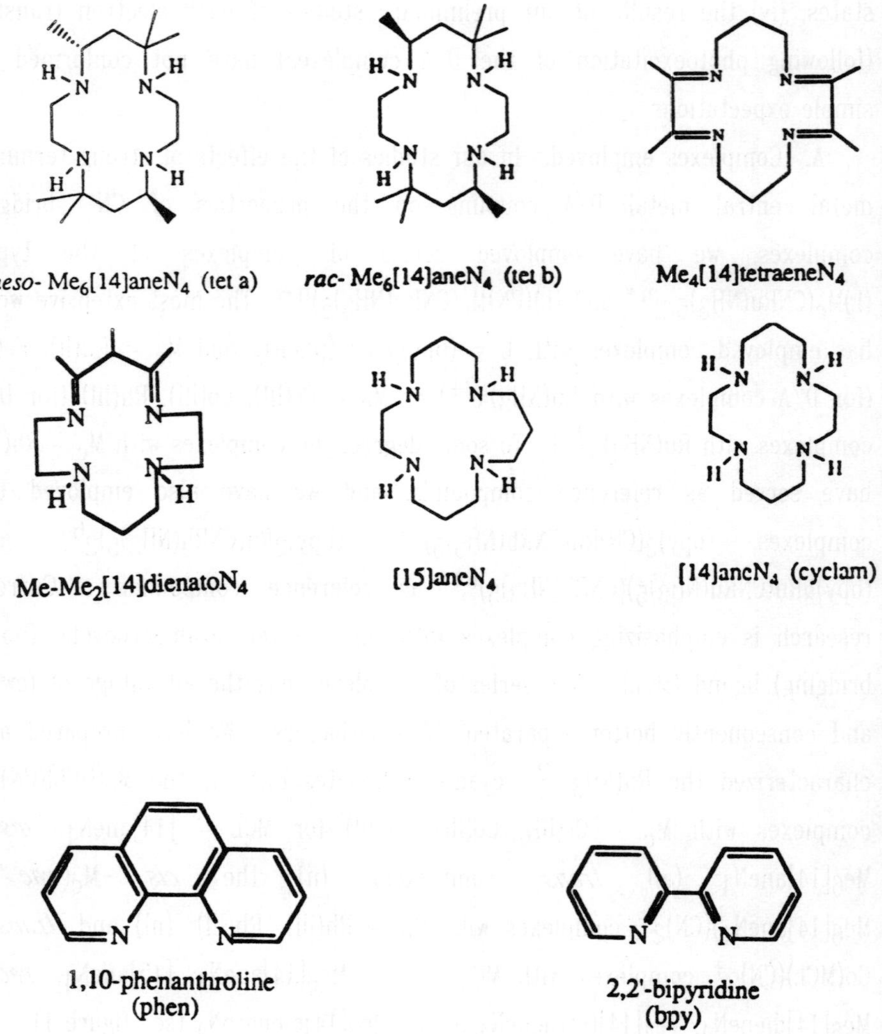

Figure 1. Skeletal structures of the ligands (L) employed.

potential variations of the $M_c(III)/M_c(II)$ couples). In many instances these properties may be evaluated by oxidation (or reduction) of donor (or acceptor) at a site different from that of the chromophore which is monitored. In other instances more elaborate comparisons are necessary.

The variations in absorption spectra with the oxidation state of transition metal donor or acceptor have been determined by the systematic oxidation (through the addition of about 1/10th of an equivalent per aliquot of a standardized Fe^{3+} or Ce^{4+} solution to the fully reduced complexes; and frequently similar additions of $Ru(NH_3)_6^{2+}$ to the substances with oxidized terminal moieties) of the terminal $Ru(NH_3)_5$ centers.

1. CT absorption spectra. Donor/acceptor complexes of the type $LM(CNRu(NH_3)_5)_2$ exhibit very rich, $Ru(NH_3)_5^{3+,2+}$ dependent CT spectra. Two examples are shown in figure 2: top, the stepwise Fe^{3+} oxidation of $(bpy)_2Ru(CNRu(NH_3)_5)_2^{4+}$; bottom, the Fe^{3+} oxidation of $(bpy)_2Fe(CNRu(NH_3)_5)_2^{4+}$. Based on comparisons between the spectra of a variety of related complexes (e.g., L = tetraazamacrocycles as well as $(bpy)_2$) and literature precedents [22] the absorption bands observed are assigned as follows:

a. The $Ru(NH_3)_5$/central metal CT absorptions. These varied in intensity (ε_{max} = (1–5) x 10^3 M^{-1} cm^{-1}) and energy (λ_{max} = 400–900 nm) depending on the central metal. The characteristics of these absorption bands are summarized in Table 1.

b. For the complexes with $Ru(bpy)_2^{2+}$ or Fe (bpy or phen)$^{2+}$ centers, a central metal to bpy, M_cLCT transition is observed (λ_{max} = 400–560 nm).

c. Absorption bands of intermediate intensity occur in the 350–500 nm region of all polypyridyl complexes containing $Ru(NH_3)_5^{2+}$. Comparable transitions do not occur when the central metal has saturated ligands such as in the $M(MCL)(Ru(NH_3)_5)^{5+}$ complexes in which the Rh(III), Co(III) or Cr(III) and

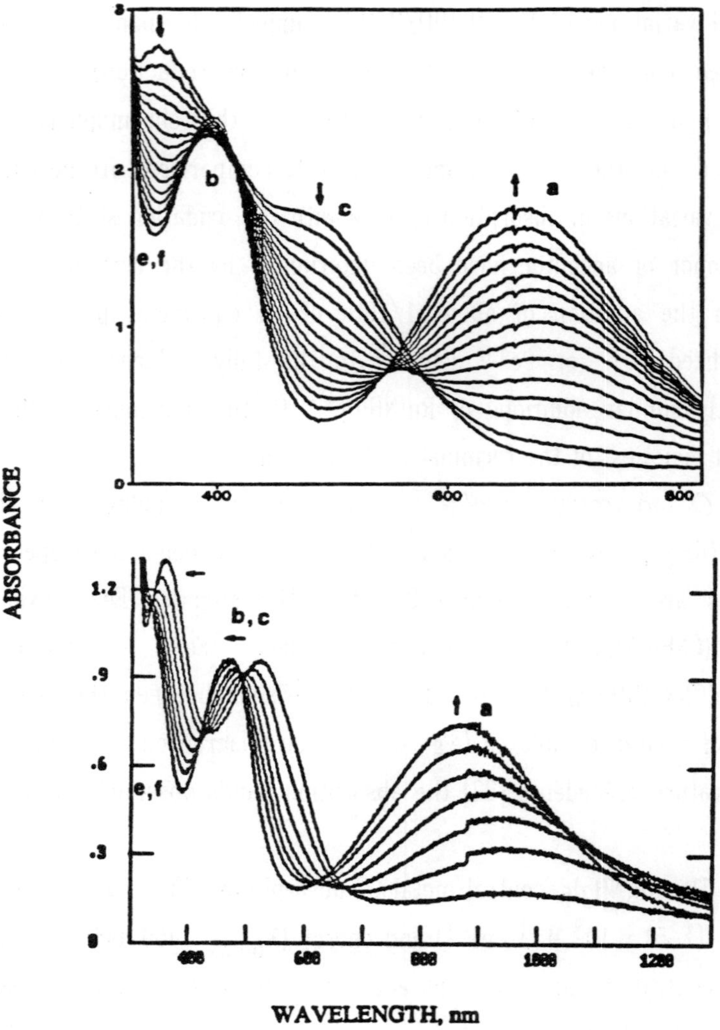

Figure 2. Spectroscopic changes which accompany the successive additions of aliquots of Fe^{3+} (aq) to aqueous solutions of $(bpy)_2M_c(CNRu^{II}(NH_3)_5)_2)^{4+}$ complexes: upper spectra for M_c = Ru(II)l lower spectra for M_c = Fe(II). Only oxidation of the terminal $Ru(NH_3)_5{}^{2+}$ moieties occurs during these titrations. Arrows indicate direction of shifts of absorption maxima which accompany oxidation.

Table 1. Ground state properties of some representative transition metal complexes: $L(CN^-)M_c(CNRu_t(NH_3)_5)^{n+}$ and $LM_c(CNRu_t(NH_3)_5)_2^{n+}$

M_c	L	n [a]	Ru Oxidation States	$E_{1/2}$ ($Ru(NH_3)_5^{3+,2+}$)[b] V *vs* SCE	DACT [c] $h\nu_{max}(\varepsilon_{max})[\Delta\nu_{1/2}]$	$\Delta\upsilon_{CN^-}$ [d] cm^{-1}
Cr(III)	$(bpy)_2$	5+	II	0.295	15.3(6.0)[5.6]	61
	$(bpy)_2$	3+	II	0.308	15.5(3.6)[6.0]	63
	[14]aneN_4	5+	II	0.321	20.0(8.0)[4.9]	120, 15
Co(III)	[14]aneN_4	5+	II	0.267	19.9(1.0)[7.0]	20
Rh(III)	$(bpy)_2$	5+	II	0.215	{31.1(2.5)[4.5]}[e]	4
Rh(III)	[14]aneN_4	5+	II	0.254	{29.2(0.8)[6.0]}[e]	
Ru(II)	$(bpy)_2$	6+	III	−0.037	15.3(7.8)[5.0]	49
	$(bpy)_2$	3+	III	−0.023	14.3(4.4)[4.5]	41

Notes:

[a] Sum of formal oxidation states of metals = n.
[b] Determined in acetonitrile with $Fe(Cp)_2^{+,0}$ internal reference.
[c] In water, energies in $cm^{-1}/10^3$; ε_{max} M^{-1} $cm^{-1}/10^3$.
[d] $\Delta\upsilon_{CN^-} = \upsilon_{CN^-}(Mc(MCL)(CN)_2^{n+}) - \upsilon_{CN^-}$(D/A complex).
[e] This absorption could also be assigned as $Ru^{II} \rightarrow CN^-$ (π^*) MLCT.

MCL = [14]aneN_4, *rac*–Me_6[14]aneN_4 and *ms*–Me_6[14]aneN_4. This supports the previous assignment [23] of these transitions as $Ru(NH_3)_5 \rightarrow$ polypyridyl M_tLCT transitions. These absorptions are almost always convoluted with MM'CT transitions or with central metal to polypyridine CT transitions. The most striking examples of this occur for the (Ru and Fe)(PP)$(CNRu(NH_3)_5)_2^{4+}$ complexes in which the $Ru(NH_3)_5^{2+} \rightarrow$ PP transitions are evidently mixed with the $M(PP)_2^{2+}$ MLCT transitions (see bottom spectrum of figure 2).

d. During the course of the redox titrations of the bis-ruthenates a weak absorption ($\varepsilon_{max} \sim 10^2$) developed in the near i.r. (800–1200 nm). The intensity of this absorption maximized when sufficient oxidant had been added to generate a 1:1 ratio of $Ru(NH_3)_5^{2+}$ to $Ru(NH_3)_5^{3+}$. This band was almost always convoluted with the more intense MM'CT transition. This absorption was most distinct in the (MCL)Rh$(CNRu(NH_3)_5)_2^{5+}$ complexes for which there were no interfering absorbances in this spectral region. The absorptions in this spectral region were sometimes complicated by impurity absorptions. This is clearly a $Ru(NH_3)_5^{2+} \rightarrow Ru(NH_3)_5^{3+}$ CT transition.

e and f. A relatively weak band can often be resolved at about 340 nm in the CN^- bridged complexes containing $Ru(NH_3)_5^{3+}$, while a band of roughly comparable intensity is almost always observed in the 350–400 nm spectral region in the CN^- bridged complexes containing $Ru(NH_3)_5^{2+}$ respectively. This assignment is consistent with the occurrence in comparable spectral regions of MLCT absorption bands in $Ru(CN)_6^{4-}$ and LMCT absorption bands in $Ru(CN)_6^{3-}$[22].

Detailed consideration of the spectroscopic evidence for the mixing of CT excited states suggests that in many cases of interest the excited state wave function must be appreciably delocalized. For example, perturbational parameters based on absorption spectra ($\beta_{ij}^2/E_{ij}^2 = \alpha_{ij}^2$, where β_{ij} is the electronic matrix element of the $i \rightarrow j$ absorption and E_{ij} is its energy) suggest that the Ru(II) → Cr(III) DACT excited state of $(bpy)_2Cr(CNRu(NH_3)_5)_2^{5+}$ may be about 50% delocalized. The extent of this delocalization appears to be much less when the central metal is coordinated to aliphatic rather than polypyridyl ligands. This delocalization becomes an issue when one attempts to evaluate matrix elements, or quantities such as stabilization energies which depend on them ($\varepsilon_S \cong \beta_{ij}^2/E_{ij}$) [15], using experimental DACT absorption parameters, as in eq. 1, [15,16,24–26] since the parameter $r_{DA}{}^c$ is the effective distance between the centers of

$$\beta_{ij} = (0.0205/r_{DA}{}^c)(\varepsilon_{max}\Delta\nu_{1/2}\nu_{max})^{1/2} \tag{1}$$

charge in the two states. A very approximate perturbational correction can be used to relate $r_{DA}{}^c$ to the geometrical separation of the donor and acceptor atomic centers, $r_{DA}{}^g$: $r_{DA}{}^c \sim r_{DA}{}^g\,(1 - \sum_{ij}\alpha_{ij}^2)$ for all the CT mixings, *ij*, which contribute to charge delocalization.

2. Infrared absorptions: variations in the cyanide stretching frequency. We have found that υ_{CN} is very sensitive to the electronic coupling of a bridged donor–acceptor pair. Many details of this behavior have been recently published [12] and may be summarized as follows:

a. In the absence of M/M' electronic coupling in a $M(CN^-)M'$ couple υ_{CN} occurs at higher energy than in the parent cyano (MCL) complex [12,27].

b. When there is appreciable M/M' electronic coupling (i.e., when CN^- bridges a D/A pair), υ_{CN} shifts to lower energy [12,27].

c. The magnitude of the shift of υ_{CN} in $D(CN^-)A$ complexes is roughly proportional to $\beta_{DA}{}^2/E_D$ [12].

d. In *trans*-$M^{III}(MCL)(CN)_2^+$ complexes, and in *trans*-$M^{III}(MCL)(CNRu^{II}(NH_3)_5)_2^{5+}$ complexes (C_{2h} symmetry), two CN^- stretching frequencies are observed. The weaker of these occurs at higher energy for the parent dicyano complexes and is assigned as the symmetric combination of CN^- stretching frequencies; the more intense absorption is assigned as the asymmetric combination.

For M = Rh(III), the ordering of υ_s and υ_a is not changed when the coordinated CN^- is ruthenated, but their energy difference is increased. For M =Cr(III), υ_s shifts about 120 cm^{-1} to lower energy while υ_a shifts (also to lower energy) only about 15 cm^{-1} (see figures 3 and 4). This is consistent with a straightforward and simple group theoretical argument: (i) only symmetric DACT excited states can mix with the symmetric (A_g in C_{2h}) ground state; (ii) any vibrational modes which facilitate this DACT excited state/ground state coupling must be totally symmetric (Ag in C_{2h}). Since the D/A coupling is very large for Cr^{III}–(CN^-)–Ru^{II}, but very small for Rh^{III}–(CN^-)–Ru^{II}, contrasting shifts of υ_s and υ_a are observed (figures 3 and 4). The much larger shift of υ_s than υ_a for *trans*-Cr(MCL)(CNRu$(NH_3)_5)_2{}^{5+}$ suggests that much of the effect is anharmonicity of the ground state P.E. surface; however, the considerable shift of the "center of gravity" of the vibrations suggests that there is also some general weakening of the C–N bond due to the D/A coupling.

Most of these observations are at least qualitatively consistent with a simple, semi-classical vibronic model (analogous to a pseudo-Jahn Teller model [28,29]) for D/A coupling [12]. This model suggests that the shifts in frequency of the vibrations which mix with H_{DA} are given by an expression of the form given in equation 2 where β°_{DA} is

$$\left(\beta_{DA} - \beta^{\circ}_{DA}\right)^2 / E_{DA} \cong a\Delta\upsilon_d/\upsilon_d \qquad (2)$$

the D/A coupling matrix element in the absence of any contributions from displacements in the nuclear coordinates, a is proportional to the nuclear reorganizational energy, λ_{CT}, associated with transfer of charge to (or from) the bridging ligand (a $\sim$ $\lambda_{CT}/2$) and it is assumed (for simplicity) that the vibronic contributions to in β_{DA} arise from a single vibrational mode, υ_d. So far, eq. 2 appears to be consistent with observations on CN^- bridged D/A

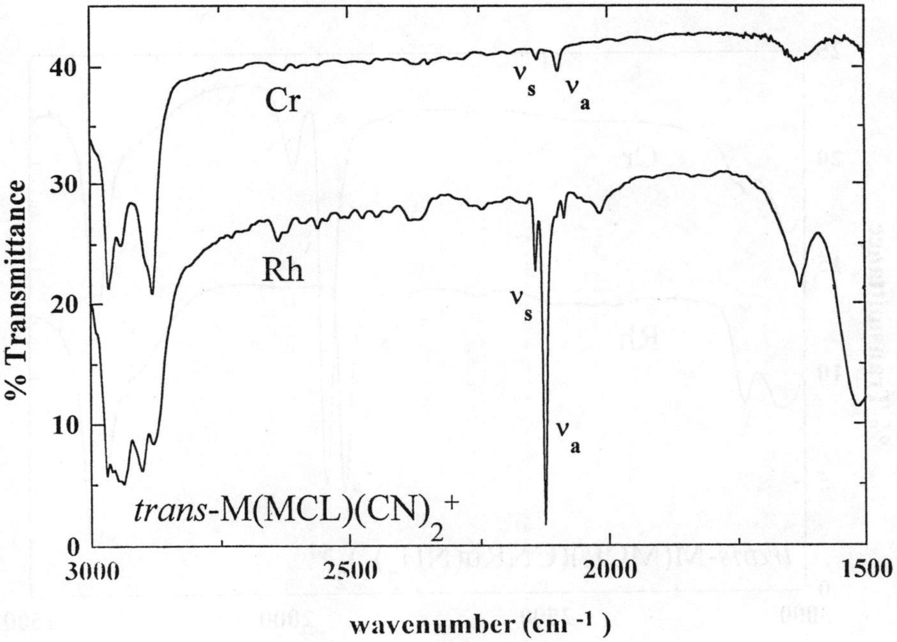

Figure 3. Infrared spectra of *trans*-([14]aneN_4)$M_c(CN)_2^+$ complexes: upper spectrum for M_c = Cr(III); lower spectrum for M_c = Rh(III). The symmetric and antisymmetric combinations of CN^- stretching frequencies are labelled υ_s and υ_a, respectively.

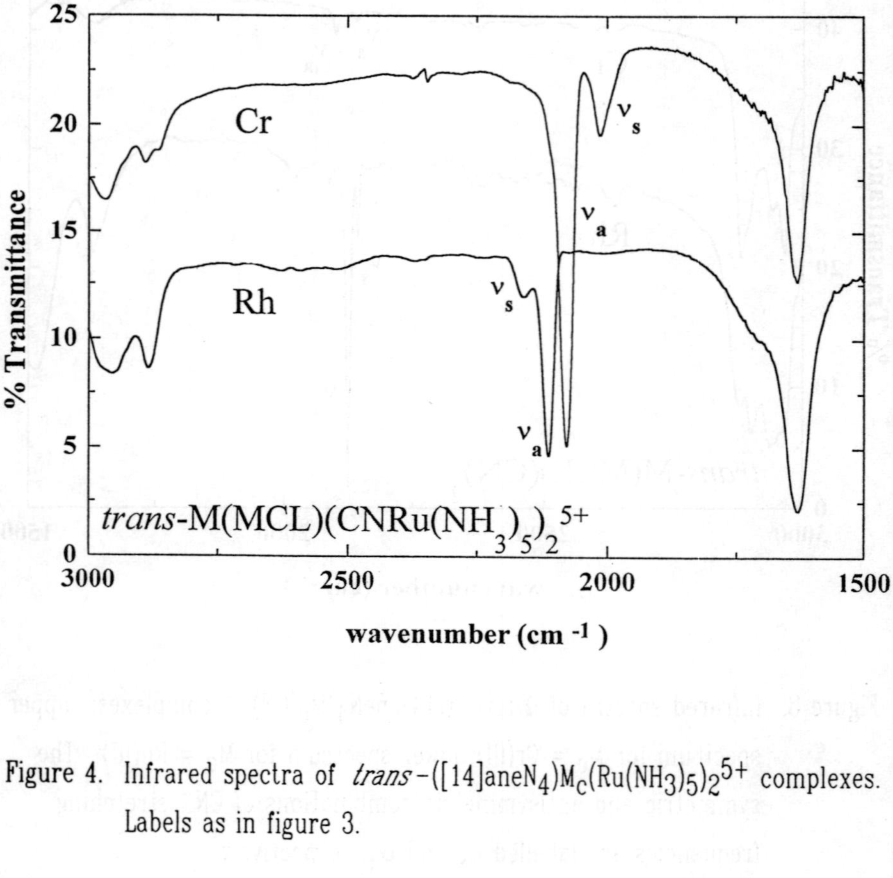

Figure 4. Infrared spectra of *trans*-([14]aneN_4)M_c($Ru(NH_3)_5)_2^{5+}$ complexes. Labels as in figure 3.

complexes.

3. Ground state stabilization energies from half-wave potentials. Since coupling to a CT excited state results in a contribution to the stability of the D/A ground state [15], and since $LRu(NH_3)_5{}^{3+,2+}$ couples are electrochemically reversible, it seemed likely that the variations of $E_{1/2}(Ru(NH_3)_5{}^{3+,2+})$ in $LM(CNRu(NH_3)_5)_2$ complexes could be also used as a measure (in addition to DACT and IR spectra) of the effective values of Ru/M electronic coupling (i.e., of β_{DA}) [11,30]. This can be expressed

$$E_{1/2}^{obsd} = E_{1/2}^{ref} \pm \varepsilon_s^{th} \qquad (3)$$

as in eq. 3 [11,30], where $E_{1/2}^{ref}$ is the appropriate reference potential in the absence of D/A coupling and the "+" sign is used if $Ru(NH_3)_5{}^{2+}$ is stabilized by D/A coupling and the "−" sign is used if $Ru(NH_3)_5{}^{3+}$ is so stabilized. Our initial results indicated that $E_{1/2}(Ru(NH_3)_5{}^{3+,2+})$ could be varied over a range of ~ 350 mV just by varying M in the $LM(CNRu(NH_3)_5)_2{}^{n+}$ complexes [11,27]. If $\varepsilon_s^{th} = \varepsilon_s^{op} = -\beta_A{}^2/E_{DA}$ (see eq. 1) then eq. 4 follows (m is a collection of constants).

$$E_{1/2}^{obsd} \cong E_{1/2}^{ref} \pm m(\varepsilon_{max}\Delta\nu_{1/2}\nu_{max})(r_{DA}{}^c)^2 \qquad (4)$$

Equation 4 is the basis for simple empirical correlations; however, a strictly literal use of eq. 3 results in a different correlation line for each class of complexes. The "slopes" of such correlations are steepest for the $(bpy)_2(CN)M(CNRu(NH_3)_5)^{n+}$ complexes (about 4-times greater than expected for $\varepsilon_s^{th} = \varepsilon_s^{op}$) and most shallow for some of the macrocyclic complexes (for *trans*-$M(Me_6[14]aneN_4)$ $(CNRu(NH_3)_5)_2{}^{5+}$ complexes the slope of this kind of, uncorrected, plot is consistent with $\varepsilon_s^{th} = \varepsilon_s^{op}$). The $(bpy)_2M(CNRu(NH_3)_5)_2{}^{n+}$ complexes are intermediate and they can best be correlated with a line whose slope is about twice as large as would be expected if $\varepsilon_s^{th} = \varepsilon_s^{op}$. The apparent contrast between the mono- and *bis* -ruthenated $(bpy)_2M(CN)_2$ complexes is removed (figure 5) if one allows for symmetry adaptation of the DACT excited states [31]: (i) the *mono* -ruthenates have C_1 symmetry and the number of excited state electronic configurations that mix with the ground state is the same as those that are allowed in absorption; (ii) the *bis* -ruthenates have nominally degenerate DACT excited states, but in the C_{2v} (or C_2) symmetry of the complexes these must split into equal numbers of symmetric and antisymmetric combinations, and optical transitions are allowed to both while only the symmetric excited state configurations can mix with the ground state (the C_{2v} symmetry is a little more restrictive than this, so that in some systems the ratio of states allowed for optical transitions to those allowed for ground state mixing is 3/1 rather than 2/1). Correction for this symmetry effect results in the correlation for $M(polypyridyl)(CN)_2$ centered complexes shown in figure 5; the solid line illustrates expectation based on $\varepsilon_s^{th} = \varepsilon_s^{op}$, the dotted line corresponds to $\varepsilon_s^{th} \sim 4.2\ \varepsilon_s^{op}$. There are a large number of issues involved in interpreting the observations summarized in figure 5. Some of these are: (i) there is a one unit difference in charge between the Rh(III) and Cr(III) centered complexes at the upper limit and the Ru(II) and Fe(II) complexes at the lower limit of the correlation, and this may alter the solvational contributions to $E_{1/2}^{obsd}$; (ii) absorptions at about $29 \times 10^3\ cm^{-1}$ in the Rh(III) centered complexes have been interpreted as Ru(II) → Rh(III) CT absorptions for purposes of this correlation, but this assignment is not certain and the absorptions could be Ru(II) → $CN^-(\pi^*)$ MLCT instead; (iii) the delocalization of electron density from (or to) the $Ru(NH_3)_5$ moieties varies across this series so that use of a constant value of $r_{DA}^c = r_{DA}^g = 5.2$. in eq. 1 is probably inappropriate; (iv) very closely related to the preceding point, the mixing of higher energy CT excited states with the DACT

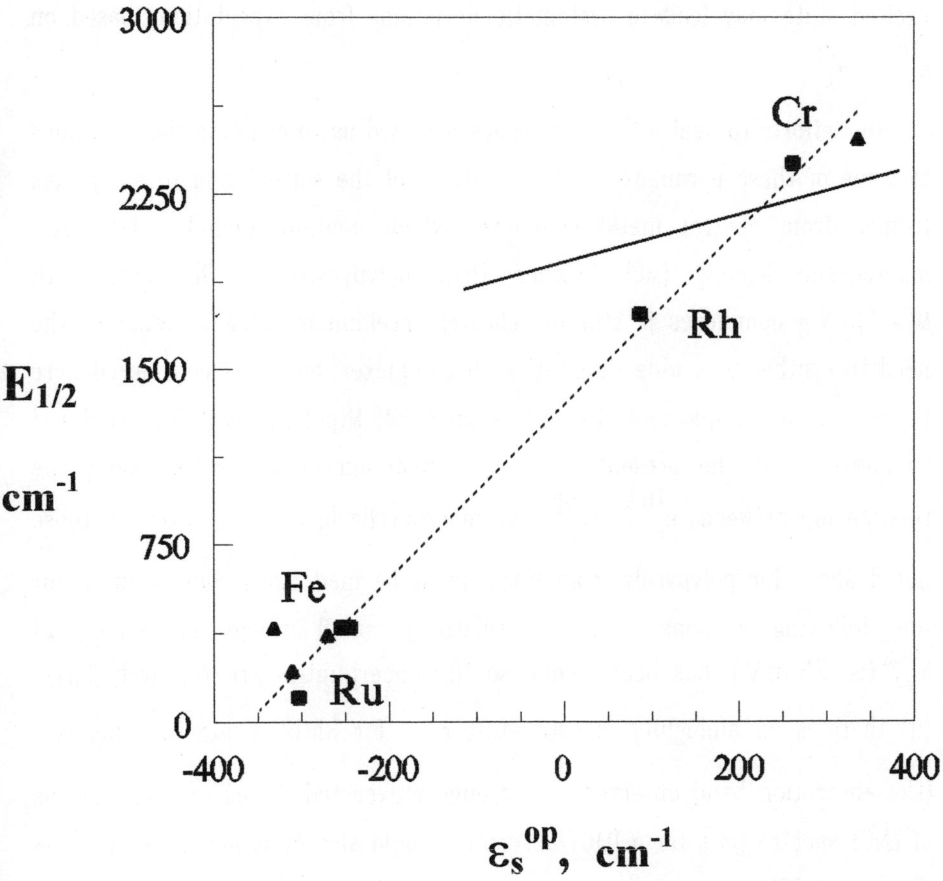

Figure 5. Correlation of $E_{1/2}(Ru(NH_3)_5{}^{3+,2+})$ with the apparent stabilization energy, $\varepsilon_s^{op} = \beta_{DA}{}^2/E_{DA}$, inferred from DACT spectra. Complexes are $(bpy)_2M_c(CNRu(NH_3)_5)_2{}^{n+}$, squares, or $(bpy)_2(CN)M_c(CNRu\,(NH_3)_5)^{n+}$, triangles. The values of ε_s^{op} for the *bis*-ruthenates have been corrected for symmetry effects (i.e.,using $\varepsilon_{max}/2$) and values of $E_{1/2}$(obsd) for M_c = Ru(II) and M_c = Fe(II) have been increased by 400 cm^{-1} (50 mV) to compensate for the variations in charge type. The points for the different central metals are noted on the figure.

excited state may lead to systematic deviations from expectations based on $\varepsilon_s^{th} = \varepsilon_s^{op}$.

Our efforts to deal with these issues have led us to examine the variations of some of these parameters with variations of the solvent and in complexes formed from central metal complexes which contain aliphatic, tetraaza-macrocyclic ligands (MCL) rather than polypyridyls. The work with $M(MCL)(CN)_2$ complexes is still in relatively preliminary stages, owing to the need to synthesize a wide range of such complexes, and it is considered here largely as a complement to our studies of $M(polypyridyl)(CN)_2$ centered complexes. At the present stage, a comparison of inferences about the relationship between $\varepsilon_s^{th} = \varepsilon_s^{op}$ for macrocyclic ligand complexes, to those noted above for polypyridyl complexes, must be made with some caution for the following reasons: (i) a relatively small range of values of $E_{1/2}^{obsd}$ **($\leq$ 75 mV)** has been found, so the uncertainties are relatively large; (ii) there is an ambiguity in estimating ε_s^{op} for Rh(III) centered complexes (the absorption band observed at the energy expected, based on comparisons of LMCT spectra [22], for a DACT absorption could also be assigned as Ru(II) $\rightarrow$ $CN^-(\pi^*)$ MLCT); (iii) changing the macrocyclic ligand appears to change the solvational contribution to $E_{1/2}^{obsd}$. We hope that work in progress will resolve the relation between ε_s^{th} and ε_s^{op} by comparing relative values of $\Delta E_{1/2}^{obsd}$ = $E_{1/2}$ (M ctr.) − $E_{1/2}$ (Rh ctr.) for a substantial number of macrocyclic ligand complexes.

Obviously, these are very complicated comparisons, and it is difficult to systematically change a single parameter in these systems without altering one or more of the others. Nevertheless, several tentative conclusions are possible: (i) ε_s^{th} is generally larger than ε_s^{op}; (ii) solvational effects are significant, but they are not the major factor contributing to the observations (see the next section); (iii) electronic delocalization is more of a factor in polypyridyl than in aliphatic macrocyclic complexes. This delocalization appears to be correlated with the higher energy (than DACT) LMCT and MLCT excited states.

4. Solvent perturbations of CT absorption spectra and $\Delta E_{1/2}^{obsd}$. We have examined the variations in Ru_c(III), Ru_t(III), DACT, Ru_c(II)/bpy, MLCT, absorption maxima and band widths and $\Delta E_{1/2}$ ($Ru(NH_3)_5^{3+,2+}$) (vs. $Fe(Cp)_2^{+,0}$) in several solvents (acetone, acetonitrile, benzonitrile, DMF, DMSO and water) [32]. We have found that $\Delta h\nu_{max} = \Delta h\nu_{max}(MLCT) - h\nu_{max}(DACT)$ correlates reasonably well with $\Delta E_{1/2}$ ($Ru(NH_3)_5^{3+,2+}/Fe(Cp)_2^{+,0}$) [32]. This correlation implies: (i) the variations of $E_{1/2}$ and CT absorption maxima with variation of the solvent can be treated perturbationally, and thus linearized; (ii) that the major contribution is localized at the cationic centers of highest charge density; (iii) that the solvational perturbations of the $Ru(NH_3)_5^{3+,2+}$ couple are about twice as large as those of $Ru(bpy)_2^{3+,2+}$ couple; and (iv) from the slope of the correlations we have inferred that the variations of the solvent reorganizational energy, for the DACT process, are comparable in magnitude to the variations in the free energy of reaction. These observations can be used to estimate the effect on $E_{1/2}$ of the one unit of charge variation among the polypyridyl/complexes discussed in the preceding section is less than or equal to about 50 mV [32]. A correction of 50 mV has been applied for Fe(II) and Ru(II) centered complexes in figure 5; i.e., $E_{1/2} = E_{1/2}^{obsd} + 400\ cm^{-1}$ for these complexes.

5. Symmetry restrictions on the observed properties. Some of the simpler aspects of symmetry related issues have been mentioned in the preceding sections. The *bis*-ruthenates of Ru(II) and Cr(III) complexes each turn out to be a little complicated. In the $(L)Ru^{II}(CNRu^{III}(NH_3)_5)_2^{6+}$ complexes, magnetic coupling between terminal configurations is possible. The ratios of the number of DACT excited state electronic configurations for which dipole allowed optical transitions are possible to those which can mix with the ground state vary somewhat from one ground state configuration to another. At the present time we are using an unbiased average over all the possible ground state configurations, and this results in about a 2-to-1 ratio of configurations contributing to absorption and stabilization.

The discussion of the $LCr^{III}(CNRu^{II}(NH_3)_5)_2^{5+}$ complexes is somewhat complicated by the three half filled d-orbitals of the Cr(III) centers. This results in relatively strong restrictions on the number of excited states which can mix appreciably with the ground state.

An additional feature which somewhat complicates the comparison of optical spectra to ground state stabilization in these complexes is the cylindrical symmetry of CN^- and the different nodal structure of its π and π^* orbitals. Owing to the cylindrical symmetry around CN^-, DACT absortions (e.g., Ru(II) $\rightarrow$ Cr(III)) are dipole allowed along the x and y as well as along the z-axis. We have no good basis for assuming significantly different intensities for these different transitions. The ambiguities of detailed assignments are similarly evident in any simple model of D/A coupling in complexes with C_{2h} symmetry since the C_2 axis, which bisects N-M_C-N (MCL) bond angles, mixes the dπ orbitals of the central metal and, probably, of the axial $Ru(NH_3)_5$ moieties. In such symmetry, all the dπ orbitals would contribute to the D/A coupling, regardless of whether they were coplanar. The symmetry adapted coupling in C_2 (or C_{2v}) complexes is more straightforward and can be adequately described in terms of coplanar interactions. The preceding discussions have not dealt with these issues. However, the variations of υ_{CN}

may turn out to be sensitive to such details. We hope that future studies with higher and with lower symmetry complexes will clarify these points.

C. Some transient studies of back electron transfer in D/A systems. Pertinent values of k_{BET}, determined by transient absorption spectroscopy following photoexcitation of some covalently linked transition metal D/A systems, are summarized in Table 2. We have included the weakly coupled $(bpy)_2Ru(bb)Co(bpy)_2{}^{5+}$ complex in Table 2 (bb = 1,2-bis(2,2'-bipyridyl-4'-yl)ethane; see figure 1) as a simple example of a weakly coupled system and for contrast with the strongly coupled, CN^- bridged complexes. The D/A coupling in this complex is estimated [33] to be $\beta_{DA} \cong 80$ cm^{-1} based on a MLCT (metal to ligand charge transfer) mediated superexchange model [5,34]. This estimate, implying $\kappa_{el} \sim 10^{-2}$, together with the standard semi-classical logic [43] account very well for the magnitude of k_{BET} (Table 3). This seems to be a reasonably typical example of the behavior of weakly coupled D/A systems, and it serves to illustrate that the semi-classical, crude Born-Oppenheimer based formalisms (i.e., based on assuming that the nuclear and electronic coordinates are independent) seem to work very well for weakly coupled D/A systems.

These semi-classical formalisms and various ground state structural and spectroscopic parameters can be used to estimate k_{BET} as in Table 3. Neither the magnitude of k_{BET}(obsd) nor the trends in back electron transfer rates are accounted for by the simple semi-classical models. Further to this point, the value of k_{BET} reported for $(NC)_5Ru(CNRu(NH_3)_5)^-$ [35-38] is much larger than the rate constants for vibrational relaxation in the photogenerated electron transfer excited state [35-38]. If the electronically excited system does not reach vibrational equilibrium within the excited state electronic manifold before electron transfer, then some of the assumptions of the semi-classical approach are invalid. At the very least, the observations on k_{BET} in

Table 3. Back Electron Transfer Parameters of Some Bridged Donor–Acceptor Complexes

Complex	λ_{DA} $cm^{-1}/10^3$	$-\Delta G^\circ_{DA}$ $cm^{-1}/10^3$	$\Delta G^{\neq a}$ $cm^{-1}/10^3$	β^{op}_{DA} $cm^{-1}/10^3$	$k_{BET}(calcd)^b$ $s^{-1}/10^{10}$
$(bpy)_2Ru(bb)Co(bpy)_2{}^{5+}$	13.8	7.7	2.2	~0.05	0.02^c
$(bpy)_2Ru(CNRu(NH_3)_5)_2{}^{6+}$	3.6	11.3	4.1	2.2	0.6
trans-$Co([14]aneN_4)(CNRu(NH_3)_5)_2{}^{6+}$	7.8	12.1	0.6	1.1	$\sim 10^3$
trans-$Cr([14]aneN_4)(CNRu(NH_3)_5)_2{}^{7+}$	~8	~14	~1	2.5	$\sim 10^3$
$(NC)_5Ru(CNRu(NH_3)_5)^-$	6	~8	0.2	1.4	$\sim 10^3$

Notes for Table 3:

a $\Delta G^{\neq} = (\lambda_{DA}/4)(1 + \Delta G^\circ_{DA}/\lambda_{DA})^2$.
b Using $k_{BET}(calc.) = 6 \times 10^{12} \times \exp[\Delta G^{\neq} - \beta_{DA})/RT]$.
c Calculated for $^4Co(II)/^2Ru(III)$ coupling from a superexchange model.

CN^- bridged D/A complexes raise questions about the validity of many of the interpretations of strongly coupled D/A complex properties in terms of the semi-classical formalisms.

D. A simple, semi-classical vibronic model to deal with the effects of CT/DACT mixing. The basic elements of a simple vibronic model have been published [12,30]. The argument is patterned after "pseudo-Jahn-Teller "approaches [19,20,28,29] in that coupling of the ground and excited electron transfer states to a higher CT excited state lead to potential energy functions as in eqs. 5 and 6 respectively (based on a Taylor's series

$$V_g = kx^2 - ax \tag{5}$$

$$V_e = E_{DA} - kx^2 + a'x \tag{6}$$

expansion around a nuclear coordinate **x**), and that the electronic coupling matrix element can be assumed to be a function of the same coordinates, eq. 7.

$$\beta_{DA} \cong \mathring{\beta}_{DA} + bx \tag{7}$$

This approach synergestically couples the upper CT and the DACT states. This approach predicts a shift of the frequencies of the coupled vibrational motions which is proportional to $\beta_{DA}{}^2/E_{DA}$ and it leads to the conclusion that $\varepsilon_s^{th} > \varepsilon_s^{op}$ in an appropriate limit, as in eq. 8

$$\varepsilon_s^{th} - \varepsilon_s^{op} \cong 2(\mathring{\alpha}_{CT})^2\lambda_{CT} - (\mathring{\alpha}_{DA})^2\lambda_{DA} \tag{8}$$

where the $\alpha_i = \mathring{\beta}_{gi}/E_{gi}$ and the λ_i are nuclear reorganizational energies associated with the respective charge transfer processes [32].

This model has been used to account for the otherwise puzzling observation [39] that the nuclear reorganizational energies for inner-sphere electron transfer reactions of *trans* $-Co(MCL)(OH_2)Cl^{2+}$/ *trans* $-Co(MCL)(OH_2)_2{}^{2+}$ couples are about one half of the reorganizational energies of the corresponding outer-sphere couples [40]. This model is based on the hypothesis that in a system in which the electronic coupling is strong, the electronic distribution adjusts very rapidly to changes in the nuclear coordinates. Most vibronic approaches to electronic coupling have focused on the skeletal modes of the donor and acceptor centers as the vibrational modes which mix with the electronic wavefunctions [17–19]. Evidence for such effects have been found by Solomon and coworkers [8] and in our work with inner-sphere electron transfer systems [39]. However, our work with CN^- bridged D/A complexes, described above, has implicated a bridging ligand vibrational mode more strongly than metal complex skeletal modes in the coupling of the D and A electronic wavefunctions. A related theory of vibronically mediated D/A coupling, which implicates bridging ligand modes has been proposed by Ferretti, Ondrechen and co-workers [20]. A theme common to these studies, and emphasized by ours, is that standard models developed for weakly coupled D/A systems [1–6] are not applicable to systems in which the electronic coupling is strong.

Conclusions.

Several properties of the CN^- bridged, transition metal donor-acceptor complexes have been found to be sensitive to the strength of the D/A electronic coupling: (a) the intensity of the DACT absorption band; (b) the magnitude of the shift to lower energies of the bridging CN^- stretching frequency; (c) the reduction potentials of $Ru(NH_3)_5^{3+}$ cyanometalates; (d) the mixing of spectroscopic CT states. These experimental observations are qualitatively, and in some cases semi-quantitatively in agreement with expectation based on a simple vibronic model for D/A electronic coupling.

The perturbational mixing considered in this work are vertical with respect to the ground state nuclear coordinates. The vibronic model's assumption that $\beta_{DA} = \beta_{DA}^{o} + bx$ implies that β_{DA} will be different at the transition state for electron transfer than it is for the vertical CT coupling interactions. Our preliminary studies of back electron transfer rates, following DACT photoexcitation of these complexes, suggests that they do not readily correlate with the vertical electronic coupling parameters or with the standard electron transfer parameters in these complexes. Vibronic coupling is likely to be a major factor in the relaxation of the DACT excited states; however, the intervention of intermediate states of different spin multiplicity is also very likely. These issues are being further investigated.

Acknowledgments

We gratefully acknowledged partial support of this research by the Division of Chemical Sciences, Office of Basic Energy Sciences, Office of Energy Research, U.S. Department of Energy. S.E.M. acknowledges support from the Conselho Nacional de Desenvolvimento Cientifico e Technológico of Brazil.

References

1. R.A. Marcus, Annu. Rev. Phys. Chem., 15 (1965)155.
2. N.S. Hush, Electrochim. Acta., 13 (1968)1005.
3. C. Creutz, Progr. Inorg. Chem., 30 (1983)1.
4. M.D. Newton and N. Sutin, Annu. Rev. Phys. Chem., 35 (1988)437.
5. M.D. Newton, Chem. Rev., 91 (1991)767.
6. N.R. Kestner; J. Logan and J. Jortner, J. Phys. Chem., 78 (1974)2148.
7. R. de la Rosa; P.J. Chang; F. Salaymeth and J.C. Curtis, Inorg. Chem., 4(1985)4229.
8. F. Saleymeth; S. Berhaue; R.Yusof; R. de la Rosa; E.X. Fung; R.Matamoros; K.W. Law; Q. Zheng; E.M. Kober and J.C. Curtis, Inorg. Chem., 32 (1993)3895.
9. R.J. Crutchley, Adv. Inorg. Chem., 41 (1994)273.
10. A.R. Rezvani; L.E.B. Evans and R.J. Crutchley, Inorg. Chem., 34 (1995)4600.
11. J.F. Endicott; X. Song; M.A.Watzky and T. Buranda, J. Photochem. Photobiol. A. Chem., 82 (1994)181.
12. M.A. Watzky; J.F. Endicott; X. Song; Y. Lei and A. Macatangay, Inorg. Chem., 35 (1996)3463.
13. D.R. Gamelin; E.L. Bominaar; C. Mathoniére; M.L. Kirk; J.-J. Girerd and E.I. Solomon, Inorg. Chem., 35 (1996)4323.
14. N.S. Hush, Progr. Inorg. Chem., 9 (1968)391.
15. R.S. Mulliken and W.B. Person, Molecular Complexes; Wiley-Interscience: New York; 1968.
16. J.R. Reimers and N.S. Hush, Chem. Phys., 208 (1996)177.
17. S.B. Piepho; E.R. Krause and P.N. Schatz, J. Am. Chem. Soc., 100 (1978)2996.
18. S.B. Piepho, J. Am. Chem. Soc., 110 (1988)6319.
19. S.B. Peipho, J. Am. Chem. Soc., 112 (1990)4197.

20. A. Ferretti; M.J. Lami; G. Ondrechen and G. Villani, J. Phys. Chem., 99 (1995)10484.
21. C. Creutz and H. Taube, J. Am. Chem. Soc., 95 (1973)1086.
22. A.P.B. Lever, Inorganic Electronic Spectroscopy; 2d ed.; Elsevier: New York, 1984.
23. C.A. Bignozzi; S. Raffia and F. Scandola, J. Am. Chem. Soc., 107 (1985)1644.
24. C. Creutz; M.D. Newton and N. Sutin, Photochem. Photobiol. A. Chem., 112 (1990) 7915.
25. R.J. Cave and M.D. Newton, Chem. Phys. Lett., 249 (1996)15.
26. J.R. Reimers and N. Hush, J. Phys. Chem., 95 (1981)9773.
27. C.A. Bignozzi; J.R. Argazzi; J.R. Schoonover; R.B. Dyer and F. Scandola, Inorg. Chem., 31 (1992)5260.
28. I.B. Bersuker, The Jahn-Teller Effect and Vibronic Interactions in Modern Chemistry; Plenum: New York, 1984.
29. C.J. Ballhausen, In Vibronic Processes in Modern Chemistry; Plenum: New York, 1984.
30. J.F. Endicott; X. Song; M.A. Watzky; T. Buranda and Y. Lei, Chem. Phys., 176 (1993) 427.
31. W.L. Parker and J.A. Crosby, Int. J. Quant. Chem., 39 (1991)299.
32. M.A. Watzky; R.A. Van Camp; X. Song; S.E. Mazzetto; A.V. Macatangay; J.F. Endicott and T. Buranda, J. Phys. Chem., submitted for publication.
33. X. Song; Y. Lei; S. Van Wallendal; M.W. Perkovic; D.C. Jackman; J.F. Endicott and D.P. Rillema, J. Phys. Chem., 97 (1993)3225.
34. H.M. McConnell, J. Chem. Phys., 35 (1961)508.
35. S.K. Doorn; D.O. Stoutland; R.B. Dyer and W.H. Woodruff, J. Am. Chem. Soc., 114 (1992)3133.

36. S.K. Doorn; R.B. Dyer; D.O. Stoutland and W.H. Woodruff, J. Am. Chem. Soc., 115 (1993)6398.

37. G.L. Walker; D.F. Barbara; S.K. Doorn and J.T. Hupp, J. Phys. Chem., 95 (1991) 5212.

38. K. Tominaga; D.A.V. Klinel; A.E. Johnson; N.E. Levinger and D.F. Barbara, J. Phys. Chem., 97 (1993)1228.

39. C.L. Schwarz and J.F. Endicott, Inorg. Chem., 34 (1995)4572.

40. B. Durham; J.F. Endicott; C.-L. Wong and D.P. Rillema, J. Am. Chem. Soc., 101 (1979)847.

41. T. Buranda; Y. Lei and J.F. Endicott, J. Am. Chem. Soc., 114 (1992)6917.

ADIABATIC AND NON-ADIABATIC ELECTRON TRANSFER IN SOLUTIONS: A SELF-CONSISTENT APPROACH BEYOND THE CONDON APPROXIMATION.

ALEXANDER M.KUZNETSOV
The A.N.Frumkin Institute of Electrochemistry, Russian Academy of Sciences, 117071 Moscow, Leninskii prospect 31, Russia

Calculational schemes enabling to go beyond crude Condon approximation in non-adiabatic electron transfer reactions are discussed with the use of continuum approximation for the solvent polarization. An algorithm for the self-consistent introduction of an effective reaction coordinate in the adiabatic transition is suggested. Effects due to deviations from the Born-Oppenheimer approximation in bridge-assisted electron transfer reactions are discussed. Interpolation formulae covering limits of coherent and sequential electron transfer in bridge-assisted processes are presented. Simple equations determining a parametric dependence of the transition probability on the reaction free energy in crude Condon approximation are included.

1 Non-adiabatic electron transfer rate in crude Condon approximation

Earlier works on the theory of non-adiabatic electron transfer reactions beginning from first papers by Marcus[1] and Levich and Dogonadze[2] used so called crude Condon approximation (CCA), i.e. the calculations of transition probability were performed neglecting possible dependence of the electron wave functions on the fluctuations of the inertial solvent polarization. Attempts to go beyond CCA were made in later works in the framework of various approaches. The interaction of the electron with local vibrational modes of the reaction complex was taken into account in[3,4]. The quantum chemical calculations were performed in [5-8]. The schemes which use continuum approach for the description of the solvent polarization will be discussed in the present paper.[9-11]

Traditional scheme of calculation of the transition probability for the non-adiabatic transition involves the introduction of the diabatic electron states φ_i and φ_f as the solutions of the Schrödinger equations

$$H_i^e \varphi_i = \varepsilon_i \varphi_i$$
$$H_f^e \varphi_f = \varepsilon_f \varphi_f \qquad (1)$$

where the Hamiltonians of the initial and final states for the electron

$$H_i^e = T_e + V_{eD} + V_{eP}$$
$$H_f^e = T_e + V_{eA} + V_{eP} \qquad (2)$$

involve the kinetic energy T_e and the interactions of the electron with the inertial polarization V_{eP} and with the donor (D) and acceptor (A) ions, V_{eD} and V_{eA}, respectively. The CCA consists of the use of fixed non-deformable basis electron wave functions $\varphi_i \to \varphi_D$ and $\varphi_f \to \varphi_A$, which are usually considered as the eigen states of the ions D and A in the gas phase or at initial and final equilibrium values of the solvent polarization, $\mathbf{P}_{0i}$ and $\mathbf{P}_{0f}$. The corresponding electron energies

$$\varepsilon_D(\mathbf{P}) = \langle \varphi_D | H_i^e | \varphi_D \rangle$$
$$\varepsilon_A(\mathbf{P}) = \langle \varphi_A | H_f^e | \varphi_A \rangle \qquad (3)$$

are therefore linear functions of **P** and being added to the Hamiltonian of the solvent result in the adiabatic Hamiltonians of the initial and final states

$$H_i^a = T_P + U_i(\mathbf{P})$$
$$H_f^a = T_P + U_f(\mathbf{P}) \qquad (4)$$

where T_P is the kinetic energy operator and the diabatic free energy surfaces of the initial and final states have the form

$$U_i = U_0(\mathbf{P}) + \varepsilon_D(\mathbf{P})$$
$$U_f = U_0(\mathbf{P}) + \varepsilon_A(\mathbf{P}) \qquad (5)$$

where

$$U_0(\mathbf{P}) = \frac{2\pi}{c}\int \mathbf{P}^2(\mathbf{r})d^3r \tag{6}$$

and $c=1/\varepsilon_\infty - 1/\varepsilon_0$ is the Pekar parameter involving the high frequency and low frequency dielectric constants of the medium.

The calculation of the transition probability for the non-adiabatic reaction may be performed with the use of the Fermi golden rule which in CCA reduces to the calculation of the averaged Franck-Condon factor. If the electron transfer is accompanied by a small change of the intramolecular structure of the reactants, harmonic approximation may be applied for the description of the intramolecular vibrations. The use of the effective Hamiltonian method for the description of the solvent polarization[12,13] reduces then the problem to the calculation of the Franck-Condon factor for the system of harmonic oscillators. The most simple general formula is obtained if the vibrational frequencies of the oscillators remain unchanged in the course of the transition[14]

$$W = \frac{2\pi}{\hbar^2}|V|^2 \exp\left\{-\frac{\Delta F}{2k_BT} - \sum_k z_k \cosh\frac{\hbar\omega_k}{2k_BT}\right\}\sum_{..l_k..}\prod_k I_{l_k}(z_k) \tag{7}$$

where $I_n(x)$ is the Bessel function of imaginary argument, $\omega_k = p_k\Omega$, the summation is performed over l_k satisfying to the condition $\Sigma_k l_k p_k = \Delta F/\hbar\Omega$, ΔF is the free energy of the transition (driving force) and

$$z_k = \frac{\left(q_{kf}^0 - q_{ki}^0\right)^2}{2}\operatorname{csc h}\frac{\hbar\omega_k}{2k_BT} \tag{8}$$

If only one intramolecular degree of freedom z is reorganized and all the medium polarization is classical, Eq.8 takes the form[15]

$$W = |V|^2 \sqrt{\frac{\pi}{\hbar^2 E_r k_B T}}$$
$$\sum_{m=-\infty}^{\infty} I_m(z) \exp\left\{ - z \cosh \frac{\hbar\omega}{2k_B T} - \frac{m\hbar\omega}{2k_B T} \right\} \quad (9)$$
$$\exp\left\{ - \frac{\left(\Delta F - m\hbar\omega + E_r\right)^2}{4E_r k_B T} \right\}$$

where E_r is the reorganization energy of the solvent polarization

$$E_r = \frac{c}{8\pi} \int d^3 r \left(\mathbf{D}_f - \mathbf{D}_i \right)^2 \quad (10)$$

and summation over m takes into account all possible transitions between different vibrational states of the local mode.

Eq.9 and its limiting forms are most popular in the literature. It should be noted however that Eq.9 is convenient when:

1. Only one intramolecular degree of freedom takes part in the transition.
2. Local vibrations are quantum ($\hbar\Omega \gg k_B T$). Then only small number of terms give considerable contribution to the sum.

More general and convenient expressions may be derived with the use of the generating function method.[13,16,17] A simple form for the case where the vibrational frequencies remain unchanged in the course of the transition is given below[18]

$$W(\Delta F) = W(\alpha) = \frac{\sqrt{2\pi}}{\hbar k_B T} V^2 \left[- \frac{d^2 G(\alpha)}{d\alpha^2} \right]^{-1/2} e^{\alpha \frac{dG(\alpha)}{d\alpha} - G(\alpha)} \quad (11)$$

$$\Delta F = \Delta F(\alpha) = -k_B T \frac{dG(\alpha)}{d\alpha} \quad (12)$$

where

$$G(\alpha) = \sum_l \frac{E_{rl}}{\frac{1}{2}\hbar\Omega_l} \frac{\sinh\frac{\hbar\Omega_l(1-\alpha)}{2k_BT}\sinh\frac{\hbar\Omega_l\alpha}{2k_BT}}{\sinh\frac{\hbar\Omega_l}{2k_BT}} + \alpha(1-\alpha)\frac{E_r^{class}}{k_BT} \quad (13)$$

where E_r^{class} is the reorganization energy of all classical degrees of freedom, E_{rl} is the reorganization energy of non-classical degrees of freedom, summation is performed over all non-classical degrees of freedom, α is the symmetry factor which is determined as the solution of Eq.12. Analytical solution of Eq.12 is possible only in some limiting cases. However Eqs.11-13 give a *parametric dependence of the transition probability on the free energy of the transition with α playing the role of running parameter.*[18] It is worth to note that Eqs.11-13 allow to calculate the dependence of the transition probability on the driving force using standard programs available for PC and incorporate possible reorganization of several local modes in a simple way (one-fold summation in Eq.13). Extensions of this type expressions to intramolecular potentials of arbitrary form in classical limit and for electrochemical processes are available.[18,19]

2 Beyond crude Condon approximation. Non-self-consistent models

Earlier attempts to improve CCA concerned either more accurate estimation of the electronic preexponential factor at the transitional configuration calculated in CCA or more exact calculation of the transitional configuration itself in the framework of Condon approximation (improved Condon approximation). The model of harmonic vibrations allows to determine the transitional configuration in CCA for both classical and quantum behaviour of the intramolecular modes.[9,10,20,21] This may be used for the calculation of the electron wave functions either by means of solution of the Schrödinger equations or with the use of a direct variational method. In the latter approximate trial forms are accepted for the electron wave functions. In the simplest case they are represented as exponentials with orbital exponents λ_i and λ_f as variational parameters. λ_i and λ_f determine the inverse radii of the distribution of the electron densities in the donor and acceptor. Their values at the transitional configuration are different from those at equilibrium configurations. Calculations in continuum model showed that the electron clouds are more diffuse at the transitional configuration as compared to those at corresponding equilibrium configurations. This results in a considerable increase of the estimated value of the electron transmission coefficient as compared to its value calculated in CCA.
An attempt to improve Condon approximation and to go beyond Condon approximation was made in classical limit in the framework of a configurational

model where certain functional form was used for the description of the fluctuation of the solvent polarization[9,10]

$$\delta \mathbf{P} = \frac{c}{4\pi}\left[\xi \mathbf{D}_{eD}\left(\varphi_D\right) + \eta \mathbf{D}_{eA}\left(\varphi_A\right)\right] \tag{14}$$

where $\delta\mathbf{P}$ is the deviation of the polarization from its initial equilibrium value, ξ and η are independent variables and $\mathbf{D}_{eD}$ and $\mathbf{D}_{eA}$ are the electric inductions created by the electronic charge distributions in the donor and acceptor ions calculated with the use of fixed basis set φ_D and φ_A . Calculations with the use of this model showed that the allowance for the variation of the electron wave functions in the process of the polarization fluctuation from the initial equilibrium value to the transitional configuration results in lower activation barrier and larger electron transmission coefficient increasing thereby the transition probability. It is clear that the models of this sort are non-self-consistent since they do not take into account a feedback between the electronic and nuclear subsystem in the fluctuation process. A self-consistent calculation of the transition probability must take into account mutual influence of the transferable electron and solvent polarization in the determining an optimum configuration. The solution of this problem was approached with the use of various models.

3 Path-integral approach

This approach exploits the integral representation of the Fermi golden rule for the transition probability of the non-adiabatic reaction

$$W = \frac{1}{i\hbar k_B T} e^{\frac{F_{i0}}{k_B T}} \int_{c-i\infty}^{c+i\infty} d\theta \, Tr\left[\hat{V}_i \rho_i (1-\theta) \hat{V}_i^{+} \rho_f(\theta)\right] \tag{15}$$

It differs from a traditional one by the fact that ρ_i and ρ_f represent here the total density matrices of the initial and final diabatic states including the electron subsystem.[22,23] They may be represented in the form of Feynman path integrals.[24] The calculation of these integrals was performed in various approximations. In the improved Condon approximation (ICA) the preexponential factor was ignored and the saddle point method was used. In classical limit for nuclear subsystem this leads in particular to two coupled non-linear Schrödinger equations for the calculation of the electron wave functions at the transitional configuration[22,23]

$$\begin{aligned}&\left[H_D - \int \mathbf{P}^*\left(\mathbf{r};\varphi_D^*,\varphi_A^*\right)\mathbf{D}_{eD}(\mathbf{r})d^3r\right]\varphi_D^* = \varepsilon_i\varphi_D^*\\&\left[H_A - \int \mathbf{P}^*\left(\mathbf{r};\varphi_D^*,\varphi_A^*\right)\mathbf{D}_{eA}(\mathbf{r})d^3r\right]\varphi_A^* = \varepsilon_f\varphi_A^*\end{aligned} \tag{16}$$

where H_D and H_A are the electron Hamiltonians of the donor and acceptor, and the solvent polarization depends on the same electron wave functions which must be found from the solution of these equations (see below). Numerical calculational schemes may be used. However much more simple and transparent *analytical* forms emerge in the framework of direct variational approach which allows also to go beyond the Condon approximation. This gives in particular for symmetric transitions the following approximate expressions for the orbital exponents of the electron wave functions and for the activation free energy[22,23]

$$\lambda = \lambda_0 \frac{\left(1 + 32z/5c\varepsilon_0 - 2R\lambda_0 k_B T/F_a^0\right)}{2\left(1 + 16z/5c\varepsilon_0\right)} \tag{17}$$

$$F_a = F_a^0 \frac{\left(3 + 64z/5c\varepsilon_s + \left[2R\lambda_0 k_B T/F_a^0\right]^2\right)}{4\left(1 + 16z/5c\varepsilon_s\right)} \tag{18}$$

where R is the transfer distance, z is the charge number of the ion and

$$F_a^0 = \frac{5}{32}ce^2\lambda_0; \qquad \lambda_0 = \frac{5mce^2}{16\hbar^2}\left(1 + 16z/5c\varepsilon_s\right) \tag{19}$$

represent the values in CCA.

Eq.18 shows that the activation barrier depends on the transfer distance increasing with R. This is the effect of the deviation from the Condon approximation. The shift of the path of the transition from the saddle point towards higher free energy values is overcompensated by the increase of the overlap of the electronic wave functions leading to an overall increase of the transition probability. The degree of this shift depends on the transfer distance.

4 Self-consistent configurational model

In classical limit we may start the calculation of the transition probability for the non-adiabatic reaction directly from the equation

$$W = Q^{-1}\frac{2\pi}{\hbar}\int V^2(\mathbf{P})e^{-\frac{U_i(\mathbf{P})}{k_B T}}\delta\left(U_i(\mathbf{P}) - U_f(\mathbf{P})\right)\delta\mathbf{P} \tag{20}$$

where Q is the configurational integral for the initial state, V is the electron matrix element and Dirac delta function means that the integration is performed over various polarization configurations along the surface separating the reactants and products regions.

In ICA the transitional configuration may be found as the point of minimum on this separating surface (i.e. the saddle point).[25] The calculational scheme differs from that in CCA by that we have to take into account the dependence of the electron wave functions on the medium polarization $\mathbf{P}$.[25] The diabatic free energy surfaces $U_i(\mathbf{P})$ and $U_f(\mathbf{P})$ have a general form

$$\begin{aligned} U_i(\mathbf{P}) &= U_0(\mathbf{P}) + \langle\varphi_D|H_D|\varphi_D\rangle - \int d^3r\mathbf{P}(\mathbf{r})\langle\varphi_D|\mathbf{D}_{eD}|\varphi_D\rangle \\ U_f(\mathbf{P}) &= U_0(\mathbf{P}) + \langle\varphi_A|H_A|\varphi_A\rangle - \int d^3r\mathbf{P}(\mathbf{r})\langle\varphi_A|\mathbf{D}_{eA}|\varphi_A\rangle \end{aligned} \tag{21}$$

where two last terms are the electronic energies. The electron wave functions are determined from the condition of the minimum of the energy functionals at fixed polarization value

$$\frac{\delta U_i}{\delta\varphi_D}\bigg|_{\mathbf{P}} = 0; \qquad \frac{\delta U_f}{\delta\varphi_A}\bigg|_{\mathbf{P}} = 0; \tag{22}$$

Eqs.22 give the electron wave functions which depend on the polarization, $\varphi_D(x;\mathbf{P})$ and $\varphi_A(x;\mathbf{P})$. The saddle point is determined as the minimum point of the function

$$U = (1-\theta)U_i(\mathbf{P}) + \theta U_f(\mathbf{P}) \tag{23}$$

over $\mathbf{P}$ and θ. This leads to two coupled equations

$$\mathbf{P} = \frac{c}{4\pi}\left\{(1-\theta)\mathbf{D}_{eA}\left[\varphi_A(\mathbf{P})\right] + \theta\mathbf{D}_{eB}\left[\varphi_B(\mathbf{P})\right] + \mathbf{D}_A + \mathbf{D}_B\right\} \tag{24}$$

$$U_i(\mathbf{P}) = U_f(\mathbf{P}) \tag{25}$$

where $\mathbf{D}_D$ and $\mathbf{D}_A$ are the electric inductions of the ionic cores in D and A.

It is noteworthy that the electric inductions in the right-hand side of Eq.24 depend on the same polarization value which is involved in its left-hand side. It is this polarization which must be used in the Schrödinger equations for the electron wave functions (see Eq.16). Eqs.16, 24 and 25 represent a set of coupled self-consistent equations for the transitional configuration in ICA.

Simple analytical expressions emerge for the activation barrier F_a and orbital exponents λ_D and λ_A if the exponential trial functions are used for $\varphi_D(x;\mathbf{P})$ and $\varphi_A(x;\mathbf{P})$[25]

$$\lambda_D = \frac{m}{\hbar^2}\left[\frac{z_D e^2}{\varepsilon_0} + \frac{5}{16} c e^2 (1-\theta)\right]$$
$$\lambda_A = \frac{m}{\hbar^2}\left[\frac{z_A e^2}{\varepsilon_0} + \frac{5}{16} c e^2 \theta\right] \qquad (26)$$

$$F_a = \theta^2 E_r = \frac{\left(E_r + \Delta F\right)^2}{4E_r} \qquad (27)$$

In this particular case the activation free energy has a familiar quadratic form, however the reorganization energy

$$E_r = \frac{5}{16}\frac{mce^2}{\hbar^2}\left(\frac{z_D e^2}{\varepsilon_0} + \frac{z_A e^2}{\varepsilon_0} + \frac{15}{32} ce^2\right) \qquad (28)$$

is smaller here than that calculated in CCA

$$E_r = \frac{5}{16}\frac{mce^2}{\hbar^2}\left(\frac{z_D e^2}{\varepsilon_0} + \frac{z_A e^2}{\varepsilon_0} + \frac{20}{32} ce^2\right) \qquad (29)$$

5 A self-consistent scheme for non-adiabatic electrochemical electron transfer

The scheme presented in Sec. 4 may be extended to electron transfer between an ion in the solution and the metallic electrode.[26] A large number of the electrons in the metal makes the calculations much more difficult. The electron density extends from the metal to the solution and is subjected to the electric field created by the ion and the solvent polarization. The electronic tail may "breath" under the action of the polarization fluctuations. Our aim below is to take into account this effect. Therefore we shall neglect possible variation of the electron density in the reacting ion.

Two main approaches exist in the electron density functional theory of metal/solution interface. One originates from Smith's work and uses the trial functions.[27] It allows to calculate the electron density and energetic characteristics of the interface. Therefore it can provide the information only about the activation barrier. The other one, Kohn - Sham method operates with the electron wave functions.[28] One may therefore hope that this method may be used for the estimation of both the activation barrier and the electron transmission coefficient. The use of the above scheme in the framework of the last method is briefly described below, although it may be equally applied to the trial function scheme.[26]

The diabatic free energy surfaces for this system have the form

$$\begin{aligned} U_i &= F_i[n_i] - \int d^3r\mathbf{P}\mathbf{D}_{ei}[n_i] - \int d^3r\mathbf{P}\mathbf{D}_A^{(z)} + \frac{2\pi}{c}\int d^3r\mathbf{P}^2 \\ U_f &= F_f[n_f] - \int d^3r\mathbf{P}\mathbf{D}_{ef}[n_f] - \int d^3r\mathbf{P}\mathbf{D}_A^{(z-1)} + \frac{2\pi}{c}\int d^3r\mathbf{P}^2 \end{aligned} \tag{30}$$

where F_i and F_f are the energies of the metal electrons which are the functionals of the initial and final electron densities n_i and n_f, z is the charge number of the ion. The second and third terms represent the energy of interaction of the slow polarization with the electric fields due to the electrons and the ion in the initial and final states.

The value of the polarization at the transitional configuration is determined by equation

$$\mathbf{P}^* = \frac{c}{4\pi}\left\{(1-\alpha)\mathbf{D}_{ei}\left[n_i(\mathbf{P}^*)\right] + (1-\alpha)\mathbf{D}_A^{(z)} + \alpha\mathbf{D}_{ef}\left[n_f(\mathbf{P}^*)\right] + \alpha\mathbf{D}_A^{(z-1)}\right\} \tag{31}$$

where α is the symmetry factor. The electron wave functions are determined in the Kohn-Sham scheme from the solution of the Schrödinger equations which at the transitional configuration are modified as follows

$$\left[T - e\Phi^i_{eff}\left(n_i^*, n_f^*\right) + \varepsilon_{xc} + V^i_{eA}\right]\varphi^i_k = \varepsilon^i_k \varphi^i_k$$
$$\left[T - e\Phi^f_{eff}\left(n_i^*, n_f^*\right) + \varepsilon_{xc} + V^f_{eA}\right]\varphi^f_k = \varepsilon^f_k \varphi^f_k \qquad (32)$$

where ε_{xc} is the exchange/correlation energy, V^i_{eA} and V^f_{eA} are the interactions of the metal electrons with the ion in the oxidized and reduced forms and the effective potentials are determined from the solution of the Poisson equations

$$\Delta\Phi^i_{eff} = -\frac{4\pi}{\varepsilon_\infty}\rho_i\left[n_i^*\right] + 4\pi c\left\{(1-\alpha)\rho_i\left[n_i^*\right] + \alpha\rho_f\left[n_f^*\right]\right\}$$
$$\Delta\Phi^f_{eff} = -\frac{4\pi}{\varepsilon_\infty}\rho_f\left[n_f^*\right] + 4\pi c\left\{(1-\alpha)\rho_i\left[n_i^*\right] + \alpha\rho_f\left[n_f^*\right]\right\} \qquad (33)$$

where

$$\rho_\gamma = n_+\theta(-x) + n_\gamma + eq_\gamma\delta(\mathbf{r} - \mathbf{R}_A); \quad \gamma = i, f; \quad q_i = z, \quad q_f = z - 1 \qquad (34)$$

and it is assumed that the metal is located in the half-space $x<0$. Eqs.31-34 represent a set of coupled equations which in principle give a self-consistent solution for the polarization and electron densities. The symmetry factor is determined then by Eq.25 and the activation barrier is equal to

$$F_a = U_i\left(\mathbf{P}^*\right) - U_i\left(\mathbf{P}_{0i}\right) \qquad (35)$$

6 New effective reaction coordinate for adiabatic electron transfer. An algorithm for the potential profile

When we go beyond CCA the effective reaction coordinate E

$$E = \int d^3r\left(\mathbf{D}_f - \mathbf{D}_i\right)\left(\mathbf{P}(\mathbf{r}) - \mathbf{P}_{0i}(\mathbf{r})\right) \qquad (36)$$

introduced in[29] for the stochastic description of the transition is inappropriate.

A new reaction coordinate is suggested in this case which is defined as follows

$$E = \varepsilon_f(\mathbf{P}) - \varepsilon_i(\mathbf{P}) \tag{37}$$

An algorithm for the calculation of the potential profile along this coordinate is given below. It involves several points.

1. First we note that the diabatic free energy surfaces may be written in the form

$$\begin{aligned} U_i(\mathbf{P}) &= U_0(\mathbf{P}) + \varepsilon_i(\mathbf{P}) \\ U_f(\mathbf{P}) &= U_0(\mathbf{P}) + \varepsilon_f(\mathbf{P}) = U_i(\mathbf{P}) + E \end{aligned} \tag{38}$$

where

$$U_0(\mathbf{P}) = \frac{2\pi}{c}\int \mathbf{P}^2 d^3r - \int \mathbf{P}(\mathbf{D}_D + \mathbf{D}_A)d^3r \tag{39}$$

2. The diabatic potential profiles $U_i(E)$ and $U_f(E)$ are defined as

$$U_i(E) = \min_{\mathbf{P}} U_i(\mathbf{P})\big|_E; \qquad U_f(E) = \min_{\mathbf{P}} U_f(\mathbf{P})\big|_E = U_i(E) + E \tag{40}$$

Thus we may restrict ourselves by the calculation of $U_i(E)$.

3. The potential profile $U_i(E)$ may be found as

$$U_i(E) = \min_{\mathbf{P},\alpha} U(\mathbf{P},\alpha) \tag{41}$$

where

$$U(\mathbf{P},\alpha) = U_i(\mathbf{P}) - \alpha\left[\varepsilon_f(\mathbf{P}) - \varepsilon_i(\mathbf{P}) - E\right] \tag{42}$$

The term in brackets in Eq.42 takes into account the constraint E=const.

4. Minimization of $U(\mathbf{P},\alpha)$ over $\mathbf{P}$ gives the electron wave functions and $\mathbf{P}(\mathbf{r};\alpha)$. This results in

$$U_i = U_0(\alpha) + \varepsilon_i(\alpha) \tag{43}$$

and

$$E = \varepsilon_f(\alpha) - \varepsilon_i(\alpha) \tag{44}$$

Eqs.43 and 44 give a *parametric* dependence of U_i on E with α *as running parameter*. This scheme involves the solution of nonlinear differential equation and is therefore rather complicated. A more simple version involves trial functions and leads to the solution of algebraic equations for the variational parameters. The adiabatic free energy surfaces are constructed from the diabatic ones in a usual way.

7 Effects of violation of the Born - Oppenheimer approximation in long-distance electron transfer and in atom group transfer

The Born - Oppenheimer approximation is one of the main tools of the theory of charge transfer processes. In electron transfer reactions it is applied usually to the diabatic electron states and is based on large separation of the electron energy levels both in the donor D and in the acceptor A. Only one, ground electron state in each reactant partner is usually involved in the calculations (two-level approximation). This approximation is quite accurate in the diabatic basis for short-distance electron transfer. It correspond physically to the fact that the interaction of the electron with the vibrational subsystem remains practically unchanged during the time which is necessary for the electron traveling along its orbits in the donor and acceptor. Being applied to the adiabatic electron states, Born - Oppenheimer approximation is also valid for these processes at all nuclear configurations except those near the transitional configuration. The non-adiabatic jumps between the adiabatic free energy surfaces may take place in this region due to violation of the Born - Oppenheimer approximation reducing thus the reaction rate constant (non-adiabatic reactions).

However the Born - Oppenheimer approximation may be insufficient for long-distance electron transfer.[30] The overlap of the diabatic states occurs in this case in the region far from the ionic cores and electron "traveling" along remote orbits takes longer time. The interaction of the electron with the vibrational subsystem may change significantly during this time. In the context of the two-level model for electron transfer reactions the violation of the Born - Oppenheimer approximation is equivalent to the break down of the two-level approximation. The delocalized or polaron-like intermediate states were incorporated into consideration in [31-34] in order to describe possible deviations from the Born - Oppenheimer approximation in the case of short-range interaction of the transferable electron with intramolecular vibrational degrees of freedom. A different solution of this problem was given earlier for the bridge assisted long-range electron transfer where the electron states of the bridge serve as natural candidates for the role of intermediate states.[35-45] The simplest extension of the two-level model is a three-state complex where one diabatic state is taken into account in each species taking part in the transition, i.e. in the electron donor D, bridge B and acceptor A. The Born - Oppenheimer approximation is used for each diabatic state leading to the introduction of three diabatic free energy surfaces, U_D, U_B, U_A (Figs.1 and 2). The transition does not require here the overlap of the electron wave functions in the remote region and in this respect is similar to the short-range electron transfer. However the transition involves now two short-range overlaps of the electron wave functions and the mechanism of the overall transition is different depending on the mutual position of the free energy surfaces (Figs.1 and 2). The situation shown in Fig.1 corresponds to so called *superexchange* transition. Fig.2 shows the arrangement of the free energy surfaces for *coherent transition through dynamically populated intermediate state (DPIS)*.

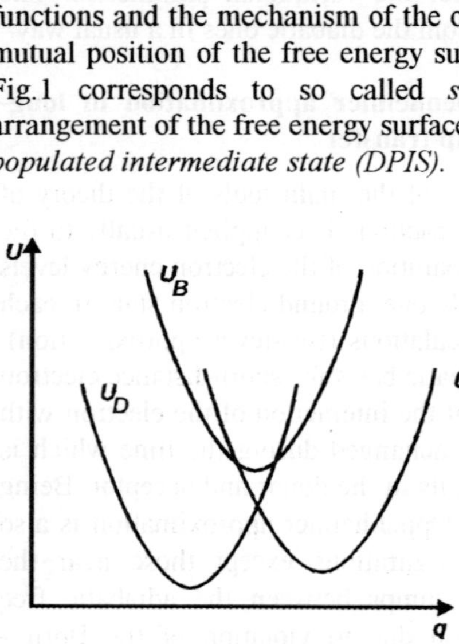

Figure 1: Diabatic free energy surfaces for superexchange transition

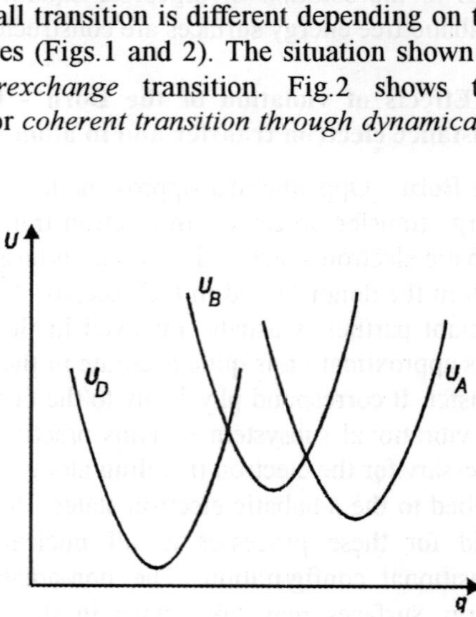

Figure 2: Diabatic free energy surfaces for coherent transition

The calculations performed in the framework of the second order quantum mechanical perturbation theory showed that in classical limit the superexchange transition involves virtual transitions between the electron states in the reactants and the electron state in the bridge molecule.[35-39,43,45] The expression for the transition probability in this case is formally similar to that for ordinary electron transfer. The activation barrier is determined by the crossing of the diabatic free energy surfaces if the initial and final states. However the electron transmission coefficient involves the second order matrix element

$$V^{(2)} = \frac{V_{DB}V_{BA}}{\Delta U} \tag{45}$$

where V_{DB} and V_{BA} are the electron matrix elements coupling the donor and acceptor with the bridge respectively and ΔU is the energy difference between the free energy surface of the intermediate state and the energy of the crossing point of the free energy surfaces of the initial and final states.

Similar calculations for the coherent mechanism[35-45] showed that in one-dimensional case the activation barrier is determined by the highest crossing point of the free energy surfaces and the preexponential factor has a more complicated form although it also involves the product of squared matrix elements V_{DB} and V_{BA}.

A landmark in the theory of bridge assisted electron transfer processes was the result obtained in[43,44]. It was shown that the results of the second order perturbation theory may be explained in the framework of a semiclassical trajectory approach. The transition in this case may be regarded as a dynamic motion along the free energy surfaces with the Landau - Zener transitions at each crossing point and without energy relaxation in the intermediate state. An extension of the theory to adiabatic bridge-assisted reactions was given.[44,46] The adiabatic transition in terms of diabatic electron states means that the Landau - Zener probabilities at each crossing points are equal to 1 leading to the overall transmission coefficient κ=1. In terms of adiabatic electron states this means that the transition occurs by means of the motion along the lower adiabatic free energy surface without violation of the Born - Oppenheimer approximation (Fig.2). The violations of the Born - Oppenheimer approximation may occur in the regions where the splitting of the free energy surfaces is small leading to the non-adiabatic transitions between the free energy surfaces and therefore to a decrease of the transition probability. The deviations from the Born - Oppenheimer approximation in this system may be thus described with the application of the Landau - Zener approach. A similar physical picture takes place for the superexchange mechanism. It corresponds to the Landau - Zener transition at the crossing point of the initial and final diabatic free energy surfaces and the probability of this transition is determined by the resonance splitting of these free energy surfaces ΔE. The latter is related to the second order electron matrix element

$$\Delta E = 2V^{(2)} = \frac{2V_{DB}V_{BA}}{\Delta U} \tag{46}$$

Interesting consequences of this approach emerged for the multidimensional systems, in particular for the enzymatic and biological reactions.[47-49]

Results obtained revealed the physical difference between the coherent and sequential mechanisms of the bridge assisted processes. Unlike the former, the latter corresponds to total energy relaxation in the intermediate state.

Further important step in the theory was the introduction of the energy relaxation into the dynamic picture.[50] It was shown that the energy relaxation in the intermediate state considerably affects the overall transition probability, especially in the systems with one accepting vibrational mode. In the case of non-adiabatic reaction the motion without energy relaxation leads for these systems to an infinite number of reactive attempts at the second crossing point. This result in a much larger transition probability with the electron transmission coefficient which is of the second order in the electron matrix element as for ordinary electron transfer rather than of the fourth order as in superexchange mechanism. The energy relaxation considerably reduces the number of the reactive attempts leading to the fourth order electron transmission coefficient. This approach allows to link the superexchange and coherent mechanisms with the sequential, step-wise one. This problem was discussed recenlty in the literature in the framework of quantum mechanical perturbation theory.[51]

The semiclassical method discussed above allows to derive interpolation formulas covering both limits using the steady state chemical kinetics approach.[52] For coherent transition through dynamically populated intermediate state the interpolation formula for the reaction rate constant in the situation, where the energy at the first crossing point is higher than that at the second one, has the form

$$k = \frac{k_{DB} k_{BA}}{k_{BD} + k_{BA}} \left(1 - p^{BA}\right) + k_{DB} p^{BA} \tag{47}$$

where k_{DB} , k_{BD} and k_{BA} are the thermal rate constants for the transitions between corresponding states, and p^{BA} is the total Landau - Zener probability at the crossing point of the intermediate and final free energy surfaces with due account of all possible multiple reactive attempts at this point[53]

$$p^{BA} = P_{LZ}^{DB} P_{LZ}^{BA} \sum_{k=0}^{n(E)} \left(1 - P_{LZ}^{DB}\right)^{k} \left(1 - P_{LZ}^{BA}\right)^{k} \tag{48}$$

and $P_{LZ}{}^{DB}$ and $P_{LZ}{}^{BA}$ are the Landau - Zener probabilities for one-fold reactive attempt

$$P_{LZ}^{MN} = \frac{1 - e^{-2\pi\gamma_{MN}}}{1 - \frac{1}{2} e^{-2\pi\gamma_{MN}}} \tag{49}$$

involving the Landau-Zener parameter γ for the corresponding crossing point. The number of reactive attempts $n(E)$ depends on the energy relaxation time. If the energy relaxation time τ_E tends to infinity, the same does $n(E)$. Therefore $p^{BA} \to 1$ and the overall transition probability is determined by the probability of the transition from the donor to the bridge, k_{DB}. If, on the contrary, the relaxation time τ_E is small, $n(E)$ is also small and the rate constant is the sum of the rates of two parallel mechanisms, sequential (the first term in Eq.47) and coherent through DPIS (the second one).

Similar expression for the situation where the energy at the first crossing is lower than at the second one has the form

$$k \approx \frac{k_{DB} k_{BA}}{k_{BD} + k_{BA}} + k_{DB}^{*} p^{BA} \tag{50}$$

where k^{*}_{DB} differs from k_{DB} by higher value of the activation energy which is determined now by the highest crossing point.

Sequential, step-wise transitions are possible also in the superexchange situation (Fig.1). The application of the semiclassical method to this case gives the following interpolation formula

$$k = k_{DA}^{SE} + \left(1 - p_{LZ}^{DA(SE)}\right) k_{DB} p_{SE}^{BA} + \left(1 - p_{LZ}^{DA(SE)}\right)\left(1 - p_{SE}^{BA}\right) \frac{k_{DB} k_{BA}}{k_{BD} + k_{BA}} \tag{51}$$

where the first term is the rate constant for the superexchange transition through the crossing point of the initial and final free energy surfaces, the second one corresponds to the coherent transition through dynamically populated intermediate state, and the last one describes the sequential mechanism with the energy relaxation in the intermediate state. The quantity $p_{SE}{}^{BA}$ has the same meaning as in the previous case and differs from p^{BA} in Eq.48 due to different arrangement of the free energy surfaces. The quantity $p_{LZ}{}^{DA(SE)}$ is the Landau - Zener factor for superexchange transition at the crossing of the initial and final free energy surfaces. Eq.51 shows that the superexchange mechanism dominates if $p_{LZ}{}^{DA(SE)} \to 1$. If the latter is small the sequential and coherent (DPIS) mechanisms can in principle compete with the superexchange one.

8 Concluding remarks

The discussion above enables one to draw to some conclusions concerning various aspects of the problem.

1. First, the crude Condon approximation remains effective for the description of the short-distance electron transfer in the systems with tightly bound electrons. The theory suggests a variety of options for the calculation of the transition probability for non-adiabatic processes. Rather simple and effective tool for the description of the dependence of the transition probability on the driving force is represented by the formulas of the type of Eqs.11-13 which give a *parametric dependence of W on ΔF*. An advantage of this equations consists of the fact that they incorporate in a rather simple form the reorganization of both the continuum polarization of the solvent and the discrete vibrational local modes. The quantum effects are also included. The extensions of these equations to the cases of changing vibrational frequencies and non-harmonic vibrations exist.[13] Standard programs available for PC may be used for routine calculations by these equations.
2. The non-Condon effects involve the mutual influence of the medium polarization fluctuations on the electron wave functions and the modulation of the latter on the free energy profiles determining the fluctuations of the polarization. A self-consistent calculational scheme is developed for the non-adiabatic reactions. The deviations from the crude Condon approximation result in a decrease of the activation barrier and to an increase of the electron transmission coefficient. The non-Condon effects may lead to a distance dependence of the activation barrier. They may be significant for the long distance electron transfer and for the electron transfer in the systems with loosely bound electrons. However they are much weaker for short-distance electron transfer between the reactants enclosed into a cavity in the solvent.[5,6]
3. A new effective reaction coordinate for the adiabatic electron transfer is introduced which takes into account possible modulation of the electron wave functions by the polarization fluctuations. This reaction coordinate represents the difference of the electron energy levels in the acceptor and donor ions. A self-consistent algorithm for the calculation of the potential profile along this reaction coordinate is suggested.
4. A violation of the Born - Oppenheimer approximation for adiabatic electron states is manifested in the non-adiabatic transitions between the diabatic free energy surfaces and leads to a decrease of the transition probability (non-adiabatic reactions). The violation of the Born - Oppenheimer approximation for

diabatic electron states is possible in long-distance electron transfer. In bridge-assisted long-distance electron transfer an intermediate electron state in the bridge molecule is introduced. The Born - Oppenheimer approximation then holds for all three diabatic states and the transition has the character of the Landau - Zener transitions in the system of three free energy surfaces. It includes two mechanisms: superexchange (i.e. the transition with virtual participation of the intermediate state) and coherent (i.e. the transition trough the dynamically populated intermediate state). Algorithms for the calculation of the transition probability were developed which cover both non-adiabatic and adiabatic limits. The former corresponds to the violations of the Born - Oppenheimer approximation in the adiabatic basis whereas the Born - Oppenheimer approximation remains valid for the latter. The coherent mechanism differs from the sequential step-wise transition by the absence of the energy relaxation in the intermediate state. Therefore the coherent transition in its pure form is possible only in the case of underdamped motion along the reactive degrees of freedom. The transition takes the character of the sequential step-wise process when the motion is overdamped. Simple interpolation formulas for the reaction rate constant are suggested which cover both limits (see Eqs.47-51).

5. Violation of the Born - Oppenheimer approximation results in a decrease of the transition probability. This is the case for both electron transfer reactions and for the reactions of atom and atom group transfer.[5]

References

1. R.A.Marcus, *J.Phys.Chem.* **24,** 966 (1956); **26,** 867 (1957)
2. V.G.Levich and R.R.Dogonadze, *Dokl.Akad.Nauk SSSR* **124,** 123 (1959)
3. L.I.Trakhtenberg, *Zh.Fiz.Khim.* **54,** 1324 (1980)
4. V.L.Klochikhin, S.Ya.Pshezhetskii and L.I.Trakhtenberg, *Dokl.Akad.Nauk SSSR* **239,** 879 (1978)
5. Y.P.Liu and M.D.Newton, *J.Phys.Chem.* **99,** 12382 (1995)
6. M.D.Newton, *J.Phys.Chem.* **79,** 2795 (1975)
7. M.V.Basilevsky, G.E.Chudinov and D.V.Napolov, *J.Phys.Chem.* **97,** 3270 (1975)
8. M.V.Basilevsky, G.E.Chudinov and M.D.Newton, *Chem.Phys.* **179,** 263 (1994)
9. A.M.Kuznetsov, *Elektrokhimiya* **18,** 594, 736 (1982)
10. A.M.Kuznetsov and J.Ulstrup, *phys.stat.sol.* (b) **114,** 673 (1982)
11. A.M.Kuznetsov, J.Ulstrup and M.V.Vigdorovich, *Chem.Phys.* **176,** 539 (1993)
12. R.R.Dogonadze and A.M.Kuznetsov, *Elektrokhimiya* **7,** 763 (1971)

13. A.M.Kuznetsov, *Charge Transfer in Physics, Chemistry and Biology* (Gordon & Breach, Amsterdam, 1995)
14. R.R.Dogonadze and A.M.Kuznetsov, *Elektrokhimiya* **3,** 1324 (1967)
15. R.R.Dogonadze, A.M.Kuznetsov and V.G.Levich, *Electrochim.Acta* **13,** 1025 (1968)
16. R.Kubo and Y.Toyozawa, *Progr.Theor.Phys.* **13,** 160 (1955)
17. R.R.Dogonadze, A.M.Kuznetsov and M.A.Vorotyntsev, *phys.stat.sol.*(b) 54, 125, 425 (1972)
18 A.M.Kuznetsov and J.Ulstrup (to be published)
19. E.D.German and A.M.Kuznetsov, *J.Phys.Chem.* **98,** 6120 (1994)
20. A.M.Kuznetsov, *Elektrokhimiya* **18,** 598 (1982)
21. A.M.Kuznetsov and J.Ulstrup, *Faraday Discussion Chem.Soc.* **74,** 31 (1982)
22. A.M.Kuznetsov, *Khim.Fiz.* **1,** 1496 (1982)
23. A.M.Kuznetsov, *Chem.Phys.Lett.* **91,** 34 (1982)
24. R.P.Feynman and A.R.Hibbs, *Quantum Mechanics and Path Integrals* (McGraw-Hill Book Company, New York, 1965)
25. A.M.Kuznetsov, J.Ulstrup and M.V.Vigdorovich, *Chem.Phys.* **176,** 539 (1993)
26. A.A.Kornyshev and A.M.Kuznetsov, in preparation
27. J.R.Smith, *Phys.Rev.* **181,** 522 (1969)
28. W.Kohn and L.J.Sham, *Phys.Rev.* **140,** A 1133 (1965)
29. L.D.Zusman, *Teor.Eksp.Khim.* **15,** 227 (1979)
30. G.K.Ivanov and M.A.Kozhushner, *Fiz.Tverd.Tela* (Solid State Phys., Russian) **20,** 9 (1978)
31. A.I.Burshtein, G.K.Ivanov and M.A.Kozhushner, *Khim.Fiz.* **1,** 195 (1982)
32. G.K.Ivanov and M.A.Kozhushner, *Khim.Fiz.* **1,** 1039 (1982); **2,** 1299 (1983)
33. V.V.Egorov, *Khim. Fiz.* **7,** 1466 (1988)
34. V.V.Egorov, *Zh.Fiz.Khim.* **64,** 2305 (1990)
35. P.P.Schmidt, *Austral.J.Chem.* **22,** 673 (1969)
36. M.V.Volkenstein, R.R.Dogonadze, A.K.Madumarov and Yu.I.Kharkats, *Dokl.Akad.Nauk SSSR* **199,** 124 (1979)
37. V.G.Levich, A.K.Madumarov and Yu.I.Kharkats, *Dokl.Akad.Nauk SSSR* **203,** 1351 (1972)
38. R.R.Dogonadze, J.Ulstrup and Yu.I.Kharkats, *Dokl.Akad.Nauk SSSR* **207,** 640 (1972)
39. Yu.I.Kharkats, *Elektrokhimiya* **8,** 1300 (1972)
40. R.R.Dogonadze, J.Ulstrup and Yu.I.Kharkats, *J.Electroanal.Chem.* **39,** 47 (1972)
41. R.R.Dogonadze, J.Ulstrup and Yu.I.Kharkats, *J.Theor.Biol.* **40,** 259, 279 (1973)

42. R.R.Dogonadze, Yu.I.Kharkats and J.Ulstrup, *J.Chem.Soc., Fraday Trans. II* **70,** 64 (1974)
43. A.M.Kuznetsov and Yu.I.Kharkats, *Elektrokhimiya* **12,** 1277 (1976)
44. A.M.Kuznetsov and Yu.I.Kharkats, *Elektrokhimiya* **13,** 1498 (1977)
45. Yu.I.Kharkats, A.K.Madumarov and M.A.Vorotyntsev, *J.Chem.Soc., Faraday Trans.II* **70,** 1578 (1974)
46. R.R.Dogonadze and A.M.Kuznetsov, *Kinetics of heterogeneous reactions in solutions* (Itogi Nauki i Tekhniki, ser. Kinetics and Catalysis, vol.5, VINITI, Moscow, 1978)
47. R.R.Dogonadze, A.M.Kuznetsov and J.Ulstrup, *J.Theor.Biol.* **69,** 239 (1977)
48. A.M.Kuznetsov and J.Ulstrup, in *Protein Structure: Molecular and Electronic Reactivity* (eds. B.Austin, E.Bukhs, B.Chance, Don DeVault, P.L.Dutton, H.Frauenfelder, V.I.Goldanskii, Springer-Verlag, New York-Berlin, 1987, p.309)
49. A.M.Kuznetsov, *J.Molecular Cat.* **47,** 219 (1988)
50. A.M.Kuznetsov and J.Ulstrup, *Chem.Phys.* **157,** 25 (1991)
51. H.Sumi and T.Kakitani, *Chem.Phys.Lett.* **252,** 85 (1996)
52. A.M.Kuznetsov, J.Ulstrup, in press
53. A.M.Kuznetsov and J.Ulstrup, *Chem.Phys.* **157,** 25 (1991)
54. A.M.Kuznetsov, *Elektrokhimiya* **22,** 291 (1986)

SURROGATE HAMILTONIAN THEORY OF CHARGE TRANSFER ENERGETICS

FERNANDO O. RAINERI, HAROLD L. FRIEDMAN,
ISABEL BRANDARIZ[†], AND BAW-CHING PERNG
Department of Chemistry, State University of New York at Stony Brook, Stony Brook, New York 11794-3400, United States
[†]*Departamento de Quimica Fundamental e Industrial, Universidade da Coruña, Campus da Zapateira, s/n 15071, A Coruña, Spain.*

ABSTRACT

In recent reports we have presented a renormalized linear response theory of equilibrium and nonequilibrium solvation, the Surrogate Hamiltonian (SH) theory, which is entirely formulated in terms of the interactions among the constituent molecules. Both solvent and solute molecules are represented by interaction site models of the sort mostly used in molecular dynamics and Monte Carlo simulation studies of solution processes, but in the SH theory the statistical mechanical averages are calculated by integral equation methods of liquid state theory rather than by simulation. The theory has been successfully applied to calculate the dynamical solvent response triggered by a sudden variation of the charge distribution of a solute molecule, and to evaluate the diabatic free energy profiles and associated solvation coefficients for electron transfer reactions in solvents of dipolar and higher order multipolar character. Here we apply three versions of the SH theory to calculate the steady state Stokes shifts of the solvatochromic probe coumarin-153 in fourteen solvents that cover a wide range of polarities. We compare the results with experimental data recently reported by Maroncelli and coworkers.

1. Introduction

The molecular details of the mechanisms of charge transfer reactions in solution continue to offer challenges to more complete understanding. We have formulated a unified *molecular* theory of equilibrium and nonequilibrium solvation that is ideally suited to the study of the structural, energetic, and dynamical aspects of the solvation process relevant to charge transfer (CT) reactions in solutions.[1−3] It is based on a simple *renormalized* linear response development that incorporates non-linear aspects of equilibrium solvation. The renormalization is carried over the solute-solvent potential energy of interaction; the outcome is a statistical mechanical theory based on a surrogate (renormalized) Hamiltonian. We refer to this approach as a surrogate Hamiltonian (SH) theory of solvation.[1−3] Both solvent and solute molecules are represented by interaction site models (ISM) of the sort mostly used in molecular dynamics and Monte Carlo simulation studies of solution processes; i. e., models in which the potential energy of interaction between two molecules is a sum of pairwise additive site-site terms including, in addition to Lennard-Jones (LJ) or similar short range interactions, Coulombic interactions between partial charges located at the molecular sites. In this way, our theory incorporates the symmetries of the molecular charge distributions and the harsh repulsive forces (i. e., the "size" and "shape" of the molecules). The statistical mechanical averages are calculated by integral equation methods of liquid state theory rather than by simulation.

In previous reports[2,3] we applied the SH theory to analyze the energetics of CT reactions in dipolar and non-dipolar solvents. The latter are solvents which, like benzene and toluene, comprise molecules with permanent dipole moment of small or zero magnitude, but with finite higher order multipoles. In these applications to CT energetics, the SH theory takes into account the electronic polarizability of the solvent molecules in an approximate way. Basically, the SH theory provides a flexible methodology for the calculation of the diabatic free energy profiles and associated solvation coefficients (such as the solvent reorganization energy and the solvation free energy change) of a CT reaction in solution. We found[3] that the solvatochromic effect on the absorption energy of betaine-30, which forms the basis of the $E_T(30)$ empirical solvent polarity scale,[4] is reproduced reasonably well by three different implementations of the SH approach: the Renormalized Site Density Theory (SH-RST), the Renormalized Dielectric Theory (SH-RDT), and the Harmonic XRISM approximation (SH-HXA). We also showed[3] that the solvent reorganization energies for the photoinduced (charge-separation) electron transfer reaction in a series of porphyrin-quinone "dyads" in dipolar and non-dipolar solvents, measured by Mataga and coworkers,[5] are correctly reproduced by the SH theory. These results clearly demonstrated that the SH theory of solvation can explain the observation that a solvent with vanishing dipole moment, like benzene, can show unmistakable polarity, as reflected by its influence on the energetics of charge transfer reactions. In another application to CT energetics, we also found how the solvent reorganization energy depends on the donor-acceptor separation distance in dipolar and non-dipolar solvents.[3,6]

In this paper, as a further application of the theory, we calculate the steady state Stokes shifts E_{ss} of the solvatochromic probe coumarin-153 (C153) in fourteen solvents that cover a wide range of polarity. We compare the SH results with the experimental data recently reported by Maroncelli and coworkers.[7] To our knowledge, this is the first report implementing an integral equation statistical mechanical theory for ISM models to calculate E_{ss} for a detailed model representation of a solute as complicated as C153 (36 atoms).

In Sec. II we give a brief outline of the statistical mechanical formulation of the solvent dependence of the energetics of CT reactions in solution. The emphasis is on the diabatic free energy profiles (constrained free energies). In Sec. III we gather the main equations of the SH theory for CT energetics. In Sec. IV we give the molecular expressions for the solvent reorganization energy under the SH-RDT, SH-RST, and SH-HXA implementations of the SH theory. Finally, in Sec. V we report the results for the steady state Stokes shift of C153.

2. Energetics of Charge Transfer Reactions in Solution

2.1. Basic Hamiltonian Description

We consider a charge transfer process that carries the solute molecule (or solute complex) from a "precursor" (P) electronic state to a "successor" (S) electronic state. When the solute is represented by an ISM model (with the interaction sites labelled with the subscript λ), the P and S states are characterized, respectively, by the sets of partial charges $\{Q_\lambda^P\}$ and $\{Q_\lambda^S\}$ at the interaction sites.

The precursor $F^P(\eta)$ and successor $F^S(\eta)$ diabatic free energy profiles are given by the formula[2,8]

$$-\beta F^D(\eta) = \ln\left\{\theta \int d\Omega \int d\Gamma\, \mathrm{e}^{-\beta\mathcal{H}^D}\, \delta(\eta - \hat{\mathcal{U}})\right\} \quad , \qquad D = P, S \tag{2.1}$$

as functions of the reaction coordinate η. As usual, $\beta = (k_B T)^{-1}$ is the inverse of the temperature in energy units (with k_B the Boltzmann constant and T the absolute temperature). The parameter θ has dimensions of energy and is independent of the solute electronic state. By Ω we indicate the set of Euler angles that describe the orientation of the solute with respect to the laboratory reference frame, while Γ indicates the set of coordinates and momenta of the solvent interaction sites in the same reference frame.

The *basic* Hamiltonian $\mathcal{H}^D$ of the system when the solute is in electronic state D is

$$\mathcal{H}^D = E^D + H_w + \hat{\Psi}^D \quad , \qquad D = P, S \ , \tag{2.2}$$

where H_w is the Hamiltonian of the homogeneous solvent in the absence of the solute, while E^D and $\hat{\Psi}^D$ are, respectively, the energy of the solute in the electronic state D in vacuum and the potential energy of interaction between the solute in electronic state D and the solvent molecules. For a given configuration of the solvent around the solute, the energy difference between the S and P states of the system is

$$\mathcal{H}^S - \mathcal{H}^P = \Delta E + \hat{\mathcal{U}} \ , \tag{2.3}$$

where $\Delta E \equiv E^S - E^P$, while

$$\hat{\mathcal{U}} = \hat{\Psi}^S - \hat{\Psi}^P \tag{2.4}$$

is the solvent contribution to the vertical energy gap. Notice that the Dirac delta function $\delta(\eta - \hat{\mathcal{U}})$ in Eq. (2.1) implicitly conveys the physical meaning of the reaction coordinate.

2.2. Generalized Characteristic Function

Equation (2.1) may be written in the alternative more convenient form[2]

$$\mathrm{e}^{-\beta F^D(\eta)} = \theta\, \mathrm{e}^{-\beta\mathcal{A}^D}\, p^D(\eta) \quad , \qquad D = P, S \ , \tag{2.5}$$

where

$$-\beta \mathcal{A}^D = \ln \left\{ \int d\Omega \int d\Gamma \, \mathrm{e}^{-\beta \mathcal{H}^D} \right\} , \tag{2.6}$$

is the Helmholtz free energy of the system when the solute is in the electronic state D, while

$$p^D(\eta) = \langle \delta(\eta - \hat{\mathcal{U}}) \rangle^D \tag{2.7}$$

is the probability density that the solvent contribution $\hat{\mathcal{U}}$ to the vertical energy gap has the numerical value η. In this equation $\langle \cdots \rangle^D$ denotes an equilibrium ensemble average with the distribution function $f^D(\Gamma) \sim \mathrm{e}^{-\beta \mathcal{H}^D}$ of the system when the solute is in the D-state.

To calculate $p^D(\eta)$ from a molecular model it is most useful[2] to introduce the generalized characteristic function $g^D(z)$ of the complex variable $z = \alpha + i\zeta$ (where α and ζ are real numbers)

$$g^D(z) \equiv \langle \mathrm{e}^{iz\hat{\mathcal{U}}} \rangle^D \quad , \quad D = P, S \, . \tag{2.8}$$

The probability density $p^D(\eta)$ can then be recovered from $g^D(z)$ by the Fourier integral

$$p^D(\eta) = \frac{1}{2\pi} \int_{-\infty}^{\infty} d\alpha \, \mathrm{e}^{-i\alpha\eta} \, g^D(\alpha) \quad , \quad D = P, S \, , \tag{2.9}$$

where the integration is carried along the real line $z = \alpha$. It follows from Eqs. (2.5) and (2.9) that $g^D(z = \alpha)$ contains all the information required to calculate the diabatic free energy profile $F^D(\eta)$. Furthermore, using Eqs. (2.2)-(2.4), (2.6) and (2.8), it is simple to express the reaction free energy $\Delta\mathcal{A} \equiv \mathcal{A}^S - \mathcal{A}^P$ in terms of $g^P(z)$ or $g^S(z)$, with z purely imaginary. We find[2]

$$\Delta\mathcal{A} = \Delta E + \Delta A \, , \tag{2.10a}$$

where the solvation free energy change in the reaction is given by

$$-\beta \Delta A = \ln g^P(z = i\beta) = -\ln g^S(z = -i\beta) \, . \tag{2.10b}$$

Unfortunately, the evaluation of $g^D(z)$ for molecular models is not straightforward, and approximations must be made. For this we turn to the surrogate Hamiltonian theory of solvation.

3. Surrogate Hamiltonian Theory for Charge Transfer Energetics

3.1. Surrogate Hamiltonian Description

We describe the system in terms of surrogate (subscript Σ) diabatic Hamiltonians

$$\mathcal{H}^D_\Sigma = E^D + H_w + \hat{\Psi}^D_\Sigma \quad , \quad D = P, S \, , \tag{3.1}$$

in which the solute-solvent potential energy of interaction $\hat{\Psi}^D$ in Eq. (2.2) is replaced by a renormalized form $\hat{\Psi}^D_\Sigma$ which needs to be specified. For a given configuration of the solvent, the energy difference between the P and S states of the system is given by

$$\mathcal{H}^S_\Sigma - \mathcal{H}^P_\Sigma = \Delta E + \hat{\mathcal{E}} \, , \tag{3.2}$$

where

$$\hat{\mathcal{E}} = \hat{\Psi}^S_\Sigma - \hat{\Psi}^P_\Sigma \tag{3.3}$$

is the solvent contribution to the vertical energy gap in the surrogate system.

In contrast with the basic description, where the equilibrium distribution function $f^D(\Gamma) \sim \mathrm{e}^{-\beta \mathcal{H}^D}$ subsumes all the information about the likehood of a particular solvent configuration around the solute in state D, the SH theory assumes[1,2] that the form of the distribution function is

$$f^D_\Sigma(\Gamma) = \frac{1}{\Omega} f^w(\Gamma) \left[1 - \beta \, \delta(\hat{\Psi}^D_\Sigma)\right] , \tag{3.4}$$

where $f^w(\Gamma) \sim \mathrm{e}^{-\beta H_w}$ is the equilibrium distribution function of the homogeneous solvent, $1/\Omega = (\int d\Omega)^{-1}$ is the probability density for a particular orientation of the solute, while $\delta(\hat{\Psi}^D_\Sigma) \equiv \hat{\Psi}^D_\Sigma - \langle \hat{\Psi}^D_\Sigma \rangle$ is the part of $\hat{\Psi}^D_\Sigma$ that fluctuates with respect to its average value $\langle \hat{\Psi}^D_\Sigma \rangle$ under the pure solvent distribution function $f^w(\Gamma)$. The renormalized potential energy of interaction $\hat{\Psi}^D_\Sigma$ in Eqs. (3.1) and (3.4) is fixed by requiring that

$$\langle\!\langle \, \hat{\rho}(\mathbf{r}) \, \rangle\!\rangle^D = \langle \, \hat{\rho}(\mathbf{r}) \, \rangle^D \tag{3.5}$$

for a certain solvent density field $\hat{\rho}(\mathbf{r})$ (where $\mathbf{r}$ locates a point in the solvent relative to the solute). Here $\langle \cdots \rangle^D$ and $\langle\!\langle \cdots \rangle\!\rangle^D$ denote ensemble averages under $f^D(\Gamma)$ and $f^D_\Sigma(\Gamma)$, respectively. Condition (3.5) implies that the solvation structure [as described by $\hat{\rho}(\mathbf{r})$] is the same in the basic and surrogate descriptions of the system, an obvious requirement if we expect the SH theory to be accurate. An important simplification of the surrogate formulation is that $f^D_\Sigma(\Gamma)$ is linear in the renormalized solute-solvent interactions $\hat{\Psi}^D_\Sigma$.

To calculate the surrogate estimates $p^D_\Sigma(\eta)$ and $\mathcal{A}^D_\Sigma$ of the probability density $p^D(\eta)$ and free energy $\mathcal{A}^D$ in Eq. (2.5), we focus on the surrogate version of Eq. (2.8)

$$g^D_\Sigma(z) \equiv \langle\!\langle \, \mathrm{e}^{iz\hat{\mathcal{E}}} \, \rangle\!\rangle^D \, . \tag{3.6}$$

Consistency between Eqs. (3.4) and (3.5) demands[1–3] that the cumulant expansion of $\langle\!\langle \mathrm{e}^{iz\hat{\mathcal{E}}} \rangle\!\rangle^D$ terminates at second order. Furthermore, in evaluating the first and second cumulants of $\hat{\mathcal{E}}$ under $f^D_\Sigma(\Gamma)$, only terms that are at most quadratic in the renormalized solute-solvent interactions are retained. This procedure gives a simple expression[3] for the surrogate characteristic function $g^D_\Sigma(z)$ that, with the surrogate forms of Eqs. (2.5), (2.9) and (2.10), allows us to calculate the surrogate estimates $F^P_\Sigma(\eta)$ and $F^S_\Sigma(\eta)$ of the diabatic free energy profiles.

3.2. Diabatic Free Energy Profiles and Solvent Reorganization Energy

As expected from a (renormalized) linear response formulation, the SH theory gives harmonic diabatic free energy profiles with the same curvature:[2,3]

$$F_{\Sigma}^{P}(\eta) = \frac{1}{4\lambda_{\Sigma}} (\eta - \eta_{\Sigma}^{P})^2 , \tag{3.7a}$$

$$F_{\Sigma}^{S}(\eta) = (\Delta E + \Delta A_{\Sigma}) + \frac{1}{4\lambda_{\Sigma}} (\eta - \eta_{\Sigma}^{S})^2 . \tag{3.7b}$$

Here

$$\eta_{\Sigma}^{P} = \Delta A_{\Sigma} + \lambda_{\Sigma} \quad , \quad \eta_{\Sigma}^{S} = \Delta A_{\Sigma} - \lambda_{\Sigma} \tag{3.8}$$

are, respectively, the values of the reaction coordinate at which $F_{\Sigma}^{P}(\eta)$ and $F_{\Sigma}^{S}(\eta)$ have their minima. Equations (3.7) correspond to the choice $\theta = (\beta/4\pi\lambda_{\Sigma})^{-1/2} e^{\beta A_{\Sigma}^{P}}$ for the scaling energy parameter, which makes $F_{\Sigma}^{P}(\eta)$ vanish at the minimum [i.e., $F_{\Sigma}^{P}(\eta_{\Sigma}^{P}) = 0$]. As is clear from Eqs. (3.7) and (3.8), $F_{\Sigma}^{P}(\eta)$ and $F_{\Sigma}^{S}(\eta)$ depend on the surrogate estimates of two solvation coefficients: the solvent reorganization energy λ_{Σ} and the change in solvation free energy ΔA_{Σ} in the $P \to S$ transition of the solute. The former is defined by either of the first two equalities in Eq. (3.9) (valid for parabolic free energy profiles)

$$\lambda_{\Sigma} = F_{\Sigma}^{S}(\eta_{\Sigma}^{P}) - F_{\Sigma}^{S}(\eta_{\Sigma}^{S}) = F_{\Sigma}^{P}(\eta_{\Sigma}^{S}) - F_{\Sigma}^{P}(\eta_{\Sigma}^{P}) = \frac{\beta}{2} \langle \delta\hat{\mathcal{E}}\, \delta\hat{\mathcal{E}} \rangle . \tag{3.9}$$

The last equality gives the expression of λ_{Σ} in terms of the fluctuations of the surrogate vertical energy gap. Convenient expressions for ΔA_{Σ} under the SH theory may be found in the original reports.[2,3]

4. Three Implementations of the Surrogate Hamiltonian Theory

4.1. Surrogate Hamiltonian-Renormalized Dielectric Theory.

When the equilibrium solvation profile $\langle \hat{\rho}(\mathbf{r}) \rangle^{D}$ in Eq. (3.5) is described by the appropriate (solute site λ)-(solvent site j) correlation functions $h_{\lambda j}^{D}(r)$ (or some combination of these functions), Eqs. (3.4) and (3.5) lead to simple expressions for the renormalized solute-solvent potential energy of interaction $\hat{\Psi}_{\Sigma}^{D}$ in terms of the RISM[9] solute-solvent site-site direct correlation functions $c_{\lambda j}^{D}(r)$.

In the Renormalized Dielectric Theory[1-3] (SH-RDT) we choose the solvent polarization charge density $\hat{\rho}_{\lambda,\mu}(\mathbf{r})$ at position $\mathbf{r}$ relative to each solute interaction site λ as the solvent density field $\hat{\rho}(\mathbf{r})$ in Eq. (3.5). As a consequence, in this theory any solvation coefficient I_{RDT} for the $P \to S$ electronic transition of the solute [solvent reorganization energy λ_{RDT}, change in solvation free energy ΔA_{RDT}] "factors" into the following generic form

$$I_{\mathrm{RDT}} \sim \int_{0}^{\infty} dk\, \psi_{\mu}(k)\, \Upsilon_{\mathrm{RDT}}(k) . \tag{4.1}$$

The function $\psi_\mu(k)$, which is a property of the pure solvent, is related to the equilibrium fluctuations of the solvent polarization charge density. The second factor $\Upsilon_{\mathrm{RDT}}(k)$ is associated with the solute-solvent interactions, which in the SH theory are described in terms of the site-site intermolecular potentials of interaction instead of a cavity construction carved out of a continuum dielectric medium.

An example of Eq. (4.1) is the formula for the solvent reorganization energy[2,3]

$$\lambda_{\mathrm{RDT}} = \frac{1}{\pi} \int_0^\infty dk\, \mathcal{C}(k)\, \Gamma_{\mathrm{RDT}}(k) \,, \tag{4.2}$$

where

$$\mathcal{C}(k) \equiv \frac{1}{\epsilon_\infty} - \frac{1}{\epsilon_L(k)} \tag{4.3}$$

is the wavevector-dependent Pekar factor,[2,3,6] defined in terms of the optical dielectric constant ϵ_∞ and the wavevector-dependent static longitudinal dielectric function $\epsilon_L(k)$ of the solvent.[6,10] Correspondingly,

$$\Gamma_{\mathrm{RDT}}(k) = \sum_{\lambda\lambda'} \omega_{\lambda\lambda'}(k)\, \Delta n_{\Sigma,\lambda}(k)\, \Delta n_{\Sigma,\lambda'}(k) \,, \tag{4.4}$$

is the SH-RDT solute-solvent coupling function for the solvent reorganization energy. In Eq. (4.4) $\omega_{\lambda\lambda'}(k)$ is the Fourier transform of the solute site-site intramolecular correlation function, while $\Delta n_{\Sigma,\lambda}(k)$ represents the Fourier component of the change in the *effective* charge distribution of the solute site λ in the $P \to S$ transition. We emphasize the distinction between $\Delta n_{\Sigma,\lambda}(k)$ and $\Delta Q_\lambda \equiv Q_\lambda^S - Q_\lambda^P$, the corresponding change of the bare or basic charge density. The difference arises from the renormalization of the solute-solvent interactions, which is the principal feature of the SH theory. More specifically,[1–3] $\Delta n_{\Sigma,\lambda}(k)$ is defined in terms of the convolution of the change $\Delta c_{\lambda j}(r)$ of the solute-solvent RISM direct correlation functions in the $P \to S$ transition with certain functions of the pure solvent that characterize its density response to an external nonuniform electrostatic potential.[2]

4.2. Surrogate Hamiltonian-Renormalized Site Density Theory

A different implementation of the SH method, the Renormalized Site Density Theory (SH-RST), is obtained when we choose for the solvent density field $\hat{\rho}(\mathbf{r})$ in Eq. (3.5) each of the microscopic solute-solvent pair densities $\hat{n}_{\lambda,j}(\mathbf{r})$; these are the dynamical variables associated with the local density of solvent sites of a given type at a point $\mathbf{r}$ measured relative to the solute interaction site λ.

The solvation coefficients I_{RST} for the $P \to S$ transition under the SH-RST theory also factor into the product of structure functions of the pure solvent times functions associated with the solute-solvent interactions. The factorization, however, is more complicated, involving the summation over each solvent-solvent pair of interaction sites. In particular, the SH-RST formula for the solvent reorganization energy is

$$\lambda_{\mathrm{RST}} = \frac{\beta}{2(2\pi)^3} \int d^3\mathbf{k} \sum_{jl} S'_{jl}(k)\, \mathcal{B}_{\mathrm{RST},jl}(k) \,, \tag{4.5}$$

where $S'_{jl}(k)$ is a modified partial structure factor of the pure solvent, while $\mathcal{B}_{\mathrm{RST},jl}(k)$ takes care of the solute-solvent coupling. The difference between $S'_{jl}(k)$ and the standard structure factors $S_{jl}(k) = \rho\,[\omega_{jl}(k) + \rho h_{jl}(k)]$ is necessary to correct for the fact that in our calculations we actually represent the solvent by a non-polarizable interaction site model.[3] The SH-RST solute-solvent coupling function

$$\mathcal{B}_{\mathrm{RST},jl}(k) = \sum_{\lambda\lambda'} \omega_{\lambda\lambda'}(k)\; \Delta u_{\Sigma,\lambda j}(k)\; \Delta u_{\Sigma,\lambda' l}(k) \;, \tag{4.6a}$$

is expressed in terms of the Fourier transform $\Delta u_{\Sigma,\lambda j}(k)$ of the change $\Delta u_{\Sigma,\lambda j}(r) \equiv u^{S}_{\Sigma,\lambda j}(r) - u^{P}_{\Sigma,\lambda j}(r)$ in the renormalized solute-solvent site-site pair potentials of the SH-RST theory. The latter are related to the RISM direct correlation functions by the equation[2,3]

$$u^{D}_{\Sigma,\lambda j}(r) = -k_B T\, c^{D}_{\lambda j}(r) \;. \tag{4.6b}$$

4.3. Surrogate Hamiltonian-Harmonic XRISM Approximation

The harmonic XRISM approximation[3] (SH-HXA) is a simple approximation to the energetics of CT reactions that is consistent with the harmonic free energy profiles of Eq. (3.7). It is derived by a harmonic approximation[3] to the energetics of CT reactions calculated with the RISM-HNC ("extended"-RISM, XRISM) integral equation.[11] The SH-HXA formulas for the solvation coefficients I_{HXA} closely resemble those of the SH-RST approximation.[3] Thus, the SH-HXA formula for the solvent reorganization energy is[3]

$$\lambda_{\mathrm{HXA}} = \frac{\beta}{2(2\pi)^3} \int d^3\mathbf{k} \sum_{jl} S'_{jl}(k)\; \mathcal{B}_{\mathrm{HXA},jl}(k) \;, \tag{4.7}$$

involving the same corrected solvent partial structure factors $S'_{jl}(k)$ that appear in Eq. (4.5), but with different solute-solvent coupling functions

$$\mathcal{B}_{\mathrm{HXA},jl}(k) = \sum_{\lambda\lambda'} \omega_{\lambda\lambda'}(k)\; \Delta u^{\mathrm{coul}}_{\lambda j}(k)\; \Delta u_{\Sigma,\lambda' l}(k) \;. \tag{4.8}$$

Here $\Delta u_{\Sigma,\lambda j}(k)$ is the same solute-solvent renormalized pair potential as in the SH-RST approximation, while $\Delta u^{\mathrm{coul}}_{\lambda j}(k)$ is the Fourier transform of $\Delta Q_\lambda q_j/r$, namely the change of the long-range contribution to the *basic* site-site potential energy of interaction between solute site λ and solvent site j.

5. Applications

In this Section we compare the predictions of the Surrogate Hamiltonian theories (SH-RDT, SH-RST, and SH-HXA) with the experimental results[7] for the Stokes shifts E_{ss} of the dye Coumarin-153 in a series of solvents covering a wide range of polarity. In view of the (albeit renormalized) linear response character of the SH theories, the Stokes shift is calculated as twice the solvent reorganization energy.

In all cases we calculated the required solvent-solvent structure factors $\mathcal{C}(k)$ and $S'_{jl}(k)$ as well as the solute-solvent direct correlation functions $c^{D}_{\lambda j}(r)$ with the RISM-HNC method introduced by Rossky *et al.*[11] For a detailed exposition of our calculational procedures we refer the reader to an earlier report.[3]

We focus on coumarin-153 (C153) because for this solute an extensive set of measurements of E_{ss} on a number of solvents has been reported.[7] Furthermore, Kumar and Maroncelli[12] have developed ISM models for C153 in both the P and S states, which we use as the basis for developing the simplified ISM models for C153 used in our calculations.

With regards to the solute and solvent ISM models, the intermolecular potential function between sites i and j of two different molecules (solute-solvent or solvent-solvent) consist of electrostatic and Lennard-Jones (LJ) terms. The LJ interaction parameters between unlike sites are evaluated with the familiar combination rules.

We constructed a 19 site model (Fig. 1 and Table I) that is a united-atom reduced version of a 36-site model used by Kumar and Maroncelli.[12] Auxiliary calculations of E_{ss} of C153 in MeCN and a few other solvents showed that our calculated Stokes shifts do not depend significantly on whether we use the 19 or 36 site C153 model. Since it takes much longer to perform the calculations for the 36 site model, here we only report results obtained with the 19 site model.

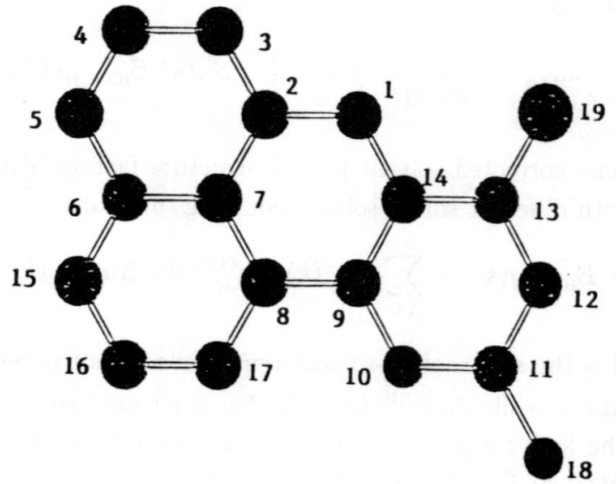

Fig. 1. Ball and stick representation of the interaction site model of C153. The ISM model has 19 interaction sites, and is constructed on the basis of the more detailed model with 36 interaction sites developed by Kumar and Maroncelli (Ref. 12). The code numbers for the interaction sites and the respective partial charges in the ground (P) and excited (S) electronic states are given in Table 1, with the Lennard-Jones parameters of the model (the same in both electronic states).

Table 1. Model Parameters of C153[a]

Site Label λ	Group	Q^P_λ/e	Q^S_λ/e	σ/Å	ε/k_B (K)
1	CH	-0.058237	-0.229037	3.75	55.3543
2	C	-0.016544	0.1544560	3.5	40.27
3	CH_2	0.085098	0.0752980	3.905	59.38
4	CH_2	0.022271	0.037371	3.905	59.38
5	CH_2	0.039630	0.037830	3.905	59.38
6	N	-0.104904	0.075496	3.25	85.62
7	C	0.170811	0.064811	3.5	40.27
8	C	-0.260409	-0.19771	3.5	40.27
9	C	0.432737	0.414537	3.5	40.27
10	O	-0.456873	-0.456373	2.96	105.8
11	C	0.970242	0.925642	3.5	40.27
12	CH	-0.398171	-0.530371	3.75	55.354
13	C	0.357045	0.0425450	3.5	40.27
14	C	-0.320389	0.010711	3.5	40.27
15	CH_2	0.054793	0.062293	3.905	59.38
16	CH_2	0.001805	0.018405	3.905	59.38
17	CH_2	0.171611	0.161511	3.905	59.38
18	O	-0.609718	-0.597918	2.96	105.8
19	CF_3	-0.080795	-0.0688950	4.5	52.8

[a] See Fig. 1 for the site labels.

The fourteen solvents considered in this study are listed in the caption of Fig. 2; six of them are "non-dipolar". Their ISM models were either taken directly from the literature, or else constructed from the recommended values in the OPLS force field.[6] We note that we have recently reported their wavevector-dependent Pekar factors.[6]

In Fig. 2 we represent E_{ss} calculated under the SH-RDT, SH-RST and SH-HXA approximations against the experimental Stokes shift for C153 in fourteen solvents. The figure shows that all three approximations basically reproduce the experimental data, with the best results given by the SH-HXA approximation. We note in particular that E_{ss} in the non-dipolar solvents 1,4-difluorobenzene, benzene and toluene is comparable to the Stokes shift in chloroform; this interesting feature is well reproduced by the SH theories. The slope $dE_{ss,\Sigma}/dE_{ss,\exp}$ is basically unity for each of the SH approximations; closer inspection reveals, however, some minor reversals. It is not clear whether these are due to imperfections of the theory or to inadequacy of the ISM models.

It is evident that in each solvent E_{ss} under SH-RDT is consistently smaller than under the other two SH approximations. This is not surprising, for it was shown

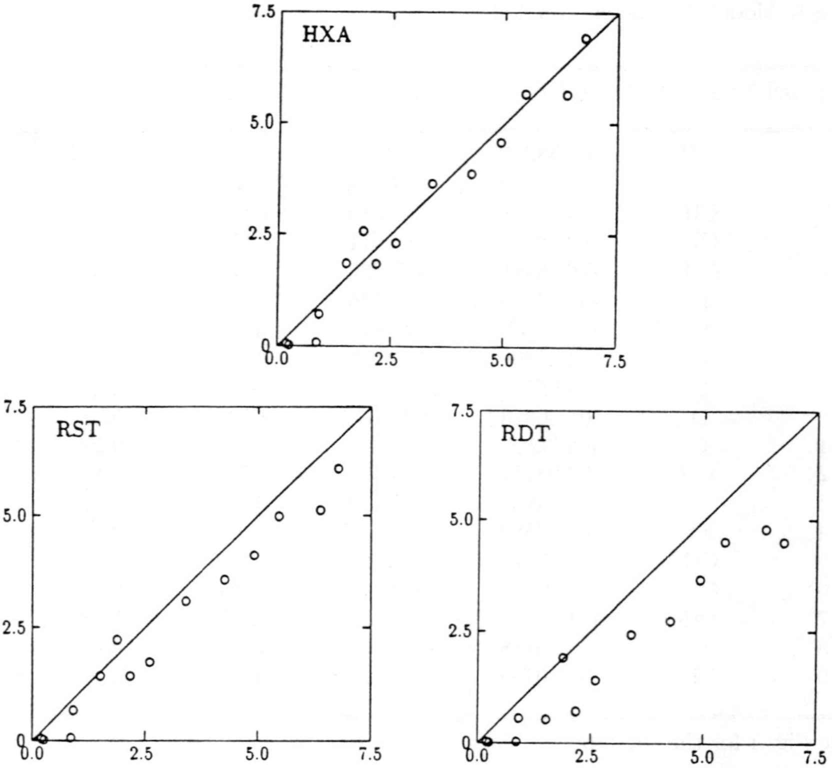

Fig. 2. Solvent-dependence of the steady state Stokes shift E_{ss} of C153: Theoretical (ordinates) versus the experimental results of Ref. 7. The solvents are in order of increasing abscissas: (1) tetrachloroethylene; (2) carbon disulfide; (3) carbon tetrachloride; (4) 1,3,5-trifluorobenzene; (5) toluene; (6) chloroform; (7) benzene; (8) 1,4-difluorobenzene; (9) tetrahydrofuran; (10) methyl acetate; (11) acetone; (12) dimethyl sulfoxide; (13) acetonitrile, and (14) methanol. E_{ss} in units of kcal mol^{-1}.

in Refs. 2 and 3 that λ_{RDT} only derives from the dielectric contribution to the solvent reorganization energy. The contributions due to the solvent number density redistribution around the solute induced by the $P \rightarrow S$ transition are underestimated in the SH-RDT approximation.

However, the overall results for the SH-RST and SH-HXA are very encouraging, considering the degree of complexity of the solute and the fact that a considerable number of solvents have large numbers of interaction sites. In view of the complexity of the solute and solvent molecules, the good agreement between the SH calculations and experiment suggests that the SH theory might well be accurate enough to study the influence of the solvent on the energetics of electron transfer reactions in solution.

6. Acknowledgements

We are very grateful to Professor M. Maroncelli for useful discussions and for providing the parameters of the 36-sites ISM model of C153. We are also grateful to Professor M. D. Newton for useful discussions. F. O. R. gratefully acknowledges a fellowship from the Consejo Nacional de Investigaciones Científicas y Técnicas (CONICET) de la República Argentina. This work was made possible by the support provided by the National Science Foundation of the United States under Grant No. CHE-9321963.

7. References

1. (a) F. O. Raineri, H. Resat, B.-C. Perng, F. Hirata, and H. L. Friedman, *J. Chem. Phys.* **100**, 1477 (1994); (b) F. O. Raineri, B.-C. Perng, and H. L. Friedman, *Chem. Phys.* **183**, 187 (1994); (c) H. L. Friedman, F. O. Raineri, F. Hirata, and B.-C. Perng, *J. Statist. Phys.* **78**, 239 (1995).
2. B.-C. Perng, M. D. Newton, F. O. Raineri, and H. L. Friedman, *J. Chem. Phys.* **104**, 7153 (1996).
3. B.-C. Perng, M. D. Newton, F. O. Raineri, and H. L. Friedman, *J. Chem. Phys.* **104**, 7177 (1996).
4. (a) C. Reichardt, *Solvents and Solvent Effects in Organic Chemistry* (Verlag Chemie, Weinheim, Germany, 1988); (b) *Chem. Rev.* **94**, 2319 (1994).
5. T. Asahi, M. Ohkohchi, R. Matsusaka, N. Mataga, R. P. Zhang, A. Osuka, and K. Maruyama, *J. Am. Chem. Soc.* **115**, 5665 (1993).
6. F. O. Raineri, B.-C. Perng, and H. L. Friedman, *Electrochim. Acta*, in press.
7. L. Reynolds, J. A. Gardecki, S. J. V. Frankland, M. L. Horng, and M. Maroncelli, *J. Phys. Chem.* **100**, 10337 (1996).
8. An incomplete list of relevant references is: A. Warshel, *J. Phys. Chem.* **86**, 2218 (1982); G. King and A. Warshel, *J. Chem. Phys.* **93**, 8682 (1990); M. Tachiya, *J. Phys. Chem.* **93**, 7050 (1989); A. Yoshimori, T. Kakitani, Y. Enomoto, and N. Mataga, *ibid.* **93**, 8316 (1989); T. Fonseca, B. M. Ladanyi, and J. T. Hynes, *J. Chem. Phys.* **96**, 4085 (1992); D. V. Matyushov, *J. Phys. Chem.* **98**, 5152 (1994); H.-X. Zhou and A. Szabo, *J. Chem. Phys.* **103**, 3481 (1995); S.-H. Chong, S.-I. Miura, G. Basu, and F. Hirata, *J. Phys. Chem.* **99**, 10526 (1995); and references therein.
9. D. Chandler, in *The Liquid State of Matter. Fluids Simple and Complex*, edited by E. Montroll and J. L. Lebowitz (North Holland, Amsterdam, 1982).
10. F. O. Raineri, H. Resat, and H. L. Friedman, *J. Chem. Phys.* **96**, 3068 (1992).
11. F. Hirata and P. J. Rossky, *Chem. Phys. Lett.* **83**, 329 (1981); F. Hirata, B. M. Pettitt, and P. J. Rossky, *J. Chem. Phys.* **77**, 509 (1982); P. J. Rossky, B. M. Pettitt, and G. Stell, *Mol. Phys.* **50**, 1263 (1983); S. J. Singer and D. Chandler, *Mol. Phys.* **55**, 621 (1985).
12. P. V. Kumar and M. Maroncelli, *J. Chem. Phys.* **103**, 3038 (1995).

HEAVY-PARTICLE CHARGE-TRANSFER REACTIONS UNDER SLOW ADJUSTMENT OF POLAR SOLVENTS

A.M. BEREZHKOVSKII
Karpov Institute of Physical Chemistry, 10 Vorontsovo Pole Str., Moscow 10364, Russia

V.Yu. ZITSERMAN
Institute for High Temperatures, Russian Academy of Sciences, Izhorskaya 13/19, Moscow 127412, Russia

D.-Y. YANG
Institute of Atomic and Molecular Sciences, Academia Sinica, Taipei, Taiwan

C.J. GUO
Department of Chemical Engineering, National Taiwan University, Taipei, Taiwan

S.H. LIN
Institute of Atomic and Molecular Sciences, Academia Sinica and Department of Chemistry, National Taiwan University, Taipei, Taiwan; Department of Chemistry and Biochemistry and Center for Study of Early Events in Photosynthesis, Arizona State University, Tempe, Arizona 85287-1604, USA

Dependence of the rate constant on the friction coefficients under slow adjustment of the solvent is studied for a heavy-particle charge-transfer reaction in a polar solvent. The consideration is performed on the basis of a model which treats the dynamics intrinsically responsible for the chemical conversion and one responsible for the solvent reorganization on an equal footing. We demonstrate that slowdown of the solvent adjustment leads to dramatic change in the Kramers-Mel'nikov-Meshkov turnover behavior of the rate constant. This effect is especially well pronounced when the coupling of the chemical and solvation modes is strong. In this case the turnover disappears and the rate constant monotonically decreases with the increase of the friction coefficient along the chemical coordinate.

1 Introduction. Stochastic Equations of Motion

In chemical reactions accompanied by redistribution of the electric charge among the reactants interaction with a polar solvent plays an important role. The solvent influence has two aspects, static and dynamical. The former consists in replacement of the gas phase potential surface by the potential of mean force which takes account of the equilibrium solvation of the reactants by the solvent.[1] In polar solvents the potential of mean force differ significantly from the gas phase one. In particular, it may have larger activation energy.[2] Certainly variation of the potential surface leads to considerable changing in the rate constant. The static solvent effect may be taken into account when the rate constant is estimated by transition state theory (TST). TST estimates the rate constant on the basis of the assumption that the chemical transformation does not disturb the thermal equilibrium of nonreactive modes of the solvent. As a result all dynamical effects of the solvent influence are neglected in this estimate.

The dynamical solvent effect in the simplest case may be taken into account in the framework of Kramers' reaction rate theory.[3] According to Kramers the chemical transformation is modeled as a transition of a Brownian particle from the potential well corresponding to the reactants to the second well corresponding to the products of the reaction. Kramers assumed that the Brownian motion responsible for the chemical transformation is one-dimensional. It is described by the Langevin equation of the form (the particle mass is taken equal to unity)

$$\ddot{x} + \gamma_x \dot{x} + \frac{dU(x)}{dx} = f_x(t) \qquad (1)$$

$U(x)$ in this equation is the potential of mean force along the "chemical" x-coordinate which has a double-well structure. The friction coefficient γ_x and the δ-correlated random force with zero mean value $f_x(t)$ are related by the fluctuation-dissipation theorem $< f_x(t) f_x(t') >= 2T\gamma_x \delta(t - t')$, where T is the temperature measured in the energy units and the angular brackets serve as a notation of averaging over realizations of the random force.

Kramers' theory does not treat the solvent dynamics explicitly. This is justified if the solvent relaxation occurs faster, in an appropriate sense, than the motion responsible for the intrinsic chemical conversation. However, the solvent dynamics may occur very slowly so that it is even possible that the solvent reorganization will determine the reaction rate.[4] So, in general it is necessary to consider the solvent reorganization dynamics on an equal footing with the dynamics of the reagent degrees of freedom responsible for the chemical conversion. Unfortunately, construction of such a theory is not available in the general case since nobody knows how to extract relevant modes of the solvent.

Several attempts to consider the solvent reorganization dynamics explicitly have been made for heavy particle charge transfer reactions in polar solvents. Following the ideas suggested by Kurz and Kurz[5], Hynes and his colleagues developed a theory[6-11] under the assumption that the solvent influence can be accounted for with the aid of one additional coordinate which is treated explicitly along with the intrinsic chemical coordinate. They, in fact, postulated that the reaction subsystem consists of two degrees of freedom: the chemical coordinate and the effective solvation coordinate s characterizing the level of solvation. Nonequilibrium solvation leads to quadratic increase in the nonequilibrium free energy of the system reactants+solvent. Therefore the potential energy for the reaction subsystem takes the form

$$V(x,s) = U(x) + \frac{1}{2}k_s(x)\left[s - s_{eq}(x)\right]^2 \tag{2}$$

where $s_{eq}(x)$ and $k_s(x)$ are the equilibrium value of the solvation coordinate at a given x-coordinate value and appropriate force constant. The two-dimensional potential surface, Eq. 2, is then used to estimate the rate constant by variational version of TST. Further development of this approach has been carried out in recent work.[12]

In fact, the theory-suggested in refs. 9-12 does not treat non-equilibrium dynamical effects of the solvent influence, since it is based on the TST estimate of the rate constant. An alternative approach which does treat the dynamical effects in non-equilibrium solvent influence on the reaction dynamics of heavy-particle charge-transfer reactions in polar solvent has been developed in[13-18]. In refs. 13, 14 and 18 it has been demonstrated how solvation coordinates may be introduced starting from the "first principles" (joint consideration of the dynamics of both the reactants and the solvent polarization at every bulk point of the solvent). Certainly reduction of the solvent (which has a continuum degree of freedom) influence on the chemical reaction dynamics to a few effective solvation coordinates is approximate. It is possible only under some simplifying assumptions. The number of effective solvation coordinates depend on the frequency dependent dielectric constant of the solvent as well as on how the charge distribution among the reactants varies in the course of the chemical conversion.

Only one extra effective solvation coordinate is required if two assumptions below are fulfilled. (1) The frequency-dependent dielectric constant of the solvent $\varepsilon(\omega)$ has a simple Debye form

$$\varepsilon(\omega) = \varepsilon_\infty + \frac{\varepsilon_0 - \varepsilon_\infty}{1 + i\omega\tau_D} \tag{3}$$

where ω is the frequency, ε_0 and ε_∞ are static and high frequency dielectric constants, respectively, and τ_D is Debye relaxation time. (2) In the course of the reaction the charge distribution among the reactants varies in such a manner that its dipole moment $\vec{d}(x)$ changes its magnitude $d(x)$ while conserving its space orientation. In such a case equations of motion for the reaction subsystem containing the chemical coordinate and the solvation coordinate y have the form[14]

$$\begin{aligned} \ddot{x} + \gamma_x \dot{x} + \frac{\partial V(x,y)}{\partial y} &= f_x(t) \\ \gamma_y \dot{y} + \frac{\partial V(x,y)}{\partial y} &= f_y(t) \end{aligned} \tag{4}$$

We denote the solvation coordinate as y to emphasize its distinction form the solvation coordinate s mentioned above. The double-well potential $V(x,y)$ here is of the form

$$V(x,y) = U(x) + \frac{1}{2}K\left[y - y_{eq}(x)\right]^2 \qquad (5)$$

The equilibrium value of the solvation coordinate is equal to the dipole moment of the chemical subsystem with a given x value: $y_{eq}(x) = d(x)$. The force constant K is determined by the expression

$$K = \frac{c}{4\pi}\int \left|3\vec{r}\left(\vec{r},\vec{e}_d\right) - r^2\vec{e}_d\right| d\vec{r} \,/\, r^8 \qquad (6)$$

where the spatial integral is over the solvent outside the reaction cage, $c = \varepsilon_\infty^{-1} - \varepsilon_0^{-1}$ is Pecar's factor and $\vec{e}_d$ is a unit vector in the direction of the dipole moment, $\vec{d}(x) = d(x)\vec{e}_d$. The friction coefficient along the solvation coordinate is given by

$$\gamma_y = K\tau_D\varepsilon_\infty / \varepsilon_0 \qquad (7)$$

It is worth emphasizing that all quantities which are applied to treat dynamical effects of the nonequilibrium solvation, i.e. $y_{eq}(x)$, K, and γ_y, are expressed in terms of well-defined characteristics of the reaction subsystem and $\varepsilon(\omega)$ of the solvent. The set, Eq. 4, describes the nonequilibrium solvation influence in the simplest case. Generalizations of this model are presented in [17,18]. Below we use the set, Eq. 4, to study the rate constant behavior as a function the friction coefficients γ_x and γ_y under the condition that γ_y is large enough and the solvation dynamics is sufficiently slow. As we will see the rate constant demonstrates a rich collection of different types of behavior.

2 Intermediate-to-high friction on the chemical coordinate

The set, Eq. 4, have been used to find the rate constant behavior under intermediate-to-high friction on the chemical coordinate and slow adjustment dynamics of the effective solvation mode in [19-21]. It has been discovered that under such conditions a nonconventional qualitative picture of the reaction dynamics (the escape scenario) may take place. Its characteristic feature is avoidance of the saddle point by the reactive flux. When it happens it leads to a collection of unusual features in the rate constant behavior such as decrease of the activation energy, fractional dependence of the rate constant Γ on γ_y, $\Gamma \sim \gamma_y^{-\alpha}$ with $\alpha < 1$, etc.

Such a phenomenon takes place in the case of moderate coupling between the chemical and the solvation modes , when $K < \omega_b^2$, where ω_b is the barrier frequency of the potential $U(x)$. In this case the potential surface $V(x,y)$ considered as a function of the chemical coordinate under a fixed value of the solvation coordinate has a double-well structure. For such potential surfaces a particle may escape the reactant domain only at the expense of motion along x. So, the escape is possible even under frozen relaxation of the solvent. The escape rate $k_{esc}(y)$ depends on y parametrically. The decay kinetics under frozen solvent is multiexponential and depends on preparation of the system. Under a given initial distribution over y $p_i(y)$ the survival probability of the chemical subsystem in the reactant state is

$$S(t) = \int p_i(y) \exp[-k_{esc}(y)t] dy \tag{8}$$

When the relaxation dynamics is unfrozen a competition between the relaxation of the distribution over y and the escape takes place. As a result a steady-state distribution is established. This distribution is close to the equilibrium one near the reactant well bottom. A particle escapes the reactant domain at the expense of the x-coordinate motion. The reactive flux passes via a point $\bar{y}$ and the rate constant reads[19-21]

$$\Gamma \cong \Gamma_{MFPT}(\bar{y}) = \frac{1}{\tau_D} \sqrt{\frac{\Delta V_{eff}(\bar{y})}{T}} \exp\left[-\frac{\Delta V_{eff}(\bar{y})}{T}\right] \tag{9}$$

where $\Gamma_{MFPT}(\bar{y})$ is the inverse mean first passage time to the point $\bar{y}$ from the well bottom. Position of the $\bar{y}$ point is determined from the condition of equality of local rates of the escape and the diffusional relaxation at $y = \bar{y}$. At $y > \bar{y}$ the escape is more efficient than the relaxation. This is why there distribution is strongly depleted compared to the equilibrium one in this region (the reactive flux escapes at $y = \bar{y}$). At $y < \bar{y}$ the relaxation wins the competition. As a result the distribution in this region is close to the Boltzmann one. The potential $V_{eff}(y)$ in Eq. 9 is the potential of mean force along the solvation coordinate. It is obtained as a result of averaging of the initial potential $V(x,y)$ over x with the Boltzmann weight:

$$\exp\left[-\frac{1}{T}V_{eff}(y)\right]=\int\exp\left[-\frac{1}{T}V(x,y)\right]dx \tag{10}$$

where the integration over x is performed over the reactant domain. $\Delta V_{eff}(\bar{y})$ in Eq. 9 is the difference of potential energies $V_{eff}(y)$ at the point $\bar{y}$ and at the well bottom. The relaxation time τ_D in Eq. 9 is the characteristic time of the diffusional relaxation of the distribution over y in the potential $V_{eff}(y)$.

Location of the $\bar{y}$ point may serve as a measure of the efficiency of relaxation of the solvent polarization. Under slow relaxation (large γ_y) $\bar{y}$ point approaches the y-coordinate of the well bottom ($y = y_w$) and the kinetics is multiexponential, Eq. 8. When the relaxation rate increases, the $\bar{y}$ point is between $y = y_w$ and the y coordinate of the saddle point ($y = y_{sp}$), $y_w < y < y_{sp}$. Under such conditions the kinetic is single exponential, $S(t) = \exp(-\Gamma t)$, with the rate constant given by Eq. 9. This case corresponds to the nontraditional escape scenario when the rate constant shows the anomalous behavior described above.

With further increase in the rate of the solvent adjustment (decrease in γ_y) $\bar{y}$ point shifts further and further from y_w. When $\bar{y}$ is located behind y_{sp}, $y_w < y_{sp} < \bar{y}$, the traditional escape scenario takes place. According to this scenario a particle escapes treactant domain passing via the saddle point on the potential surface $V(x,y)$. In such a case the rate constant is given by the Langer theory[22] which is a generalization of the Kramers intermediate-to-high friction solution to many dimensions. Langer's expression for the rate constant is

$$\Gamma_L = (H_L/\omega_b)\Gamma_{TST} \tag{11}$$

Where $\Gamma_{TST} = (\omega_0/2\pi)\exp[-\Delta U/T]$ is TST estimate of the rate constant in the potential $U(x)$ with the activation energy ΔU and the reactant well frequency ω_0 and H_L is a single positive root of the equation

$$(H^2+\gamma_x H-\omega_b^2+K)(H\gamma_y+K)=K^2 \tag{12}$$

Note that variation of the escape scenario induced by changing in the solvent relaxation rate takes place only in the case of moderate mode coupling ($K < \omega_b^2$). When the mode coupling is strong ($\omega_b^2 < K$) the escape follows the traditional scenario over the whole range of the relaxation rate. As a consequence, the Langer theory is applicable in the whole range when the friction along both coordinates is intermediate-to-high.

It is worth nothing that there is an alternative way how to take slow dynamics of the solvent into account being in the framework of the one-dimensional picture of the reaction dynamics. According to this way one has to replace the ordinary Langevin equation, Eq. 1, by the generalized one

$$\ddot{x} + \int_0^t \gamma(t-t')\dot{x}(t')dt' + \frac{dU(x)}{dx} = f(t) \tag{13}$$

All the information about the solvent relaxation in this approach is incorporated into the memory kernel $\gamma(t)$. When the underlying reaction dynamics is described by Eq. 13 the rate constant under intermediate-to-high friction may be found by the Grote-Hynes theory.[23-25] This theory is equivalent to the Langer theory.[22] Therefore the anomalous behavior of the rate constant (which takes place under slow relaxation of the solvent when the nontraditional escape scenario is realized) is lost by this theory.

Our consideration above has been restricted to situations when the friction along the chemical coordinate is intermediate-to-high . The case of low x-coordinate friction as well as the turnover phenomenon under slow adjustment of the solvent are considered in the rest of this article.

3 Low friction on the chemical coordinate (Energy diffusion)

When constructing the reaction rate theory Kramers took account of the fact that in the low friction regime the friction and random force are small perturbations and, crudely spacing, a particle moves along its purely ballistic trajectory in the potential $U(x)$ (see Eq. 1). The particle energy $E = \frac{1}{2}\dot{x}^2 + U(x)$ smoothly changes on a time scale which is much longer than the period of the particle motion in the reactant well of the potential $U(x)$. This is why the low friction regime is also referred to as energy diffusion. The Fokker-Planck equation describing the time evolution of the distribution $f(E,t) = f$ over the energy coordinate reads

$$\frac{\partial f}{\partial t} = \frac{\partial}{\partial E}\left[D(E)\left(\frac{\partial}{\partial E} + \frac{1}{T}\right)(\omega(E)f)\right] \tag{14}$$

where $\omega(E)$ is the frequency of the particle vibration, $D(E) = \gamma_x TJ(E)$ is the diffusion coefficient along the energy coordinate and $J(E)$ is the action variable

$$J(E) = \frac{1}{\pi} \int_{U(x)<E} \sqrt{2[E - U(x)]}dx \tag{15}$$

When γ_x is small enough, the rate constant is proportional to γ_x, $\Gamma \sim \gamma_x$, since $D(E)$ is proportional to γ_x. So, according to Kramers when $\gamma_x \to 0$ both Γ and $D(E)$ tend to zero also.

When the energy diffusion of the chemical mode occurs against the background of slow adjustment of the solvent polarization, two slow variables should be taken into account when describing the reaction dynamics. They are the y-coordinate (its slowness is due to large values of γ_y) and the total energy $E = \frac{1}{2}\dot{x}^2 + V(x,y)$ of the chemical subsystem (its slowness is due to large values of γ_y and small values of γ_x). The effective Fokker-Planck equation describing the time evolution of the distribution over the slow variables may be derived.[26] In [26] this equation is then used to estimate the rate constant and its dependence on the two friction coefficients γ_x and γ_y. The results are discussed in the rest of this section.

The analysis in its spirit is similar to one which has been used to study the case of intermediate to high friction on the chemical coordinate.[19-21] The rate constant behavior is quite different depending on the strength of the mode coupling. When the coupling is strong ($\omega_b^2 < K$) the particles escape reactant domain moving via the saddle point on the potential surface. So, they follow the Langer scenario even in the regime of energy diffusion for the chemical mode. This is a consequence of the fact that the path via the saddle is a unique way to escape the reactant domain. The rate constant behavior in this case is given by the Langer theory, Eq. 11 and 12, with $\gamma_x = 0$ and sufficiently large values of γ_y

$$\Gamma \cong \frac{1}{\gamma_y} \frac{\omega_b K}{K - \omega_b} \Gamma_{TST} \tag{16}$$

Two points are worth emphasizing: (1) The rate constant Eq. (16) is independent on γ_x. This is contrast to the Kramers result obtained in the case of sufficiently fast adjustment of the solvent. According to Kramers in the energy diffusion regime the rate constant linearly increases with γ_x and tends to zero when $\gamma_x \to 0$. (2) The rate constant decreases with γ_y as $1/\gamma_y$. This dependence on γ_y and independence of the rate constant of γ_x show that the dynamics along the solvation coordinate (but not the chemical one) plays a decisive role. So, the solvation coordinate may be interpreted as a reaction coordinate under such conditions.

When the mode coupling is moderate ($\omega_b^2 > K$) the escape scenario is quite different. Particles escape the reactant domain at the expense of their motion along the chemical coordinate. The effective evolution for the distribution function over the solvation coordinate $f(y,t) = f$ has the form

$$\frac{\partial f}{\partial t} = D_y\left[\frac{\partial}{\partial y}\left(\frac{\partial f}{\partial y} + \frac{1}{T}V'_{eff}(y)f\right)\right] - k_{low}(y)f \tag{17}$$

where $D_y = T/\gamma_y$ is diffusion coefficient and $V_{eff}(y)$ is the potential of mean force defined by Eq. 10. The sink term $k_{low}(y)f$ describes the chemical conversation due to the x -coordinate motion. It is obtained by solution of an effective escape problem under a fixed value of the solvation coordinate (1-d Kramers problem with the potential $V(x, y = const)$). To indicate that the effective escape problem is solved in the low friction (energy diffusion) regime we use the subscript *"low"* at the local escape rate.

Eq. 17 describes the competition between the relaxation and the escape discussed in Section 2. So, the way how to estimate the rate constant discussed in that section may be applied here also. When the sink intensity is small enough, the $\bar{y}$ point is located behind y_{sp}, $y_w < y_{sp} < \bar{y}$ and the rate constant is given by the perturbation theory estimate

$$\Gamma = \int k_{low}(y)f_{eq}(y)dy \tag{18}$$

Estimates show that the application of the perturbation theory is justified when the friction along the chemical coordinate is sufficiently small and γ_x satisfies the inequality $\gamma_x << 1/\gamma_y$ The rate constant, Eq. 18, monotonically decreases when $\gamma_x \to 0$. However, in contrast to the Kramers solution, it does not tend to zero. It tends to small, but finite value proportional to γ_y^{-1}

When γ_x is larger than $1/\gamma_y$, but sufficiently small so that the chemical mode motion is the energy diffusion in character the perturbation theory becomes inapplicable. Under such condition the $\bar{y}$ point is located between y_w and y_{sp}, $y_w < \bar{y} < y_{sp}$, not too close to well bottom. Particles escape the reactant domain at $y = \bar{y}$ before they rich the saddle point. The rate constant in this case is given by Eq. 9. The important distinction of the situations considered here and in Section 2 concerns the dependence of the sink intensity $k(y)$ on γ_x. Here we consider the energy diffusion while in Section 2 the case of intermediate-to-high friction is considered. This distinction is also manifested when determining the $\bar{y}$-point position as a function of γ_x and γ_y. This position is determined from the requirement of the equality of local rates of the relaxation and the escape. When the solvent relaxation occurs so slowly that on the characteristic time scale of the escape the polarization may be considered as frozen, the decay kinetics is multiexponential and given by Eq. 8 with the replacement of $k_{esc}(y)$ by $k_{low}(y)$.

4 Turnover behavior of the rate constant under slow adjustment of the solvent

In the two proceeding sections we have described the behavior of the rate constant under weak and intermediate-to-strong friction on the chemical coordinate when the dynamics of the polarization adjustment is sufficiently slow. Here we will discuss the rate constant behavior over the whole range of the friction of the chemical coordinate.

Under fast adjustment of the solvent when $\gamma_y \to 0$ (the classical Kramers problem)[3] the rate constant considered as function of γ_x demonstrates a turnover behavior. It linearly increases with γ_x for small γ_x, reaches its maximal value and then decreases with further increase in γ_x showing $1/\gamma_x$ behavior when γ_x is large enough. While the solution for low and intermediate-to-high friction was obtained by Kramers,[3] the whole solution of the turnover problem have been obtained by Mel'nikov and Meshkov only recently.[27] Below we will discuss how the turnover behavior is modified under slow adjustment of the solvation coordinate. Our consideration is based on the results presented in.[28]

Again the behavior is strongly dependent on the strenght of the mode coupling. When the mode coupling is strong ($K < \omega_b^2$) a single possibility for a particle to escape the reactant domain is to go out via the saddle point. We have shown[26] that in the energy diffusion regime the escape scenario is exactly the same as that accepted by the Langer theory. As a consequence the rate constant, Eq. 16, coincides with one predicted by Eq. 11 and 12 with $\gamma_x = 0$. It turns out that the Langer expression, Eq. 11 and 12, works well in the whole range of the friction on the chemical coordinate. This leads us to a very interesting qualitative result. When the mode coupling is strong and the solvent relaxation is sufficiently slow the turnover behavior of the rate constant disappears. It is replaced by monotonous decrease of this quantity considered as a function of γ_x under a fixed value of γ_y given by the Langer theory.

The situation is quite different under the moderate mode coupling. A characteristics feature of the escape scenario in this case lies in the fact that particles escape at the expense of their motion along the chemical coordinate under a fixed value of the solvation coordinate. This is why we analyze the turnover problem using here Eq. 17 in which the sink intensity $k_{low}(y)$ should be modified to make this equation applicable not only in the case of low friction. Recall that $k_{low}(y)$ is obtained by solution of the 1-d Kramers problem in the potential $V(x, y = const)$ in the low friction regime. To extend the range of applicability of Eq. 17 we replace $k_{low}(y)$ in this equation by a solution of the 1-d Kramers problem in the potential $V(x, y = const)$ which works in the whole range of γ_x from the energy diffusion ($\gamma_x \to 0$) till the high friction regime $\gamma_x \to \infty$.

The modified sink intensity is of the form

$$k(y) = A_{MM}(\Delta(y))k_{Kr}(y) \tag{19}$$

The factor $k_{Kr}(y)$ describes Kramers' intermediate-to-high friction solution of the 1-d problem under a fixed value of the y-coordinate

$$k_{Kr}(y) = \left(\sqrt{1 + \frac{\gamma_x}{4\Omega_b^{\,2}}} - \frac{\gamma_x}{2\Omega_b}\right)\frac{\Omega_0}{2\pi}\exp\left[-\frac{\Delta E(y)}{T}\right] \tag{20}$$

where Ω_0, Ω_b and $\Delta E(y)$ are the well and barrier frequencies and the activation energy of the potential $V(x, y = const)$. $A_{MM}(z)$ is a function introduced by Mel'nikov and Meshkov[27] as an integral which parametrically depends on z. $A_{MM}(z)$ is a monotone function varying from zero at $z = 0$ to unity at $z >> 1$. $A_{MM}(z) \cong z$, for $z << 1$. The argument of A_{MM} in Eq. 19 is given by $\Delta(y) = 2\pi\gamma_x J_b(y)/T$, where $J_b(y)$ is the action variable found for the potential profile $V(x, y = const)$ at the energy of the chemical mode equal to the local activation energy $\Delta E(y)$.

The escape scenario described above is not universal. It takes place when the chemical coordinate friction γ_x is sufficiently small compared to γ_y. When γ_x approaches γ_y the scenario (and as a consequence Eq. 17) fails. In this range the rate constant is correctly described by the Langer theory, Eq. 11 and 12. It is important that the rate constant obtained from solution of Eq. 17 with modified sink term, Eq. 19, smoothly turns into the Langer solution with the increase in γ_x.[19-21]

Now we trace the turnover behavior with the increase in γ_x from zero to infinity (under a fixed sufficiently large value of γ_y). When γ_y is not too large the perturbation theory solution is applicable in the whole range of applicability of Eq. 17 with the modified sink term. Under such values of γ_y the rate constant considered as a function of γ_x is given by

$$\Gamma(\gamma_x,\gamma_y) = A_{MM}(\Delta(y_{sp}))\Gamma_L(\gamma_x,\gamma_y) \tag{21}$$

When γ_x tends to zero, $A_{MM}(\Delta y_{sp}) \to \Delta y_{sp} \cong \gamma_x J_b(y_{sp})/T$ while $\Gamma_L(\gamma_x,\gamma_y) \to \Gamma_L(0,\gamma_y) = \sqrt{1-(K/\omega_b^2)} \cdot \Gamma_{TST}$. As a result Eq. 21 takes the form

$$\Gamma(\gamma_x\gamma_y) \cong \gamma_x J_b(y_{sp}) T^{-1}\sqrt{1-(K/\omega_b^2)} \cdot \Gamma_{TST}, \gamma_x \to 0 \tag{22}$$

With the increase in γ_x the function $A_{MM}(\Delta(y_{sp}))$ quickly approaches to unity. In this range $\Gamma(\gamma_x,\gamma_y) = \Gamma_L(\gamma_x,\gamma_y)$ and decreases with the increase in γ_x. When γ_x becomes larger than γ_y the behavior of the rate constant, Eq. 21, is given by

$$\Gamma(\gamma_x,\gamma_y) = (\omega_b/\gamma_x)\Gamma_{TST}, \gamma_x >> \gamma_y \tag{23}$$

Eq. 21-23 describe the turnover behavior when the perturbation theory solution of Eq. 17 with the modified sink term is justified. A more general case when for some domain of values of the friction coefficients a nonperturbative treatment is required is considered in.[28] The key point of the general case is an overlapping of the domains where the rate constant $\Gamma_{eff1d}(\gamma_x,\gamma_y)$ found from solution of Eq. 17 and $\Gamma_L(\gamma_x,\gamma_y)$ loose their dependence on γ_y and becomes equal to each other. In this domain their dependence on γ_x is given by $\Gamma_L(\gamma_x,\gamma_y = \infty)$. The domain where $\Gamma_L(\gamma_x,\gamma_y)$ becomes equal to $\Gamma_L(\gamma_x,\gamma_y = \infty)$ is determined by the condition $\gamma_x < a\gamma_y$ (for sufficiently large γ_y), where a is a constant. At the same time $\Gamma_{eff1d}(\gamma_x,\gamma_y)$ is equal to $\Gamma_L(\gamma_x,\gamma_y = \infty)$ for $\gamma_x > b\gamma_y$, where b is another constant. It can be shown that $a > b$. These allows us to write down an interpolating formula which picks up the rate constant behavior over the whole range of γ_x

$$\Gamma(\gamma_x,\gamma_y)=\frac{\Gamma_{eff1d}(\gamma_x,\gamma_y)\Gamma_L(\gamma_x,\gamma_y)}{\Gamma_L(\gamma_x,\infty)} \tag{24}$$

When the nonperturbative solution of Eq. 17 is of importance Eq. 24 correctly describes the whole bunch of such unconventional phenomena like saddle point avoidance, fractional behavior of the rate constant on the friction coefficients, ect.

Note that the description above is rather crude. So, numerous subtleties are missed. For example, when $\gamma_x \to 0$, the rate constant decreases and tends to a finite value proportional to $1/\gamma_y$, which is small, but not equal to zero as it is described above. One can find a detailed theory in [26,28].

5 Concluding remarks

Realistic theory of condensed phase chemical reaction dynamics has to treat dynamics of slow modes of the solvent on an equal footing with the reactant modes intrinsically responsible for the chemical conversion. This is especially important for reactions accompanied by redisiribution of electric charge occurring in polar solvents. A simplest theory of such a type is considered in the present note. The theory treats a heavy-particle charge-transfer reaction in a simple Debye solvent. When adjustment of the solvent polarization is sufficiently fast, the problem is reduced to the standard Kramers problem in one dimension. When the adjustment is slow enough the dependence of the rate constant on the friction coefficient along the chemical coordinate may be considerably modified compared to the case of fast adjustment.

The difference between cases of fast and slow adjustment of the solvent is especially well pronounced when the mode coupling is strong. Here the turnover behavior in the case of fast solvent is replaced by monotonous decrease of the rate constant with γ_x when the solvent is slow. The turnover behavior of the rate constant considered as a function of γ_x under slow relaxation of the solvent polarization (sufficiently large value of γ_y) takes place in the case of moderate coupling. However, the situation here is much more complicated than that for the fast solvent. For example such phenomena like the saddle point avoidance, decrease of the activation energy, fractional dependence of the rate constant on the friction coefficients, ect. may take place.

Acknowledgments

This work was supported in part by Academia Sinica and in part by NSC of Taiwan under the contract NSC 86-2811-M001-0021. Two of us (A.M.B. and V.Yu.Z.) are grateful to the Russian Foundation of Fundamental Researches for partial support of this work (Grants Nos. 95-03-09133a and 96-03-32733a).

References

1. C. Reichardt, *Solvents and Solvent Effects in Organic Chemistry* (VCH, Weinheim, 1988).
2. J. Chandrasekhar, S.F. Smith, and W.L Jordensen, *JACS* **107**, 154 (1985)
3. H. Kramers, *Physica* 7, 284 (1940).
4. D.H. Waldeck, *J. Mol. Liquids* **57**, 127 (1993).
5. J.L. Kurz and L.C. Kurz, *JACS* **94**, 4451 (1972).
6. G. van der Zwan and J.T. Hynes, *J. Chem. Phys.* **76**, 2993 (1982).
7. G. van der Zwan and J.T. Hynes, *J. Chem. Phys.* **78**, 4174 (1983).
8. G. van der Zwan and J.T. Hynes, *Chem. Phys.* **90**, 21 (1984)
9. S. Lee and J.T. Hynes, *J. Chem. Phys.* **88**, 6853 (1988).
10. S. Lee and J.T. Hynes, *J. Chem. Phys.* **88**, 6863 (1988).
11. J.T. Hynes, H.J. Kim, J.R. Mathins, and J.J. Timoneda, *J. Mol. Liquids* **57**, 53 (1993).
12. D.G. Truhlar, G.K. Schenter, and B.C. Garret, *J. Chem. Phys.* **98**, 5756 (1993).
13. A.M. Berezhkovskii, *Dokl. Akad. Nauk SSSR* **293**, 880 (1988).
14. A.M. Berezhkovskii, *Chem. Phys.* **164**, 331(1992).
15. A.M. Berezhkovskii and V.Yu. Zitserman, *Chem. Phys.* **164**, 341 (1992).
16. A.M. Berezhkovskii, S.A. Dudko, and V.Yu.Zitserman, *Chem. Phys.* **187**, 275 (1994).
17. A.M. Berezhkovskii and S.A. Dudko, *Dokl. Russ. Akad. Nauk* **327**, 228 (1992).
18. A.M. Berezhkovskii and S.A. Dudko, *J. Chem. Phys.* **100**, 5949 (1994).
19. A.M. Berezhkovskii and V.Yu. Zitserman, *Physica* A **166**, 585 (1990).
20. A.M. Berezhkovskii and V.Yu. Zitserman, *Chem. Phys.* **157**, 141 (1991).
21. A.M. Berezhkovskii and V.Yu. Zitserman, *J. Phys.* **187**, 519 (1992).
22. J.S. Langer, *Ann. Phys.* **54**, 258 (1969).
23. R.F. Grote and J.T. Hynes, *J. Chem. Phys.* **73**, 2715 (1980).
24. P. Hanggi and F. Mojtabai, *Phys. Rev.* A **26**, 1168 (1982).
25. P. Hanggi, P. Talker, and M. Borkovec, *Rev. Mod. Phys.* **62**, 251 (1990).
26. A.M. Berezhkovskii, V.Yu. Zitserman, D.-Y. Yang, C.J. Guo, and S.H. Lin, (the paper under preparation).
27. V.I. Mel'nikov and Meshkov, *J. Chem. Phys.* **85**, 1018 (1986).
28. A.M. Berezhkovskii, V.Yu. Zitserman, D.-Y. Yang, C.J. Guo, and S.H. Lin, (the paper under preparation).

MOLECULAR DYNAMICS CALCULATION OF THE CUPROUS-CUPRIC ELECTRON TRANSFER RATE AT A COPPER ELECTRODE

J. W. HALLEY, S. WALBRAN
School of Physics and Astronomy, University of Minnesota, Minneapolis, MN 55455, USA

B. B. SMITH
National Renewable Energy Laboratory, Golden, CO 80401, USA

We describe preliminary results of a calculation of the cuprous-cupric electron transfer rate in aqueous solution at a copper electrode. The methods are similar to those we reported earlier for the ferrous-ferric rate. In this case, it was necessary to make a model of the differential capacitance in order to fix the point of zero charge for the copper electrode.

1 Introduction

Electron transfer at electrode-electrolyte interfaces is a key step in a wide variety of chemical processes, including for example those occurring in stress corrosion cracking[1,2]photelectrochemical reactions of various sorts[3] and in fuel cells and batteries[4] . The general framework for considering such questions set out by Marcus[5–12] is likely to be correct. The objective of our collaboration with Argonne National Laboratory is to obtain detailed models and quantitative predictions for these rates using modern computational capabilities and techniques and to compare these predictions with new, detailed experiments on selected reaction rates.

The first reaction selected for study was the ferrous-ferric electron transfer at a gold electrode[13] , [14] In the model reported in reference 14 we took acount of the effects of the electric field arising from the charge on the electrode using a method reported earlier[15]. Entropic effects in the free energy were fully included by use of several importance sampling techniques. The same methods permitted an evaluation of the assumption of quadratic free energy surfaces. A full study of the dependence of the free energies on the distance of the ion from the electrode and the charge on the electrode was carried out. We found in the study of the ferric-ferric reaction that a quantitative model of the process required a more detailed consideration of the dependence of the reactant concentration, the electron transfer matrix element, and the work function on distance of the ion from the electrode than had heretofore been included in such studies. Including these effects, we were able to obtain quantitative estimates for the activation energy, the transfer coefficient and the standard potential for the reaction which were in reasonable quantitative agreement with experimental results obtained by our collaborators at Argonne.

In this paper we present results of a similar study of the cuprous-cupric electron transfer, which is believed to play a role in limiting rates of stress corrosion cracking in some engineering environments. Experimental studies of the temperature dependence of this rate at a copper electrode are currently under way at Argonne. We found that the literature was unclear concerning the potential of zero charge at a copper electrode. In the first part of this paper we describe a model of the differential capacitance which provides some clarification of this issue, which is important for the quantitative application of our model for the electron transfer. The third section describes a calculation of the relevant free energy for the electron transfer and provides a preliminary estimate of the standard potential for this reaction.

2 Model for the differential capacitance and determination of the point of zero charge.

Traditionally, the point of zero charge is deduced from the experimental differential capacitance using the Gouy-Chapman theory (sometimes, modified in various ways to take account of effects inside the diffuse double layer). In that theory, the point of zero charge occurs for low electrolyte concentrations at the minimum in the differential capacitance as a function of the electrostatic potential drop across the interface. Unfortunately, in experiments[16, 17] on the copper water interface, *two* minima occur in the differential capacitance in the relevant potential and electrolyte concentration region. Which, if either, of these minima represents the point of zero charge of the interface has been a point of some uncertainty and dispute in the experimental literature. We have considered three contributions to the capacitance: 'electronic' [18–25] (meaning contributions arising from the rearrangements of electrons in the metal electrode in response to changes in the electrostatic potential drop), dielectric (meaning contributions from the polarization of the water molecules near the interface where screening of the electric field arising from the electrode by mobile ions is not expected to be important) and double layer (meaning contributions of the ions of the electrolyte to the capacitance). It is well known that all three contributions can be significant in determining the differential capacitance. Here we attempt to model each contribution with sufficient realism and accuracy to permit the issue of the location of the point of zero charge at the copper water interface to be addressed quantitatively. In our model the contribution from electronic effects near the metal has been simply added in series to the capacitance arising from the other two effects (though this is an oversimplification) but we have used a slightly more general approach to combining the contributions from Stern and double layer effects. The form which we have chosen for the 'electronic' contribution , which has features consistent with the discussion in references 18-25.

$$C_{me}(\sigma) = A_{me}(1. + exp((\sigma - B_{me})/\sigma^*_{me})) \quad (1)$$

The parameters A_{me}, B_{me}, and σ^*_{me} were varied in fitting the model for the capacitance to experimental data.

To describe the solvent near the interface we used a model for water [26]near the interface which is the same as one which we use in our studies of the cuprous-cupric electron transfer rate. In our simulation, we place model water molecules between semi-infinite slabs of ideal classical metal (regions of infinite dielectric constant), as described in reference 15 where our methods for handling Ewald sums associated with long range parts of the coulomb interaction are described. In the studies reported here we have used a different short range water-wall potential than that used in reference 14 . The water-wall potential used here was obtained by fitting the interaction potential of a water molecule with the electrode to the results of (nearly) first principles calculations of water at a copper electrode as described in reference 27. In the calculation described in reference 27 the electronic structure of a model of the copper electrode was calculated at each time step of a molecular dynamics simulation in which model water molecules move in a force field provided by the electrons of the electrode and the other water molecules. We fit the interaction energy of a single water molecule with the electrode determined as described in reference 27 to a sum of electrostatic images and single-particle potentials, with the image plane at 8.5 au from the center of the copper slab . The single- particle potentials used for the fit are of the form

$$(z) = A/z^9 - Be^{-z/a} \quad (2)$$

The results of reference 27 showed that the wall induced a significant water-water repulsive force when two particles were simultaneously near the wall. To reproduce this effect in our classical simulation , we included an extra repulsion from the wall for oxygens which are close to one another. The parameters of this multibody force were adjusted so that the classical simulation reproduced the oscillations in the oxygen density which were obtained from the less phenomenological model of reference 27. The multibody potential used in the classical simulation for oxygen centers 1 and 2 was of form

$$U(\vec{r}_1, \vec{r}_2) = Ce^{-r_\perp^2/(r_o/2)^2}\left(\frac{1}{z_1^9} + \frac{1}{z_2^9}\right) \quad (3)$$

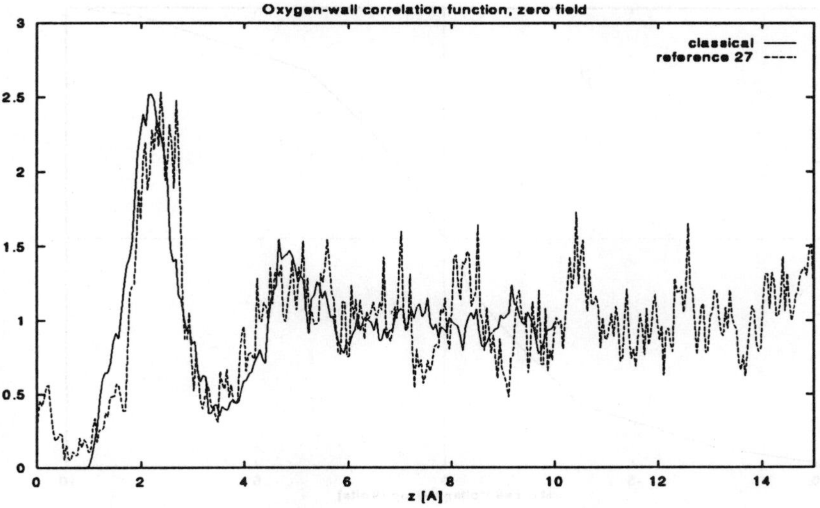

Figure 1: Density of oxygen as a function of distance z from the electrode in the model.

Here $r_\perp$ is the magnitude of the projection of the vector $\vec{r}_1 - \vec{r}_2$ between oxygens onto a plane parallel to the metal walls and z_1 and z_2 are the distances of oxyens 1 and 2 from the wall. We used $C = 0.3 * A_{oxy}$ where A_{oxy} is the coefficient of the $1/z^9$ term in the oxygen-wall interaction (see Eq. (2))

The simulation sample consisted of 216 water molecules plus their images in one of the perfect metal walls. This sample was periodically repeated an infinite number of times in all three dimensions as described in reference 15. As described there, one can simulate a slab of water between opppositely charged ideal metal walls using the same code, simply by adding a constant force $q_i E_0$ in the direction perpendicular to the walls to each center with charge q_i in the simulation. Then the electrostatic potential drop V_0 from one electrode to the other is $V_0 = -E_0 L$ where L is the water sample size in the direction normal to the walls. It is not hard to show[15] that this simulation then corresponds to placing a charge per unit area σ of

$$\sigma = \frac{E_0}{4\pi} + < P_z > \tag{4}$$

on one metal wall (and $-\sigma$ on the other wall). $< P_z >$ is the component of the water polarization perpendicular to the walls, averaged over the sample and over the simulation time. In Figure 1, we compare the density of oxygen centers in zero field as calculated in the classical simulation and in reference 27 .

In Figure 2, we show σ calculated as in 4 as a function of the poten-

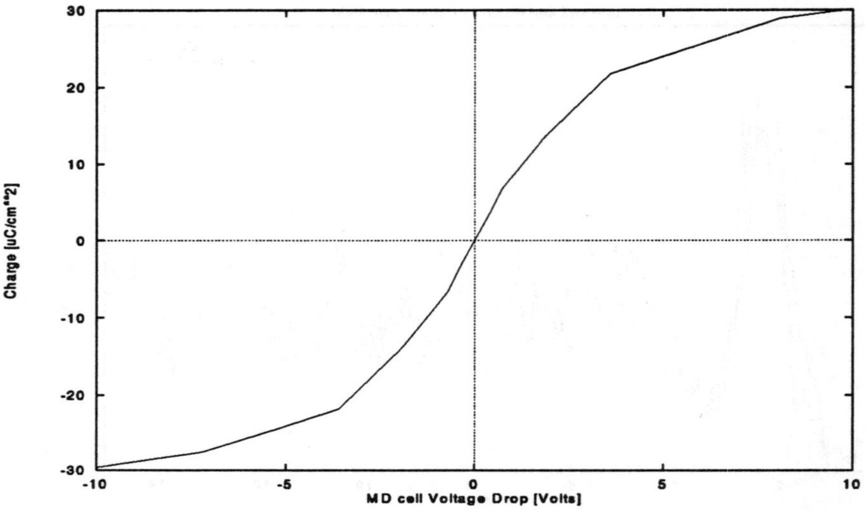

Figure 2: Calculated charge/ area vs of the potential drop across the Stern layer.

tial drop V_0 across the sample. From this curve one could easily determine the differential capacitance of the Stern layer alone in this model. Note that, as expected, there is a linear region at small potential drops corresponding to easy reorientation of the water dipoles in the field, followed by saturation at larger potentials of either sign. Thus, if one were to take a derivative of this curve, to obtain the capacitance of the Stern layer alone, one would find that the capacitance had a simple peak at the point of zero charge and died away symmetrically to each side of it. However, we are not interested in the capacitance of the Stern layer alone

We next discuss how we interface this model for the Stern layer to the standard Gouy Chapman model for the double layer region in which the ions of the electrolyte screen out the field. One approach is to simply add the Gouy-Chapman capacitance in series with the capacitance of the Stern layer. This approach implictly and unrealistically postulates a sharp interface between the Stern layer and the region in which the electrolyte can be treated by the Gouy-Chapman model. If such a sharp interface existed, then the boundary condition on the macroscopic electric field in the Gouy Chapman theory would be simply 4π times the charge σ applied to the electrode, leading to the usual series expression for the lumped capacitances. However, when the interface is not sharp, the boundary condition for the Gouy Chapman model can be estimated from the molecular dynamics simulation of the Stern layer region by calculating the electric field at the plane $z = z_c$ at which the model begins to represent the

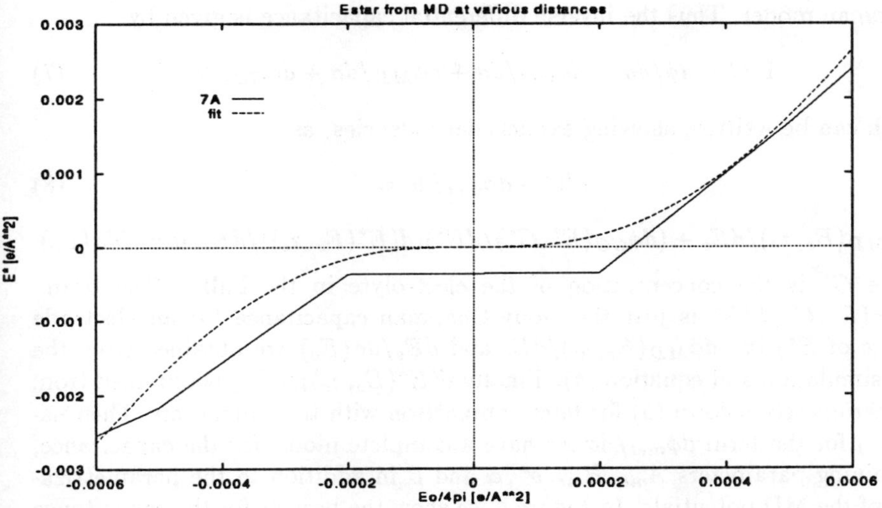

Figure 3: The average macroscopic field E^* as calculated in the simulation model at 7 Åfrom the electrode. Dashed line is field needed in Gouy Chapman theory to fit the capacitance.

system by the Gouy-Chapman model. The appropriate value of z_c is not known a priori. We show smoothed values of the electric field as calculated from the simulations at z= 7 Å in Figure 3 as a function of the external field E_0. (See equation 4 for the relationship of E_0 to σ.) The smoothing was accomplished by averaging the calculated fields around each z value with a Gaussian weight of width 3 Åin the z direction. This removes irrelevant fluctuations due to detailed molecular structure. To determine an appropriate z_c for representing the experimental capacitance in the model, we represent these data by the form

$$E^* = E_0 - \alpha \tanh(\beta E_0) \tag{5}$$

which performs adequately as a representation of the data in Figure 3. We use (5) to determine the boundary condition on the field for the Gouy Chapman theory and regard α and β as fitting parameters. After making the fit we will compare (5) with the molecular dynamics values of E^* at various z to determine the effective value of z_c at which the match is made.

With these preliminaries in hand we can represent the potential drop from electrode to bulk solution as

$$\phi = \phi_{met} + \phi_{MD} + \phi_{GC} \tag{6}$$

where ϕ_{MD} is the potential drop from z=0 to z_c in the molecular dynamics cell, and ϕ_{GC} is the potential drop from z_c to $z \to \infty$ determined by the Gouy-

Chapman model. Thus the inverse differential capacitance is given by

$$1/C = d\phi/d\sigma = d\phi_{met}/d\sigma + d\phi_{MD}/d\sigma + d\phi_{GC}/d\sigma \quad (7)$$

which can be written, showing explicit dependencies, as

$$1/C - d\phi_{met}/d\sigma = \quad (8)$$

$$(d\phi_{MD}(E_o, z_c)/dE_o + (d\phi_{GC}(E^*, C^*)/dE^*)(d(E^*(E_o, z_c))/dE_o))dE_o/d\sigma(E_o)$$

where C^* is the concentration of the electrolyte in the bulk. The term $d\phi_{GC}(E^*, C^*)/dE^*$ is just the Gouy-Chapman capacitance for an electrode charge of $E^*/4\pi$. $d\phi_{MD}(E_o, z_c)/dE_o$ and $dE_o/d\sigma(E_o)$ are obtained from the MD simulation and equation (4). Finally $d(E^*(E_o, z_c))/dE_o)$ is obtained from the parametrized form (5) for later comparison with the simulation. Then using (1) for the term $d\phi_{met}/d\sigma$ we have a complete model for the capacitance, containing parameters $A_{me}, B_{me}, \sigma^*, \alpha$ and β in addition to the parametrization of the MD potentials. In Figure 4 we show the best fit for the capacitance compared with the experimental results of reference for MD simulations in the simulation. The point of zero charge occurs at the peak. The metal contribution is somewhat larger than suggested by the calculations of references 18-25 , but does not affect the predicted experimental capacitance significantly. The dominant new physical effect in the model arises because the field E^* experienced as a boundary condition by the Gouy Chapman part of the model is less than the total charge on the electrode at small imposed potentials. As a result, the Gouy Chapman capacitance does not contribute significantly to the changes in the capacitance and one observes mainly the Stern layer peak around the point of zero charge. The model quite unambiguously indicates that the point of zero charge is at the position of the peak in the observed capacitance, and not at one of the two minima as previously suggested. This is quite similar to suggestions of Guidelli and others that saturation of the water in the Stern layer accounts for the peaks in the capacitance seen in the differential capacitance curves of several sd metals.

3 Calculation of the free energy surface for the cuprous-cupric electron transfer.

We use the same approach which was fully described in reference 14 . This requires a description of the interaction of the cuprous and cupric ions with the solvent (water). As in our ferrous- ferric work, these were obtained from ab initio calculation by L. Curtiss and coworkers [28] . Particularly for the cupric ion, many body effects are significant, so it is necessary to take some account

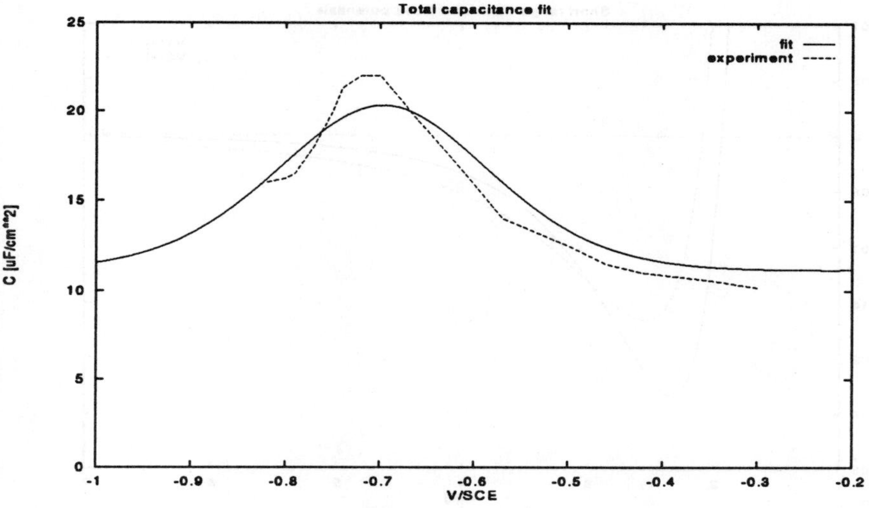

Figure 4: omparison of the experimental capacitance with the fit to the model described here.

of these in an empirical two body copper-oxygen potential used to describe the interactions. Here we have used the form

$$V_{Cu-O} = Ae^{-Br} + \frac{C}{r^4} + \frac{D}{r^6} + \frac{F}{r^{12}} \tag{9}$$

Values for the parameters were respectively for the cuprous and cupric ions A=236.389,64.051 $e^2/\text{Å}$,B=3.6763,3.626 Å^{-1},C=.30224, 1.6300 $e^2\text{Å}^3$, D=9.4561,-4.3257 $e^2\text{Å}^5$,F=-24.0109, -1.313 A $e^2\text{Å}^{11}$. The potentials are shown in Figure 5.

We calculate the difference in the free energy of solvation of the two ions to be 0.92 $e^2/\text{Å}$(1276 kJ/mol) using these potentials. This is to be compared with a reported experimental value[29] of 1.06 $e^2/\text{Å}$(1471 kJ/mol).

The radial distribution functions for the cupric ion in water for these potentials are shown in Figure 6. For the cupric case, they are consistent with neutron scattering experimental results by Enderby and coworkers [30, 31] Our estimates of the Jahn Teller effect [32] indicate that it is not large in the cupric ion, contrary to references 30-31 and we have not included it in the present study. (We do not regard the question of whether the Jahn Teller affect might effect the electron transfer rate as closed.) We use the following expression for the transfer rate, which follows from the Fermi Golden rule of quantum mechanics, in a classical approximation for the solvent and assuming the electronically

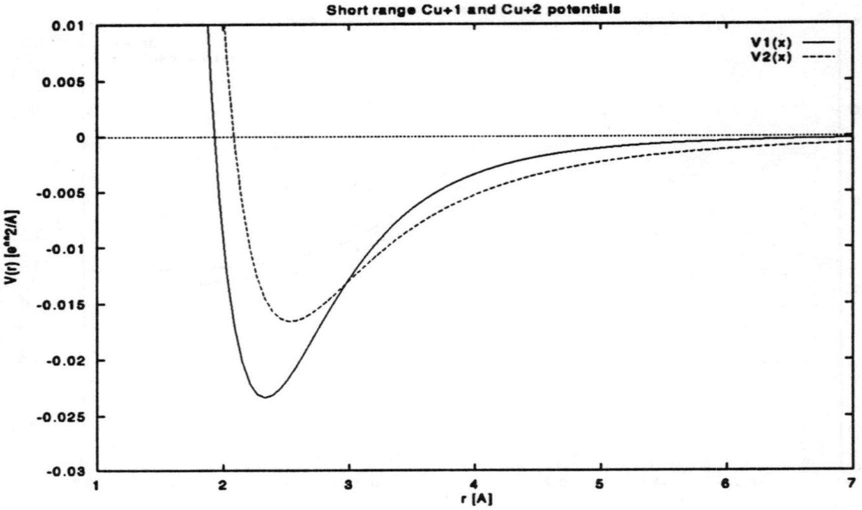

Figure 5: Copper-oxygen potentials used in the simulation for the free energy surface determination.

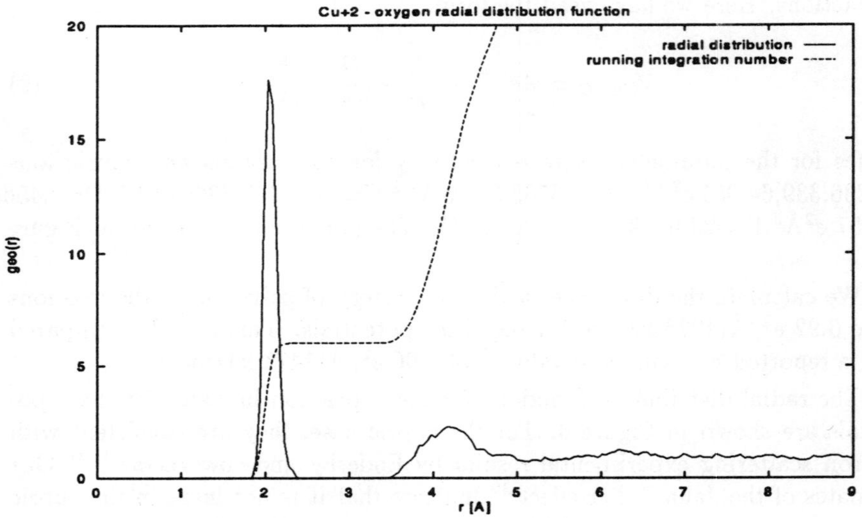

Figure 6: Radial distribution function of oxygen around the cupric ion in the model.

diabatic (non-adiabatic) limit:

$$rate = \frac{2\pi}{\hbar} <| \Lambda(\{R\}) |^2 \delta(V_i'(\{R\}) - V_f'(\{R\})) >_i \tag{10}$$

Here $\Lambda(\{R\})$ is the electronic matrix element for the electron transfer rate. The dependence of this matrix element on nuclear positions $\{R\}$ is an important feature of the present problem. Unlike the ferrous-ferric case, an electronic structure calculation of this matrix element is not yet available and we have simply used a functional form for the matrix element which decays exponetially as the distance z of the ion from the electrode increases, as found by L. Curtiss for the ferrous-ferric case [13].

V_i' is the potential energy of the system when the ion is in its initial state and V_f' is the potential energy of the system at the same set of nuclear coordinates $\{R\}$ but when the ion is in its final state. $(V_i'(\{R\}) - V_f'(\{R\}))$ contains a term $E_e(z)$ depending on the distance z of the ion from the electrode and containing terms in the energy difference between reactant and product which are not included explicitly in our molecular dynamics model. As in reference 14, we separate this part of the energy difference out, writing $(V_i'(\{R\}) - V_f'(\{R\})) = (V_i(\{R\}) - V_f(\{R\})) - E_e(z)$. The average $< ... >_i$ in 10 is an equilibrium average in which the system is described by $V_i'(\{R\})$, but in our application, the average will is described by $V_i(\{R\})$. It is convenient to separate the coordinate z describing the distance of the ion from the electrode out of the integrals in 10 which then becomes

$$rate = \int dz\kappa(z)C(z) \tag{11}$$

in which

$$\kappa(z) = \frac{2\pi}{\hbar} | \Lambda(z) |^2 < \delta(V_i'(\{R\}) - V_f'(\{R\})) >_{i,z} \tag{12}$$

where $< ... >_{i,z}$ is an equilibrium average taken with the constraint that z is fixed. $C(z)dz$ is the probability of finding the ion in the range of distances z to $z + dz$ from the electrode. Equation (11) will, of course arise on general phenomenological grounds in both the adiabatic and diabatic limits. In the canonical ensemble $C(z)$ takes the form

$$C(z) = \frac{\int d\Omega e^{-\beta V_i}}{\int dR e^{-\beta V_i}} \tag{13}$$

where the integral $\int d\Omega ...$ is over all coordinates $\{R\}$ except z . Noting that $C(z)$ will vanish as $1/L$ where L is the sample depth in the z-direction, we can produce a finite quantity in the limit $L \to \infty$ and make contact with the experimentally measured electrochemical rate constant κ_{ec} (which has units of velocity) by rewriting the previous expressions as

$$\kappa_{ec} = \int dz \kappa(z) C'(z) \tag{14}$$

in which

$$C'(z) = \lim_{L\to\infty} LC(z) \tag{15}$$

can be interpreted as the ratio of the reactant ion density at z to the reactant ion density in the bulk of the electrolyte.

It is convenient to rewrite (15) in terms of a sum of terms over slices of the sample volume taken parallel to the electrode surface. We denote the center of each such slice by z_α and its width by Δ. Then the expression for the rate becomes

$$\kappa_{ec} = \sum_\alpha \kappa_\alpha C'_\alpha \tag{16}$$

in which

$$\kappa_\alpha = \frac{\frac{2\pi}{\hbar^2}\int_{z_\alpha-\Delta/2}^{z_\alpha+\Delta/2} dz \int d\Omega \mid \Lambda(z) \mid^2 \delta(V'_i(\{R\}) - V'_f(\{R\}))}{\int_{z_\alpha-\Delta/2}^{z_\alpha+\Delta/2} dz \int d\Omega e^{-\beta V_i}} \tag{17}$$

and C'_α is the number of ions in slice α divided by the bulk number of ions per unit distance normal to the electrolyte surface.

We write κ_α in terms of a free energy at fixed value ΔE of the difference $V_i - V_f$: We define:

$$F_{i,\alpha}(\Delta E) = \tag{18}$$

$$-ln\left[\frac{k_BT\int_{z_\alpha-\Delta/2}^{z_\alpha+\Delta/2} dz \frac{|\Lambda(z)|^2}{k_BT}\int d\Omega e^{-\beta V_i}\delta(\Delta E - E_{e,\alpha} - \delta E_{e,\alpha}(z) - V_i + V_f)}{\int_{z_\alpha-\Delta/2}^{z_\alpha+\Delta/2} dz \int d\Omega e^{-\beta V_i}}\right]$$

$E_{e,\alpha} = \frac{1}{\Delta}\int_{z_\alpha-\Delta/2}^{z_\alpha+\Delta/2} dz E_e(z)$, $\delta E_e(z) = E_e(z) - E_{e,\alpha}$ (The term $\delta E_e(z)$ is set to zero here.) An arbitrary factor of k_BT is inserted inside the ln in the definition in order to make the units of $F_i(\Delta, z)$ those of energy. (The result in the next equation is independent of this factor.) Then the rate constant can be written as

$$\kappa_{ec} = \sum_\alpha \frac{2\pi}{\hbar k_B T} e^{-\beta F_{i,\alpha}(\Delta E = E_{e,\alpha})} C'_\alpha \tag{19}$$

In practise we will take the matrix element $\Lambda(z)$ out of the integral in the above definition and replace it by its average, Λ_α over a slice giving

$$\kappa_{ec} = \sum_\alpha \frac{|\Lambda_\alpha|^2}{k_B T} \frac{2\pi}{\hbar} e^{-\beta F'_{i,\alpha}(\Delta E = E_{e,\alpha})} \tag{20}$$

in which

$$F'_{i,\alpha}(\Delta E) = -k_B T \ln k_B T < \delta(\Delta E - V_i - V_f) >_{i,\alpha} \tag{21}$$

Here the average is taken over the ensemble in which the forces are defined by the potential V_i and the ion is confined to slice α.

The only difference between this formulation and the one in reference is that we have formulated the problem in terms of averages over ensembles in which the ion is confined to slices of space normal to the electrode, rather than being fixed in space. This difference is not essential for the results reported here, though we hope to make use of it for improved computations of the factors C'_α in the future. The confinement is accomplished by the imposition of reflecting 'walls' which reflect the ion but not the solvent in an energy conserving way. There is a description of some aspects of the algorithm[35] for accomplishing the confinement in reference 35.

The umbrella sampling techniques used to calculate the free energies $F'_{i,\alpha}$ were essentially the same as those described in reference 14. We show some results in Figure 7. These curves have been shifted in a way which permits them to be interpreted in the usual way as describing the barriers for the forward and reverse reactions. A full analysis, requiring a computation of the sum of forward and reverse currents is not yet complete, but we can extract a rough estimate of the standard potential for this reaction from these curves. The value of $E_0/4\pi$ at equilibrium is estimated from the data of Figure 8. to be about $0.0018e/\text{Å}^2$ (The 'selection factors' $C'(z)$ and $\Lambda(z)$ do not appear to be strongly affecting this in this case.) This is translated to a charge on the electrode by use of equation (4). This yields $\sigma \approx 26\ \mu\text{C/cm}^2$. Then, integrating the experimental capacitance we obtain the relation shown in Figure 9 between charge and voltage, permitting the estimate $\approx 1.5 \pm 1.0$ V *vs* SCE for the equilibrium potential , This is roughly a volt above the experimental standard potential[34] . The error estimate is based on the long extrapolation required to convert the charge to a potential as shown in Figure 9.

4 Discussion

We can trace the error in the calculated standard potential to the incorrect value of the solvation energy difference which was found using our force fields

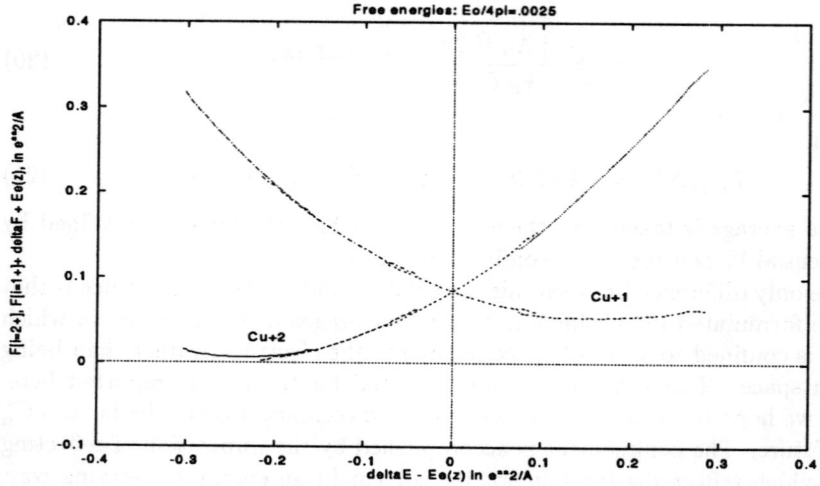

Figure 7: Free energies for two fields for z between 6 Å and 7 Å.

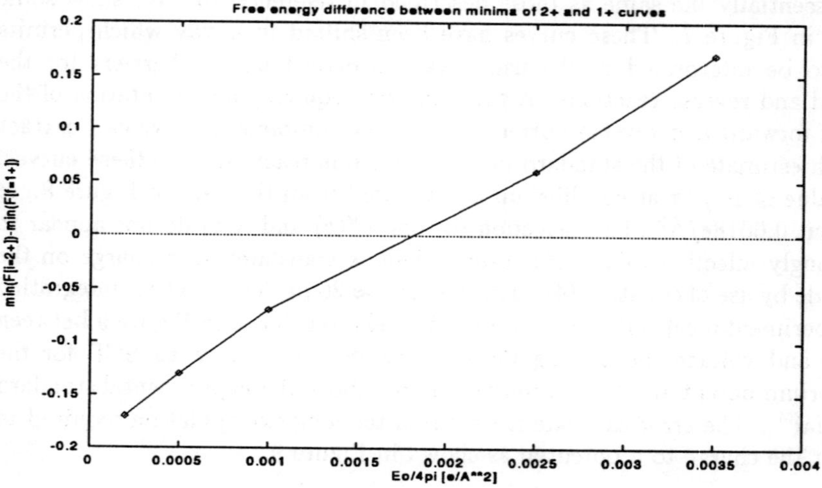

Figure 8: Free energy difference between the cuprous and cupric states in equilibrium as a function of field E_0.

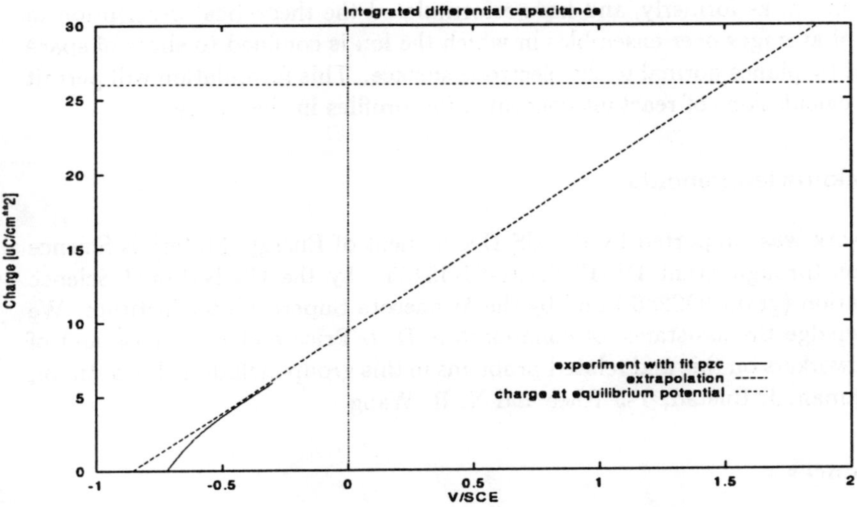

Figure 9: Integrated capacitance and extrapolation for estimate of the equilibrium potential.

for description of the water copper ion interactions. Indeed, by simply shifting the constant part of the offset $E_e(z)$ by a fixed amount to compensate for this error we would get a much better prediction of the equilibrium potential. This, however, is not a correct procedure. What is required is a revision of the force fields to obtain the correct solvation energy and a repetition of the calculation. This is under way.

There is evidence that our values for the many body repulsion (Equation (3)) are somewhat too high. This is both because the repulsion is somewhat too successful in depopulating the bound states near the surface relative to the result of the first principles calculations as shown in Figure 1 and because we get a better fit to the capacitance (not shown here) than the one shown in Figure 4 with a smaller many body repulsion. Accordingly we plan to repeat these analyses with such a smaller many body term.

In conclusion we have successfully modeled the differential capacitance of a copper electrode and used it to determine the point of zero charge which was in experimental dispute. The model includes all the known factors contributing to the capacitance and uses a more accurate method for combining them to obtain the total capacitance than the usual series sum. This is essential to obtaining the successful fit. We have used the information obtained in this exercise in our calculations of the free energy surfaces associated with the cuprous-cupric electron transfer at a copper electrode. The methods used for these calculations were the same as those used in reference 15 except that we have not fixed the ion position, as formerly, and have reformulated the theoretical description in terms of averages over ensembles in which the ion is confined to slices of space defined by planes normal to the electrode surface. This formulation will permit better calculations of reactant concentration profiles in the future.

5 Acknowledgements

This work was supported by the US Department of Energy, Materials Science Division through grant DE-FG02-91-ER45455 , by the US National Science Foundation (grant 952286) and by the Minnesota Supercomuter institute. We acknowledge the assistance of collaborators D. L Price and L. Curtiss and of earlier workers on this and related problems in this group including J. Hautman, A. Rahman, J. Rustad, Y-J Rhee and X. R. Wang.

References

1. W. J. Shack, T. F. Kassner, P. S. Maiya , J. Y. Park, and W. E. Ruther, Nucl. Eng. Des.108, 199 (1988).
2. W. E. Ruther, W. K. Soppet, and T. F. Kassner, Corrosion 44, 791 (1988).
3. F. Williams and A. Nozik, Nature 311, 21 (1984).
4. Proc. Symp. on Batteries and Fuel Cells for Stationary and Electric VehicleA. R. Landgrebe, Z. Takehara eds, Pub. Electrochem. Soc., Pennington, NJ, (1993).
5. R. Marcus, J. Chem Phys. 24, 966 , 979(1956)
6. R. Marcus, J. Chem. Phys. 26, 867, 872 (1957)
7. R. Marcus, Can J. Chem 37, 155 (1959)
8. R. Marcus, Trans. Symp. Electrode Processes, Phila .Pa. 1959, 239 (1957)
9. R. Marcus, Trans. N. Y.Acad. Sci. 19, 423 (1957)
10. R. A. Marcus, Dis. Fara. Soc. 29, 21 (1960)
11. R. Marcus, J. Phys. Chem. 67, 853, 2889 (1963)
12. R. Marcus, J. Chem. Phys,43, 679 (1965)
13. L. A. Curtiss, J. W. Halley, J. Hautman, Z. Nagy, Y.-J. Rhee and R. M Yonco, Journ. Electrochem. Soc. 138, 2032 (1991).
14. B. Smith and J. W. Halley, J. Chem. Phys. 101, 10915 (1994)
15. J. Hautman, J. W. Halley and Y-J. Rhee, J. Chem. Phys. 91, 467 (1989)
16. U. Bertocci and D. R. Turner in Encyclopedia of electrochemistry of the elementsAllen J. Bard, ed., M. Dekker, N Vol 2, sec. 6, pp. 427-30
17. S. Haertinger, K. Doblhofer, Journal of Electroanalytical Chemistry 380, 185 (1995)

18. D. Price and J.W. Halley, J. Electroanal. Chem. 150 , 347 (1983)
19. J.W. Halley, B. Johnson, D. Price and M. Schwalm, Phys. Rev. B31, 7695 (1985)
20. J. W. Halley and D. Price, Phys. Rev. B35, 9095 (1987)
21. D. Price and J. W. Halley, Phys. Rev B38, 9357 (1988)
22. W. Schmickler and D. Henderson, J. Chem. Phys. 85, 165
23. J. Goodisman, J. Chem. Phys. 90, 5756 (1989)
24. S. Amokrane, V. Russier, J.P. Badiali, Surf. Sci. 217, 425 (1989)
25. A.A. Kornyshev, Electrochim. Acta 34, 1829 (1989).
26. K. Toukan and A. Rahman, Physical Review B 31, 2643 (1985)
27. D. L. Price and J. W. Halley, J. Chem. Phys. 102, 6603(1995)
28. L. A. Curtiss and R. Jurgens, Journ. of Physical Chemistry 94, 5509 (1990)
29. Y. Marcus, Ion Solvation, J. Wiley and Sons, NY (1985)p.107
30. P. S. Salmon, G. W. Nielson and J. E. Enderby, J. Phys. C21, 1335 (1988)
31. P. S. Salmon and G. W. Nielson, J. Phys. Cond. Matter 1, 529 (1989)
32. J. W. Halley, Modern Physics Letters B8, 1319 (1994)
33. P. T. Boinske, L. Curtiss, J. W. Halley, B. Lin and A. Sutjianto,Journal of Computer Aided Design (in press)
34. A. J. Bard and L. R. Faulkner, Electrochemical Methods, J. Wiley and Sons, NY (1980),p. 700
35. P. T. Boinske, L. Curtiss, J. W. Halley, B. Lin and A. Sutjianto, Journal of Computer Aided Design (in press)

Non-adiabaticity in proton tunnelling at the electrochemical interface

K.L. Sebastian
Department of Inorganic and Physical Chemistry
Indian Institute of Science
Bangalore 560012
India

Abstract : We consider the effects of non-adiabaticity on the tunnelling of a proton from an H_3O^+ in water to the surface of a metal electrode. We use a simple model hamiltonian for the description of the process and calculate the tunnelling probability using the semi-classical approaches of Miller[8]. The continuum of electron-hole excitations is accounted for by treating them as bosons. We show that typically, tunnelling probability can be decreased by one to two order of magnitude by non-adiabaticity.

1. Introduction

In this paper we consider the effects of non-adiabaticity on tunnelling probability of a proton through a barrier. On one side of the barrier the proton is in the ionic state, while on the other side of the barrier, it exists as a hydrogen atom, adsorbed on the surface of a metal. So the tunnelling process has to be accompanied by the transfer of an electron from the metal to the proton. Because of this the probability of tunnelling will be reduced and the question is by how much? We show that it can decrease the tunnelling probability by orders of magnitude.

The above question arises in connection with electrochemical hydrogen evolution reaction, in which the proton transfer occurs at the interface between an electrolyte and an electrode in contact with it. Éven though the calculation of the rate of the process has attracted quite a bit of attention [1-5], the above question has been addressed only in the papers of Dogonadze et al [3,4]. All other calculations simply take the reaction to proceed on an adiabatic potential energy surface and the proton to tunnel through the adiabatic barrier. We consider a model which mimics the important aspects of the tunnelling process and show that non-adiabaticity effects are important. After presenting our method, we also discuss the advantages of our method, in comparison to that of Dogonadze et al.

In the initial state, the proton is away from the surface of the electrode and is hydrated so that one has an H_3O^+. The distance of the proton from the metal surface is roughly 4 - 6 angstroms [1,2]. Here the acceptor orbital on H_3O^+ is above the Fermi level of the metal. It is slightly broadened by its interaction with the continuum of levels in the metal. In the reaction, solvent polarisation around the H_3O^+ changes and the proton moves towards the electrode. As it moves, the energy of the affinity orbital changes, so that eventually it falls below the Fermi level of the metal, as a result of which the orbital gets occupied, resulting in a hydrogen atom, which is adsorbed on the surface of the metal. In the adsorbed state, the interaction of the 1s atomic orbital of the hydrogen with the metal is fairly strong and therefore it is broadened considerably. We assume that there is no net charge on the hydrogen atom in the adsorbed state so that the 1s orbital is well below the Fermi level (see figure 1). If the time scale for tunnelling is smaller than that for electronic motion, then the electronic system can respond to the tunnelling process and hence the tunnelling would be adiabatic. However, the electronic system of a metal has a continuum of excited states with a high density of states and hence it is never able to adjust fully to the new position of the proton. Hence, after the tunnelling, it will not be in the ground state. One says that the process is then non-adiabatic or equivalently, electron-hole pairs are created in the process. This also means that the barrier for tunnelling is never the adiabatic one and that one has to account for the corresponding change in the tunnelling probability.

2. The hamiltonian and its bosonized version

The reaction may be pictured in the following fashion: the solvent polarisation around H_3O^+ fluctuates as a function of time. The time scale for this is larger than for actual proton transfer. Therefore, when the solvent modes acquire the right configuration, the proton tunnels through the barrier. A detailed calculation of the rate should take both solvent motion and tunnelling into account. However, our aim in this paper is only to demonstrate that non-adiabaticity can play an important role, and for this we make a simple model calculation. We consider only the tunnelling at the solvent orientation that is most favourable for proton transfer. For this solvent polarisation, we take the proton and the electronic system to be described by the hamiltonian

$$H = p^2 / (2m) + V_0(q) + H_e(q) \tag{1}$$

with

$$H_e(q) = \varepsilon_a(q) n_a + \sum_k \varepsilon_k n_k + \sum_k V_k(q)\,(c_a^+ c_k + c_k^+ c_a) \tag{2}$$

H is the total hamiltonian, p the momentum operator for the proton, whose position co-ordinate is q. $V_0(q)$ is a potential energy term, coming from the interaction of the proton with everything other than the electronic system (nuclei in the electrode and solvent molecules). H_e is a hamiltonian describing the electronic system. $\varepsilon_a(q)$ is the energy of the affinity orbital $|1s>$ on the proton which changes as a function of distance of the proton from the electrode surface [1,2]. n_a is the occupation number operator for this orbital (we write the hamiltonian for the electronic part in second quantized notation) ε_k represent the energy of the orbital $|k\rangle$ on the metal, and n_k is the associated occupation number operator. $V_k(q)$ represents the hopping interaction between $|1s\rangle$ and $|k\rangle$ c_a (c_a^+) is the annihilation (creation) operator for $|1s\rangle$ and c_k (c_k^+) for $|k\rangle$. We take $V_k(q) = A_k V(q)$, with A_k independent of q, so that the position dependence of all $V_k(q)$'s are the same. Note also that to keep things simple we have assumed a spin-less hamiltonian [6]. We can now follow Schönhammer and Gunnarsson [6,7] and change over to a bosonized version of this hamiltonian, which is rather easy to analyse. The bosonized hamiltonian is

$$H = p^2/(2m) + V_a(q) + H_b(q) \tag{3}$$

with

$$H_b(q) = \int_0^\infty d\varepsilon\; \varepsilon\, b_\varepsilon^+ b_\varepsilon - \Delta(q)\int_0^\infty d\varepsilon\; \sqrt{\varepsilon}\, e^{-\varepsilon/(2B)}(b_\varepsilon^+ + b_\varepsilon) + \Delta(q)^2 \int_0^\infty d\varepsilon\; e^{-\varepsilon/B} \tag{4}$$

In the above, b_ε (b_ε^+) is the boson annihilation (creation) operator, having an energy ε, obeying the commutation relation $[b_{\varepsilon'}, b_\varepsilon^+] = \delta(\varepsilon - \varepsilon)$. $\Delta(q) = \delta_F(q) - \delta_F(q_0)$ where $\delta_F(q)$ is the phase shift at the Fermi level, defined below [6]. The above hamiltonian differs from that of [6] in that we have taken the bosons to have any energy between zero and infinity, but the interaction with them is cut off in a smooth fashion using the energy scale B. In contrast, SG have an abrupt cut-off in the energy of the bosons at the value B. $V(q)$ is the adiabatic potential energy for proton motion and is equal to $V_0(q)+E_0(q)$ where $E_0(q)$ is the energy of the electronic system in its ground state, with the proton kept fixed at q. The phase-shift $\delta_F(q)$ is given by [6]

$$\delta_F(q) = \tan^{-1}\left((\varepsilon_F - \varepsilon_a(q) - \mathrm{Re}\{\Gamma(q)\})/\mathrm{Im}\{\Gamma(q)\}\right)/\pi \tag{5}$$

$\Gamma(q) = \sum_k \frac{V_k(q)^2}{(\varepsilon_F - i0 - \varepsilon_k)}$. q_0 denotes the initial position of the proton. In the figure 2, we give a typical plot of $\Delta(q)$ as a function of q which shows that $\Delta(q) \approx 0$ for $q \approx q_0$, and $\Delta(q_f) \approx 1$, where q_f is the position co-ordinate of the proton in the product state - i.e.

adsorbed state of the proton. It is also clear that the change over from *0* to *1* of $\Delta(q)$ occurs over a rather small range of q and therefore, we make the approximation $\Delta(q) \approx \theta(q-q_c)$, where q_c is the value of q at which the orbital crosses the Fermi level and θ is the Heaviside step function, defined by $\theta(q-q_c) = 1$ if $q>q_c$ and $\theta(q-q_c) = 0$ if $q<q_c$.

3. The rate

To calculate the rate, we follow the techniques of Miller [8]. As we are interested only in estimates, we adopt the simplest possible technique - tunnelling probability at a temperature T is determined by the periodic orbit having period $\beta = 1/kT$ (we put $\hbar = 1$). In fact the tunnelling probability $P(T) \sim exp[-S_{classical}]$ where the exponent $S_{classical}$ is just the Euclidean action S evaluated along the (classical) periodic orbit having period β in the inverted multidimensional potential, which minimises the action. The probability has a pre-exponential factor which we shall neglect in this rough estimate of non-adiabaticity effects. In order to evaluate $S_{classical}$ we proceed as follows: we first introduce position and momentum co-ordinates for the bosons by $x = (b_\varepsilon^+ + b_\varepsilon)/\sqrt{2\varepsilon}$ and $p = (b_\varepsilon^+ - b_\varepsilon)i\sqrt{\varepsilon/2}$ and write the above in terms of these to get the Euclidean action

$$S_E = \int_0^\beta dt\left(m\dot{q}^2/2 + \dot{V}_a(q)\right) + \int_0^\beta dt\left[\int_0^\infty d\varepsilon\left((\dot{x}_\varepsilon^2 + \varepsilon^2 x_\varepsilon^2)/2\right) - \sqrt{2}\Delta(q)\int_0^\infty d\varepsilon\; x_\varepsilon \varepsilon\, e^{-\varepsilon/(2B)} + \Delta(q)^2 \int_0^\infty d\varepsilon\; e^{-\varepsilon/B}\right] \tag{6}$$

In the above, we have neglected an infinite "zero point energy like term" which makes no contribution whatsoever to the dynamics. Now we have to find the multidimensional periodic classical path having period β that minimises the action in the equation (6). We do this in two steps. First we find the equation of motion for x_ε.

$$\ddot{x}_\varepsilon = \varepsilon^2 x_\varepsilon - \sqrt{2}\Delta(q)\, e^{-\varepsilon/(2B)}\, \varepsilon \tag{7}$$

The solution satisfying the periodicity condition is:

$$x_{\varepsilon}(t) = \sqrt{2}\varepsilon\, e^{-\varepsilon/(2B)} \sum_{n=-\infty}^{\infty} \frac{1}{\beta} \int_0^{\beta} ds \frac{Exp\left(\frac{2i\pi\, n(t-s)}{\beta}\right)}{\left(\frac{2\pi\, n}{\beta}\right)^2 + \varepsilon^2} \Delta(q(s)) \tag{8}$$

Using this equation back in S_E, we get the partially minimised action, which will determine the motion of the q co-ordinate. It is

$$S_E = \int_0^{\beta} dt \left(m\dot{q}^2/2 + V_a(q) + \Delta(q)^2 \int_0^{\infty} d\varepsilon\, e^{-\varepsilon/B} \right) - \frac{1}{2} \int_0^{\beta} dt \Delta(q(t)) \int_0^{\beta} ds \Delta(q(s)) K(t-s) \quad . \tag{9}$$

$$K(t-s) = \int_0^{\infty} d\varepsilon\; 2\varepsilon^2\, e^{-\varepsilon/B} \sum_{n=-\infty}^{\infty} \frac{1}{\beta} \frac{Exp\left(\frac{2i\pi\, n(t-s)}{\beta}\right)}{\left(\frac{2\pi\, n}{\beta}\right)^2 + \varepsilon^2}$$

$$= \int_0^{\infty} d\varepsilon\, e^{-\varepsilon/B}\, \varepsilon \left[Coth(\beta\varepsilon) Cosh(\varepsilon|t-s|) - Sinh(\varepsilon|t-s|) \right] \tag{10}$$

The above action is just what is obtained in the imaginary free energy approach, by integrating out the harmonic oscillators [8]. Now we look at the equation of motion for q. Up to now, our approach has been quite general. Because of the special nature of our $\Delta(q)$ shown in the figure 2 we make the assumption that $\Delta(q) = \theta(q-q_c)$. Further, to make the analysis simple, we shall take the potential $V_a(q)$ to be an Eckart barrier ($V_a(q) = V_b/Cosh(ax)^2$), as this makes the problem analytically solvable. The equation of motion for q may be obtained by optimising the action in equation (9) to be

$$m\ddot{q} = \frac{\partial V_a(q)}{\partial q} + F(t_+)\delta(q - q_c) \tag{11}$$

with

$$F(t_+) = B - \int_0^{\beta} ds K(|t-s|)\theta(q(s)-q_c) \tag{12}$$

Since the periodic orbit $q(t)$ obeys $q(t+\beta)=q(t)$ and as $K(t-s+\beta)=K(t-s)$, *F is independent of the time t*, even though it appears on the right hand side of the equation (12). The above equations show that the tunnelling particle executes oscillations in a modified upside down Eckart potential (see figure 2). In each oscillation, it would spend some time in the region $q > q_c$, which we denote as t_+. F depends on t_+ and may be evaluated to be

$$F(t_+) = \frac{B}{(1+Bt_+)} - \sum_{n=1}^{\infty} \frac{2B^2 t_+}{(n\beta B+1)^2-(Bt_+)^2} \tag{13}$$

The equation (11), while it has the appearance of a Newton's equation of motion, is slightly more complex, as the equation of motion itself depends upon t_+, a global characteristic of the trajectory $q(t)$. Hence t_+ has to be found self-consistently. We shall take q_c to be the origin $(q_c=0)$. We denote the time that the particle spends in the $q_c< q$ region as t_-. Obviously, $t_+ + t_- = \beta$. The equation (11) corresponds to a classical particle moving in a potential $-V_0(q) - F(t_+)\theta(q)$. For such a particle, the energy is conserved and as it is negative, we denote it as $-E$. Then one has

$$m\dot{q}^2/2 - V(q) - F(t_+)\theta(q) = -E \tag{14}$$

For the Eckart potential, the above equation can be easily analysed to find t_+ and t_-. This leads to: $E = \left(\frac{\pi}{\alpha t_-}\right)^2 \frac{m}{2}$ and $E - F(t_+) = \left(\frac{\pi}{\alpha t_+}\right)^2 \frac{m}{2}$. Eliminating E and t_- from these two equations, we get the self-consistency condition for t_+

$$F(t_+) = \left(\frac{\pi}{\alpha}\right)^2 \frac{(2t_+ - \beta)\beta}{[t_+(\beta - t_+)]^2} \quad . \tag{15}$$

This equation can be numerically solved to find t_+. Using t_+, E can be calculated. Thus all the parameters needed to know the periodic orbit are known and the value of S_E can be determined along it. This gives

$$S_E = \frac{\pi}{\alpha}\left[2\sqrt{2mV_b} - \sqrt{2m(E-F(t_+))} - \sqrt{2mE}\right] + E\beta - F(t_+)t_+ - \ln A \tag{16}$$

where

$$A = \frac{1}{(1+Bt_+)} \prod_{n=1}^{\infty} \frac{(1+nB\beta)^2}{(1+nB\beta)^2 - (Bt_+)^2}$$

$$= \frac{1}{(1+Bt_+)} \frac{\Gamma(1+(1+Bt_+)/(B\beta))\Gamma(1+(1-Bt_+)/(B\beta))}{\Gamma(1+1/(B\beta))^2} \quad \textbf{(17)}$$

where the infinite product has been expressed in terms of Gamma functions Hence the tunnelling probability at the temperature T, is given by $P(T) = exp(-S_E)$ to be

$$P(T) = A \ Exp\left[-\left(\frac{\pi}{\alpha}\left[2\sqrt{2mV_0} - \sqrt{2m(E - F(t_+)} - \sqrt{2mE}\right] + E\beta - F(t_+)t_+\right)\right] \quad \textbf{(18)}$$

Using the above formula, we can calculate the probability of tunnelling $P(T)$ and compare it with the probability of tunnelling through the adiabatic potential, $P_a(T)$. What is of greatest interest is the ratio between the two, $r(T) = P(T)/P_a(T)$, which is plotted in the figure 3. The barrier height and width of the potential chosen are typically what are expected for proton transfer. The results show that the tunnelling probability is decreased roughly by a factor of ≈ 0.02 This arises from the term $1/(1+Bt_+)$ which is contained in A. As $B \approx$ 1-2 eV and as $t_+ \approx \beta$, clearly $1/(1+Bt_+) \approx 0.02$ to 0.05. Even though we have considered a rather simplified model, it is obvious that non-adiabaticity makes an important contribution to the rate, and hence must be included in any realistic calculation of proton transfer rates in electrochemistry.

4. Comparison with the work of Dogonadze et al [3,4]

Dogonadze et al [3,4] have addressed the problem of non-adiabaticity in electron transfer reactions. They consider a two continua of intersecting diabatic curves. As the motion takes place along any particular curve of the initial (final) state, it can go over to any curve of the final (initial) state at the point of intersection between the two. The probability of this is determined by the usual Landau-Zener factor. Successive transitions are assumed to be independent of one another. The Landau-Zener factor has the velocity of the nuclear motion in it and the velocity that is to be used is apparently the average thermal velocity at the temperature T. Actually, it is possible for the transition to occur one after the other, and energy conservation requires that the velocity at the second intersection point should be different from the velocity at the first intersection, and this is neglected in the approach. As the transitions are assumed to be independent of one another, a master equation type equation is written for the probabilities. This procedure of Dogonadze et al can be justified in the limit of small transition probabilities at each intersection. However, in the case of proton transfer reaction the interaction of the atomic orbital on the proton with the orbitals of the metal is strong. As a result the reaction is likely to be adiabatic and such an approach would be difficult to justify. To make this statement clearer, consider the case of adiabatic reactions discussed in section 7B (ii) in reference 3. In the limit of small temperatures, the effective potential energy curve given in equation (7.4b) should reduce to the adiabatic surface. Instead, it becomes (in their notation)

$$U = \min imum\ of (\bar{U}_i(Q), \bar{U}_f(Q))$$

This cusped curve is not the proper adiabatic limit and hence it would not be proper to use the approach in a case where the interaction between the orbitals is strong. On the other hand, our approach accounts for the transition amplitudes properly, and the lowest surface on which the reaction can happen is the adiabatic surface. Further, the change of velocity of the tunnelling particle as one goes from curve to curve is automatically accounted for. The only assumption made in the bosonization approach is that the probability of exciting any particular boson is small, which seems to be quite reasonable for the problem, as there is no particular electronic excitation which has a large probability of getting excited. Moreover, as has been done in other related problems, it should be possible to derive the results of this paper using approaches which do not make use of bosonization [9]. Work in this direction is in progress and shall be reported later. Thus our method provides a much better method than that of Dogonadze et al, in the limit where one starts with the adiabatic limit and tries to take non-adiabaticity into account.

Acknowledgements I wish to thank all my colleagues, particularly Prof. K. Kishore at the Department of Inorganic and Physical Chemistry, for the wonderful atmosphere that exists.

1. T.R. Knowles, Surface Science, 101, 224 (1980).

2. S.U.M. Khan and J.O'M. Bockris, in Comprehensive Treatise of Electrochemistry, edited by B.E. Conway et al. (Plenum, New York, 1983), Vol. 7, p41 and the references therein.

3. R.R. Dogonadze and A.M. Kuznetsov, in Progress in Surface Science, edited by S.G. Davison (Pergamon, Oxford, 1975), Vol 6, p 2.

4. R.R. Dogonadze, A.M. Kuznetsov and M.A. Vorotyntsev, Croatica Chemica Acta, 44, 257 (1972).

5. J. Ulstrup, Charge transfer processes in condensed media, Springer, New York, 1979 and the references therein.

6. K. Schonhammer and O. Gunnarsson, Physical Review B22, 1629 (1980).

7. K.L. Sebastian, J. Chem. Phys. 90, 5056 (1989).

8. W.H. Miller, J. Chem. Phys. 62, 1899 (1975).

9. A. Yoshimori and K. Makoshi, in Progress in Surface Science, edited by S.G. Davison (Pergamon, Oxford, 1986), Vol 21, p 251.

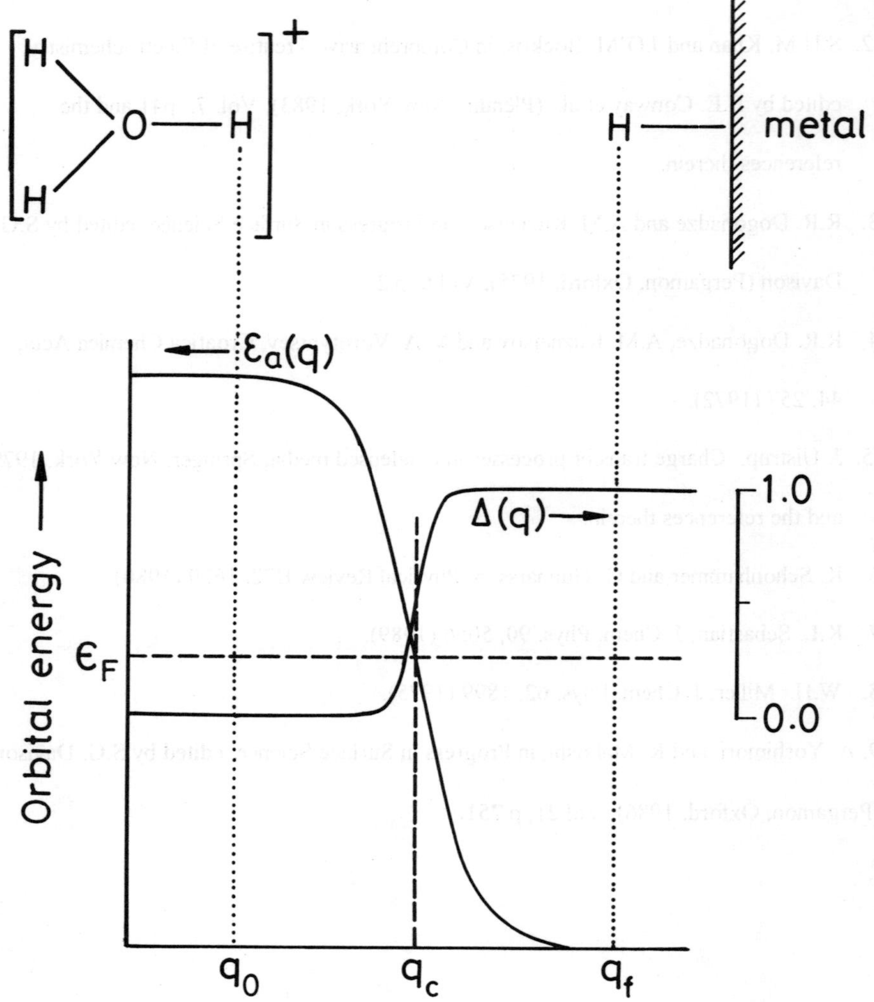

Figure 1. The upper part of the figure shows H_3O^+ in the vicinity of the metal (the initial state in the reaction). The H atom enclosed within the dotted circle shows the proton's final position after the reaction. The lower part of the figure shows the variation of $\varepsilon_a(q)$ and $\Delta(q)$ as functions of q .

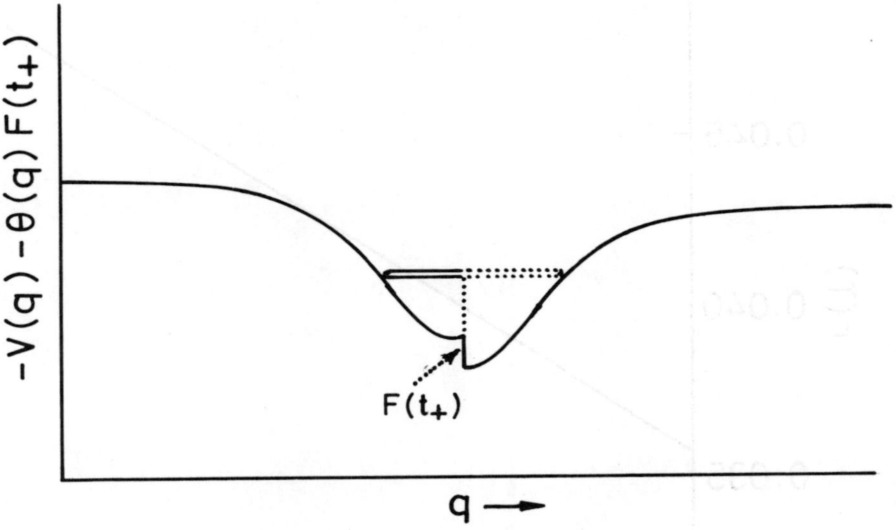

Figure. 2. The upside down Eckart potential modified due to interaction with e-h pairs [$= -V(q)-\theta(q)F(t_+)$], in which the tunnelling particle moves. The closed path followed by the particle is shown in the figure. The path is divided into two portions. The dotted portion, with $q > 0$ and the portion which is not dotted having $q < 0$. The time spent in traversing these portions are denoted t_+ and t_- respectively.

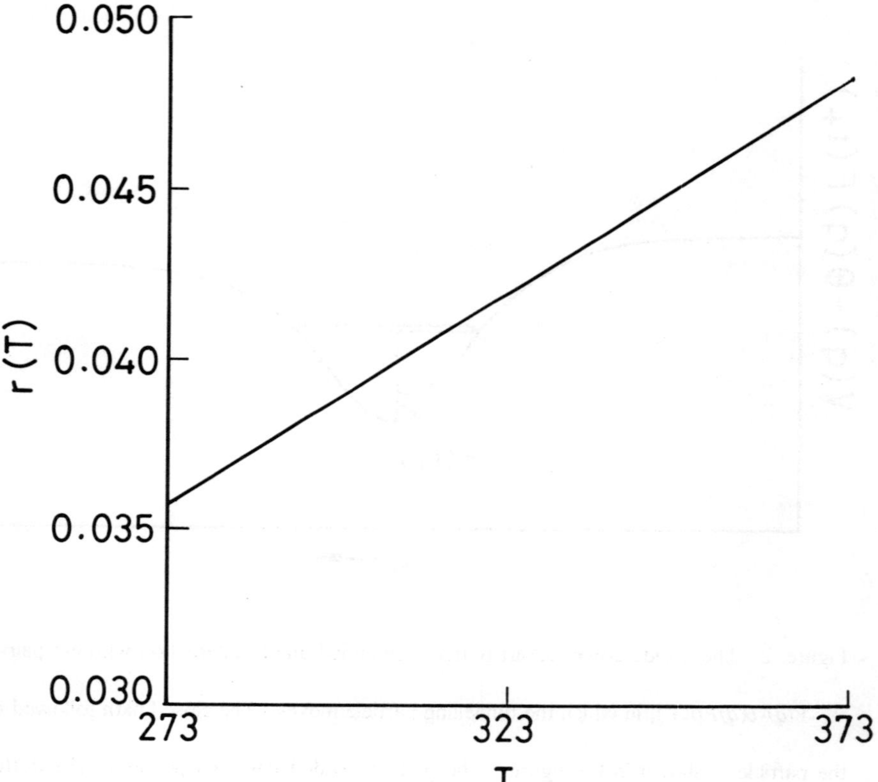

Figure 3. Plot of the ratio $r(T)$ as a function T. The parameters used are: V_0 =0.715 eV, a = 1.0/angstrom, m = mass of proton, B = 1.0 eV

Electrochemical Stepwise Formation of Metal Phases of Different Dimensionality

W. J. Lorenz and G. Staikov

Institute of Physical Chemistry and Electrochemistry, University of Karlsruhe, Karlsruhe, Germany

SUMMARY

Metal deposition on foreign substrate occurs stepwise as found by in situ local probe measurements. On the basis of structural experimental results, thermodynamics and kinetics of the initial steps of electrocrystallization forming 0D, 1D, and 2D Me phases in the undersaturation range and 3D Me phases in the supersaturation range are discussed.

INTRODUCTION

Electrochemical Phase Formation is now well understood in the case of Metal (Me) deposition which can take place either on foreign substrate (S) or on native (Me) substrate. The second case involves 3D Me phase formation and growth phenomena in the supersaturation range and represents only an advanced stage of the first more general case. The initial stage of the metal ion charge transfer process is the formation of Me adatoms (Me_{ads}) on S in the undersaturation range. Such an adsorption process is caused by the Me-S interaction energy (1).

The most important parameters determining the mechanism of Me deposition on a foreign substrate are the Me_{ads}-S binding energy and the crystallographic misfit between S and a 3D Me-deposit. Considering the deposition process at nearly equilibrium conditions, i.e. small supersaturation and negligible kinetic influences, and disregarding Me-S alloy formation, two different growth modes can be distinguished as schematically illustrated in Fig. 1 (1-7).

i) In the first case (Fig. 1a), the binding energy of Me_{ads} on foreign substrate S, $\Psi_{Meads\text{-}S}$, is lower than that of Me_{ads} on native substrate Me, $\Psi_{Meads\text{-}Me}$. Consequently, the surface concentration of Me_{ads} at the Nernst equilibrium potential of the 3D Me bulk phase, $E_{3D\ Me}$, is small and deposition of a 3D Me bulk phase takes place at supersaturation (overpotential deposition, OPD at $\eta = E - E_{3D\ Me} < 0$) on an unmodified substrate surface according to the "Volmer-Weber" or 3D island growth mode independent of crystallographic Me-S misfit.

ii) In the second case (Fig. 1b and 1c), the binding energy of Me_{ads} on the foreign substrate S, $\Psi_{Meads\text{-}S}$, is higher than that of Me_{ads} on native substrate Me, $\Psi_{Meads\text{-}Me}$. Then, 2D Me_{ads} phases can be formed at undersaturation (underpotential deposition, UPD at $\Delta E = E - E_{3D\ Me} > 0$) and the surface concentration of Me_{ads} at $E_{3D\ Me}$ can reach one or more Me_{ads}-monolayers depending on $\Psi_{Meads\text{-}S}$. Two different subcases can be distinguished:

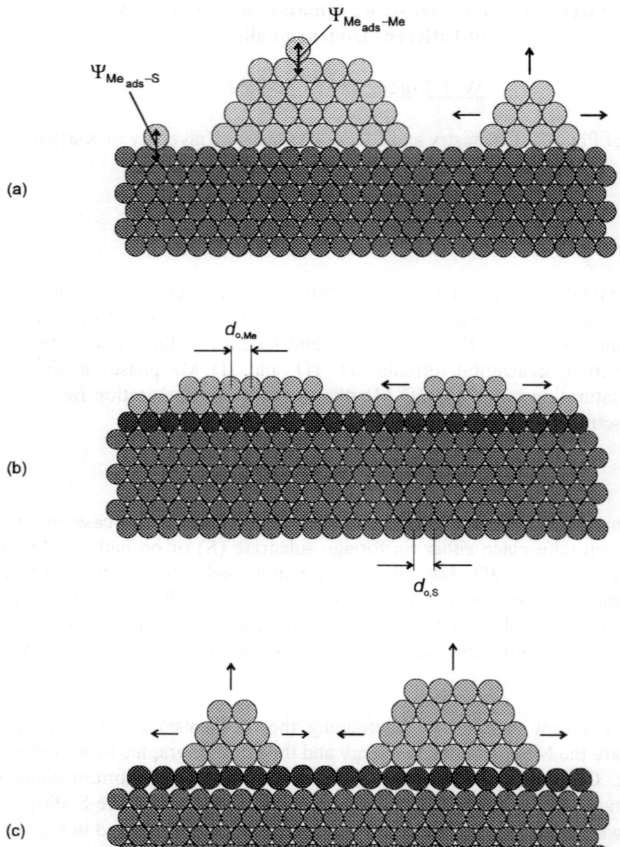

Figure 1: Schematic representation of different growth modes in metal (Me) deposition on foreign substrate (S) depending on the binding energy of Me_{ads} on S, $\Psi_{Me_{ads}-S}$, compared to that of Me_{ads} on native substrate Me, $\Psi_{Me_{ads}-Me}$, and on the crystallographic misfit characterized by the interatomic distances $d_{0,Me}$ and $d_{0,S}$ of 3D Me and S bulk phases, respectively. (a) "Volmer–Weber" growth mode (3D Me island formation) for $\Psi_{Me_{ads}-S} \ll \Psi_{Me_{ads}-Me}$ independent of the ratio $(d_{0,Me} - d_{0,S})/d_{0,S}$. (b) "Frank–van der Merwe" growth mode (Me layer-by-layer formation) for $\Psi_{Me_{ads}-S} \gg \Psi_{Me_{ads}-Me}$ and ratio $(d_{0,Me} - d_{0,S})/d_{0,S} \approx 0$. (c) "Stranski–Krastanov" growth mode (3D Me island formation on top of predeposited 2D Me_{ads} overlayers on S for $\Psi_{Me_{ads}-S} \gg \Psi_{Me_{ads}-Me}$ and $(d_{0,Me} - d_{0,S})/d_{0,S} > 0$ (positive misfit) or $(d_{0,Me} - d_{0,S})/d_{0,S} < 0$ (negative misfit).

ii_1) In systems with negligibly small crystallographic Me-S misfit (Fig. 1b), 2D Me UPD overlayers and 3D Me OPD crystallites are epitaxially oriented following the "Franck-Van der Merwe" or layer-by-layer growth mode up to the disappearance of the strong Me-S attraction. Further Me growth continues in a layer growth fashion as on native Me substrate.

ii_2) In systems with significant crystallographic Me-S misfit (Fig. 1c), the 2D Me_{ads} phase formed in the UPD range has a different structure compared to that of the 3D Me bulk phase and contains considerable internal strain. The formation and growth of unstrained 3D Me crystallites (islands) on top of strained 2D Me overlayers is energetically favored following the "Stranski-Krastanov" growth mode. The misfit between the 2D Me_{ads} phase and the 3D Me bulk phase is mainly removed by misfit dislocations (5-8).

Generally, in Me UPD systems with non-vanishing solubility of Me in S, the formation of a 2D Me-S surface alloy and/or a 3D Me-S bulk alloy becomes energetically possible even in the UPD range. A 2D Me-S surface alloy is considered if site exchange processes between Me and S occur across a limited number of surface layers (1).

2D and 3D Me-S alloy formation processes are often kinetically hindered and difficult to be studied. However, in some Me UPD systems clear evidence was found experimentally for such 2D and 3D Me-S alloy formation processes complicating Me UPD phenomena (1).

2D Me_{ads} overlayers, 2D Me-S surface alloys and 3D Me-S bulk alloys, if present in the UPD range, can act as precursors for the nucleation and growth of the 3D Me bulk phase in the OPD range (1).

STRUCTURE AND EPITAXY

The process of Me deposition on S was found to occur stepwise by in situ STM and AFM measurements. Now, thermodynamics, structure and kinetics of the stepwise formation of Me phases of different dimensionality (0D, 1D, 2D, 3D) can be treated on an atomic level which is the aim of this paper.

Me phase formation starts at extremely high undersaturation (underpotential) by a decoration of 0D surface point defects. Unfortunately, single Me adatoms cannot be detected by local probe techniques up to now. A relatively great mobility of Me_{ads} (surface diffusion, exchange with the electrolyte, tip or cantilever influence) seems to be the reason for this lack of information. However, single inorganic large ions such as iodide or organic molecules adsorbed on metal single crystal faces have already been imaged (9-11). Fig. 2 shows an example for an expanded and ordered 2D adlayer of organic molecules on an Au(111) surface (11). The imaging of the decoration of different kinds of surface point effects such as emergence points of step and screw dislocations, kink sites, atomic disorder etc.) certainly will follow in the next future. The adsorption of single adatoms, molecules or small clusters at a surface point defect can be considered as a 0D phase formation process. The influence of a small amount of adsorbate with vanishing degree of coverage on electron or ion charge transfer reactions is well known in the fields of electrocatalysis, batteries, fuel cells, and corrosion. In the field of corrosion inhibition, such an effect was previously denoted as "deactivating coverage" (12,13).

The adsorption process of single Me adatoms is followed by a decoration of monatomic steps at relatively high undersaturation. This process corresponds to a 1D Me phase formation. In various Me UPD systems, such a preferential decoration of monatomic steps was directly observed using in situ local probe measurements. As typical examples, step decoration by Pb

(a)

(b)

(c)

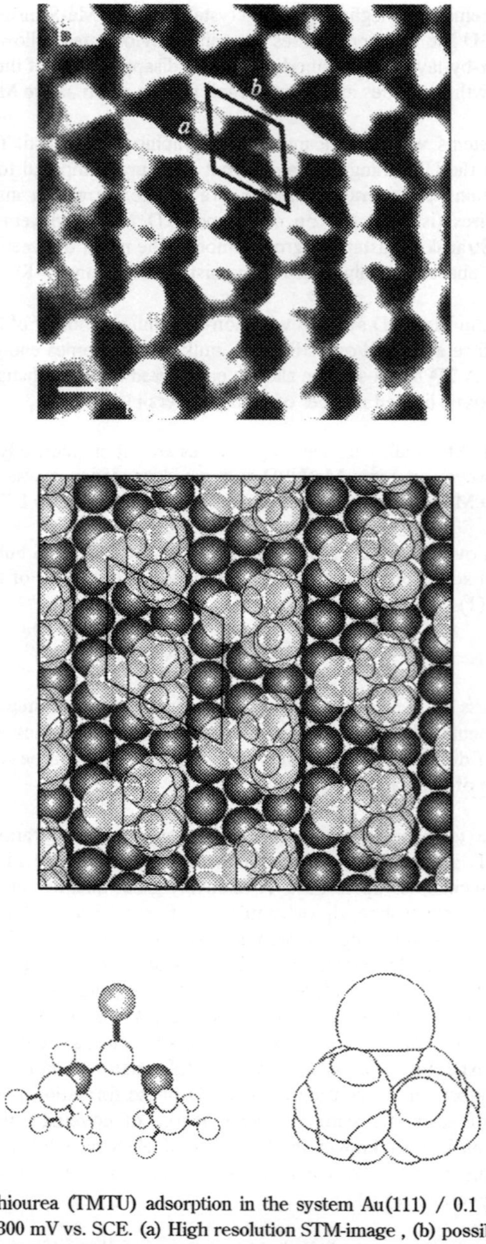

Figure 2: Tetramethylthiourea (TMTU) adsorption in the system Au(111) / 0.1 M H_2SO_4 + 10^{-5} M TMTU at 298 K and E = 300 mV vs. SCE. (a) High resolution STM-image , (b) possible conformation of TMTU molecules lying slightly tilted on the surface with the sulphur atoms being placed over a threefold hollow site, and (c) model of TMTU molecules (by courtesy of E. Bunge et. al. [11]).

UPD on Ag(hkl) with (hkl) = (100), (111) single crystal surfaces is shown in Figs. 3 and 4. Fig. 3a represents the bare Ag(100) substrate, Fig. 3b the step decoration after Pb UPD at relatively high underpotentials (14). Fig. 4 shows an in situ STM line scan for Pb UPD on Ag(111) (15).

The Me adsorption process proceeds at high and medium undersaturation by the formation of expanded commensurate 2D Me adlayers on atomically flat terraces as shown in Fig. 5. At lower undersaturation, expanded 2D Me adlayers are transformed to condensed 2D Me monolayers via a first order phase transition which is demonstrated in Fig. 6 for Pb UPD on Au(111) (1). Condensed 2D Me_{ads} overlayers are either commensurate, higher order commensurate or even incommensurate depending on the crystallographic Me-S misfit. The condensed 2D Me_{ads} overlayers act as precursors for the subsequent formation of 3D Me bulk phases (3D) in the supersaturation (OPD) range and determine the structure, epitaxy, and morphology of compact thin Me films and heterostructures. As an example, the structure of a close-packed Pb UPD overlayer on Ag(111) at low undersaturation is shown in Fig. 7 (1). The significant crystallographic Pb-Ag misfit leads to an isotropically internal strained overlayer which is incommensurate and rotated by 4.5° with respect to the crystallographic close-packed [110] directions of the silver substrate.

Previous morphological studies of Pb deposition in the OPD range on quasi-perfect silver single crystal faces (100) and (111) showed an epitaxial growth of isolated 3D Pb crystallites as shown in Fig. 8 (16-22). A strong correlation between the close packed crystallographic directions [110] of substrate and 3D Pb_{ads} crystallites is detectable. In the case of Ag(100), two different azimutal orientations of the 3D Pb crystallites are seen. In the case of Ag(111), the rotation of about 4.5° of the 3D Pb crystallites can be detected. Consequently, the epitaxy of the 2D Pb_{ads} overlayer

Ag(100) [110] || compressed 2D hcp Pb [110]
Ag(111) [110] || compressed 2D hcp Pb [110] R 4.5°

is reflected in the epitaxy of the 3D Pb crystallites:

Ag(100) [110] || 3D Pb (111) [110]
Ag(111) [110] || 3D Pb (111) [110] R 4.5°

This indicates that the UPD-OPD transition obviously proceeds via the Stranski-Krastanov mechanism (Fig. 1c) involving the formation and growth of 3D Pb crystallites on top of the 2D internally strained Pb UPD adlayers which act as precursors for the nucleation and growth process in the OPD range. The unstrained 2D hcp surface structure of a 3D Pb(111) crystal face is reached after deposition of about 10 Pb monolayers as shown in Fig. 9.

THERMODYNAMICS

The thermodynamic stability condition for a 3D Me phase is clearly given by the Nernst equation:

$$E_{3D\,Me} = E^{o}_{3D\,Me} + \frac{RT}{zF} \ln\left(\frac{a_{Me^{z+}}}{a_{3D\,Me}}\right) \qquad [1]$$

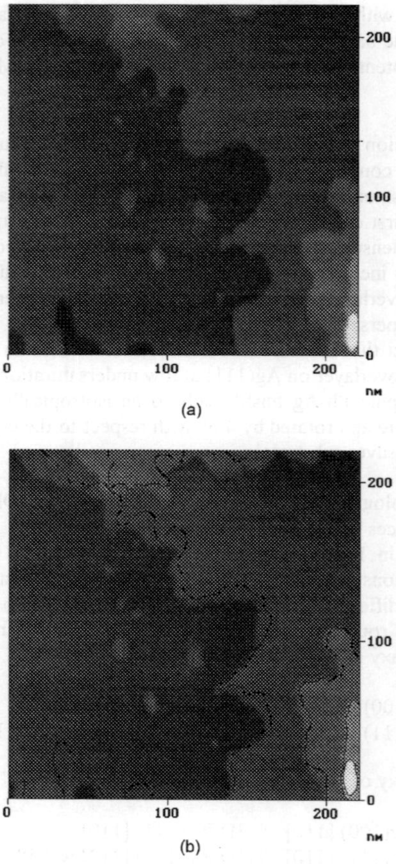

Figure 3: In situ STM images showing the decoration of monatomic steps in the system Ag(100)/5 x 10^{-3} M $Pb(ClO_4)_2$ + 10^{-2} M $HClO_4$ at T = 298 K with I_T = 40 nA and Pt –Ir tip. (a): Bare Ag(100) substrate at ΔE = 300 mV; (b): stepdecoration at ΔE = 200 mV; the lines indicate the location of step edges in (a).

where $E_{3D\,Me}$ and $E^{o}_{3D\,Me}$ denote the equilibrium and standard potentials of the 3D Me/Me^{z+} electrode, respectively, and $a_{3D\,Me}$ is the activity of 3D Me phase, which is equal to the activity of a Me atom in a kink position being unity for a pure 3D Me bulk phase.

The thermodynamic stability condition for a condensed 2D Me phase is given by a quasi-Nernst equation:

$$E_{2D\,Me} = E^{o}_{2D\,Me} + \frac{RT}{zF} \ln\left(\frac{a_{Me^{z+}}}{a_{2D\,Me}}\right) \quad [2]$$

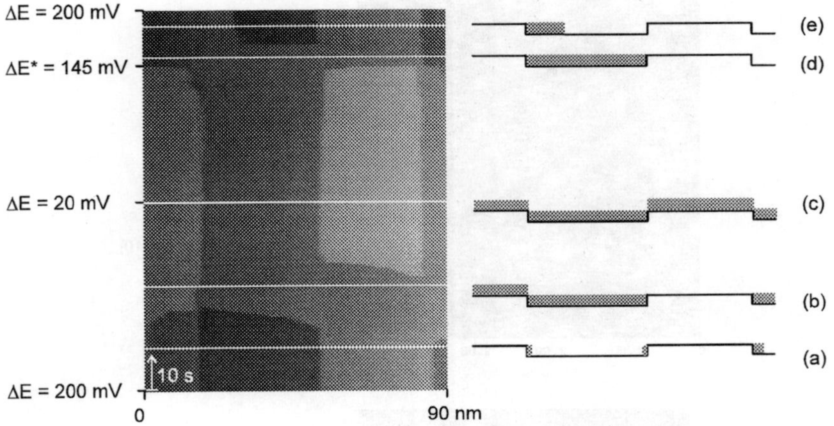

Figure 4: In situ STM line scan plot showing the formation and dissolution of a 2D Pb_{ads} overlayer on Ag(111) during a potential scan in the UPD range 20 mV $\leq \Delta E \leq$ 200 mV in the system Ag(111)/4 x 10^{-3} M $Pb(ClO_4)_2$ + 10^{-2} M $HClO_4$ with $|dE/dt|$ = 5 mV/s at T = 298 K and I_T = 10 nA. The cross sections represent: (a) step decoration at $\Delta E > \Delta E^*$; (b) formation of a condensed 2D Pb_{ads} overlayer on substrate terraces and pits starting at monatomic steps ($\Delta E < \Delta E^*$); (c) formation of a condensed 2D Pb_{ads} overlayer on top of substrate islands starting at more negative potentials ($\Delta E < \Delta E^*$); (d) dissolution of a condensed 2D Pb_{ads} overlayer on substrate terraces and islands at $\Delta E > \Delta E^*$; (e) dissolution of a condensed 2D Pb_{ads} overlayer in substrate pits starting at more positive potentials $\Delta E > \Delta E^*$. $\Delta E^*$is defined as $E_{2D\,Me} - E_{3D\,Me}$ (cf. eqs. [1] and [2]).

where $E_{2D\,Me}$ and $E^{o}_{2D\,Me}$ denote the equilibrium and standard potentials of a 2D Me phase in contact with the electrolyte, respectively, and $a_{2D\,Me}$ is the activity of the 2D Me phase, which is equal to the activity of an adatom in a kink-like position of a 2D Me_{ads} overlayer as illustrated in Fig. 9a.

Analogously, the thermodynamic stability condition for a condensed 1D Me phase is also given by a quasi-Nernst equation:

$$E_{1D\,Me} = E^{o}_{1D\,Me} + \frac{RT}{zF} \ln\left(\frac{a_{Me^{z+}}}{a_{1D\,Me}}\right) \qquad [3]$$

where $E_{1D\,Me}$ and $E^{o}_{1D\,Me}$ denote the equilibrium and standard potentials of a 1D Me phase in contact with the electrolyte, respectively, and $a_{1D\,Me}$ is the activity of the 1D Me phase, which is equal to the activity of an adatom in a kink-like position at a monatomically decorated 1D monatomic step as illustrated in Fig. 10c.

Formally, a similar treatment is also possible for the 0D decoration of surface point defects.

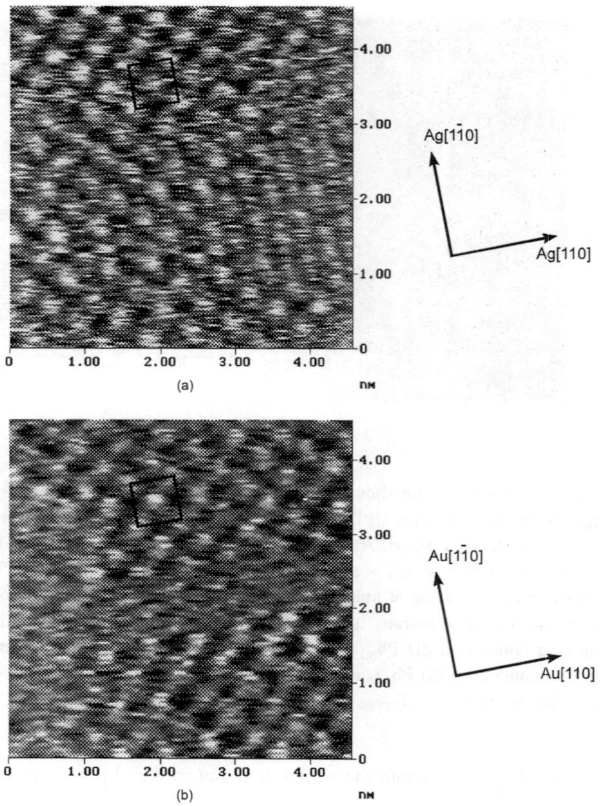

Figure 5: In situ STM images of S(100) - c(2x2) Pb_{ads} overlayer domains. (a): system Ag(100)/5 x 10^{-3} M $Pb(ClO_4)_2$ + 10^{-2} M $HClO_4$ at T = 298 K and ΔE = 175 mV with I_T = 40 nA and Pt –Ir tip; (b): system Au(100)/5 x 10^{-3} M $Pb(ClO_4)_2$ + 10^{-2} M $HClO_4$ at T = 298 K and ΔE = 550 mV with I_T = 40 nA and Pt –Ir tip.

According to the classical nucleation theory, the Gibbs energy for the formation of a critical 2D nucleus of a 2D Me_{ads} phase on an atomically flat terrace (T), $\Delta G_{crit,T}$, is given by (1):

$$\Delta G_{crit,\,T} = \frac{N_A b\Omega\varepsilon^2}{zF\,|E - E_{2DMe}|} = \frac{N_A b\Omega\varepsilon^2}{zF\,|\Delta E - \Delta E^*|} \qquad [4]$$

where $\Delta E^* = E_{2D\,Me} - E_{3D\,Me}$, b is a geometric factor depending on the shape of the critical nucleus, Ω represents the area of Me_{ads} in a condensed Me_{ads} overlayer, ε denotes the specific edge energy, and $zF|E - E_{2D\,Me}| = zF|\Delta E - \Delta E^*|$ corresponds to supersaturation with respect to the condensed 2D Me_{ads} phase.

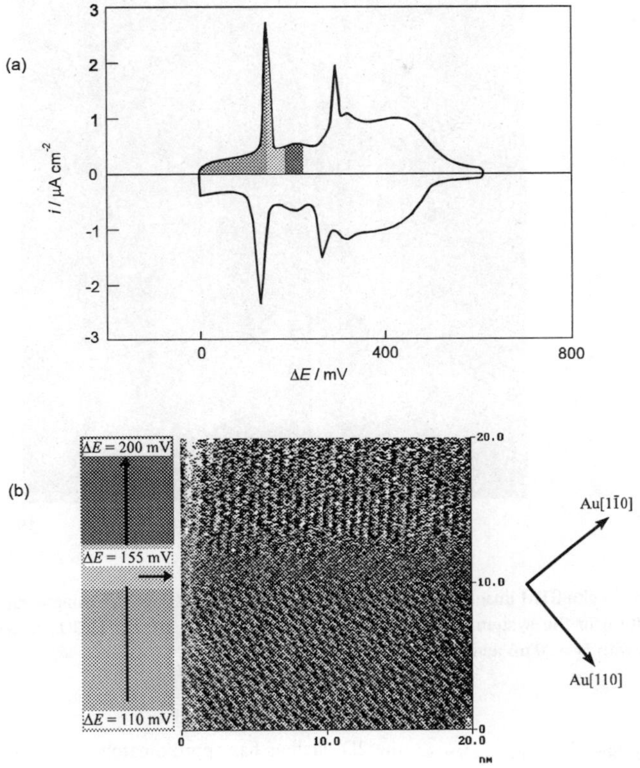

Figure 6: In situ STM image of a first–order phase transformation of a Pb_{ads} overlayer on Au (100) during anodic polarization in the system Au(100)/ 5 x 10^{-3} M $Pb(ClO_4)_2$ + 10^{-2} M $HClO_4$ at T = 298 K. (a): cyclic voltammogram $|dE/dt|$ = 10 mVs^{-1}; (b): in situ STM image during cyclic voltammetric sweeps: I_T = 40 nA, Pt–Ir tip, ▩: Au (100)–c($3\sqrt{2} \times \sqrt{2}$) R 45° Pb, ▩: bare Au (100) substrate, ▩: Au (100)–hcp Pb_{ads} overlayer; (c): in situ STM image at constant ΔE = 155 mV showing the Au(100) substrate structure. I_T = 40 nA, Pt–Ir tip.

However, the Gibbs energy for the formation of a critical 2D Me cluster at a monoatomic step (S), $\Delta G_{crit,S}$, is reduced by an additional term including the lateral interaction energy between Me_{ads} and a step, $\Psi_{Me_{ads}\text{-Step}}$. For a nucleus of a 2D Me_{ads} phase with a square lattice structure (b = 4, c = 4) assuming only first nearest neighbor interactions ($\psi_{Me_{ads}\text{-}Me_{ads}}$) one obtains for $\Delta G_{crit,S}$ (1):

$$\Delta G_{crit,S} = \Delta G_{crit,T} \left(1 - \frac{\Psi_{Me_{ads}\text{-Step}}}{\psi_{Me_{ads}\text{-}Me_{ads}}}\right) \qquad [5]$$

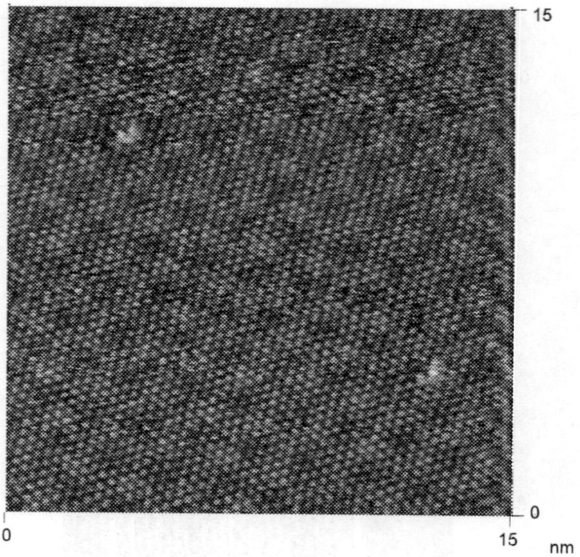

Figure 7: In situ STM image of a compressed 2D hcp Pb_{ads} overlayer showing a superstructure with moiré pattern in the system Ag(111)/5 x 10^{-3} M $Pb(ClO_4)_2$ + 10^{-2} M $HClO_4$ at ΔE = 30 mV and T = 298 K with I_T = 20 nA and Pt –Ir tip .

If $\Psi_{Me_{ads}\text{-Step}} << \psi_{Me_{ads}\text{-}Me_{ads}}$, the 2D nucleus has approximately the same size, form, and Gibbs energy of formation ($\Delta G_{crit,S}$ _ $\Delta G_{crit,T}$) a nucleus formed at a flat terrace (Fig. 10a). In the case of $\Psi_{Me_{ads}\text{-Step}} < \psi_{Me_{ads}\text{-}Me_{ads}}$ follows $\Delta G_{crit,S} < \Delta G_{crit,T}$ and 2D Me nucleation takes place preferentially at monoatomic steps (Fig. 10b). For $\Psi_{Me_{ads}\text{-Step}}$ _ $\psi_{Me_{ads}\text{-}Me_{ads}}$ follows $\Delta G_{crit,S}$ _ 0 and the formation of a condensed 2D Me_{ads} phase occurs at steps without activation energy for 2D nucleation. If $\Psi_{Me_{ads}\text{-Step}} >> \psi_{Me_{ads}\text{-}Me_{ads}}$, the formation of a 1D Me_{ads} phase along monoatomic steps becomes possible in the undersaturation range $\Delta E > E_{2D\ Me}$ with respect to the condensed 2D Me_{ads} phase (Fig. 10c). The width of such a 1D Me_{ads} phase at monoatomic step is not necessarily restricted to a monoatomic dimension, but depends on the step activity.

KINETICS

The influence of monoatomic steps on $\Delta G_{crit,i}$ with i = T,S in eq. [6] is also reflected in the nucleation rate, J_i, per nucleation site by (1):

$$J_i = K_i \exp\left(-\frac{\Delta G_{crit,i}}{kT}\right) \qquad [6]$$

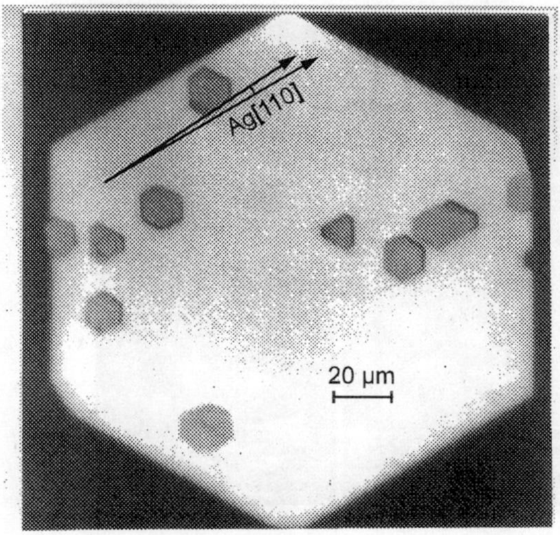

(a)

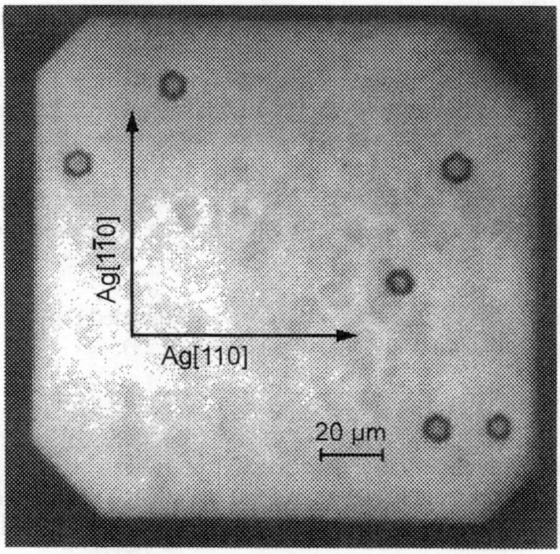

(b)

Figure 8: Epitaxy of 3D Pb crystallites deposited in the system Ag(*hkl*)/10^{-2} M $Pb(ClO_4)_2$ + 5 x 10^{-1} M $NaClO_4$ + 5 x 10^{-3} M $HClO_4$ at T = 298 K. (a): Ag(111); (b): Ag(100).

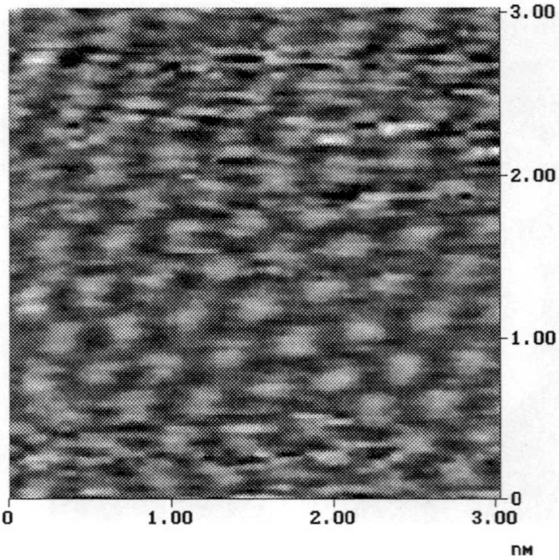

Figure 9: In situ STM image of the surface of ca. 1 μm thick Pb epitaxial film deposited on Ag(100) in the system Ag(100)/5 x 10^{-3} M $Pb(ClO_4)_2$ + 10^{-2} M $HClO_4$ at T = 298 K. I_T = 2 nA, Pt–Ir tip.

where the preexponential factor K_i and $\Delta G_{crit,i}$ depend on the supersaturation $zF|E - E_{2D\ Me}|$. However, the supersaturation dependence of K_i, which also includes the density of nucleation sites, is relatively weak and can be disregarded. The relationship [6] is characterized by a critical value, $|E - E_{2D\ Me}|_{crit}$, for a detectable 2D nucleation rate J_i. A schematic representation of the nucleation rates J_T and J_S as a function of $|E - E_{2D\ Me}|$ is shown in Fig. 11. It is seen that the critical threshold value $|E - E_{2D\ Me}|_{crit}$ is reduced for 2D nucleation processes at monoatomic steps in accordance with experimental results (1).

Generally, the kinetics of a phase transition from an expanded (ep) to a condensed (cd) 2D Me_{ads} phase under potentiostatic conditions is given by (1,23):

$$-i(t) * 1 = q_{cd}\, S(t) + [1 - S(t)]\, q_{ep}(t) - q_{ep}(t{=}0) \qquad [7]$$

where i(t) is the overall cathodic current density, S(t) represents the part of the substrate surface covered by the condensed 2D Me_{ads} phase, and q_{ep} and q_{cd} denote the charge amounts of expanded and condensed 2D Me_{ads} overlayers, respectively. The symbol * represents the convolution product.

Usually, S(t) is described by the well-known Avrami theorem (1,23) supposing multiple nucleation on a quasi-homogeneous substrate surface with a sufficient density of nuclei, statistically local distribution of nuclei, and overlapping of growing 2D islands:

$$S(t) = 1 - \exp\left[-1 * \frac{j_{free}(t)}{\Gamma_{cd}}\right] \qquad [8]$$

where $i_{free}(t)$ describes the extended growth of 2D islands without overlapping. It should be noted, however, that i(t) contains nucleative and sorption components which depend on each other and cannot be easily separated (1). Furthermore, real surfaces are always inhomogeneous consisting of flat terraces separated by monoatomic steps which act as active sites for 2D Me nucleation and restrict the growth of 2D islands. Therefore, the Avrami theorem is not strictly valid for real surfaces with a finite step density.

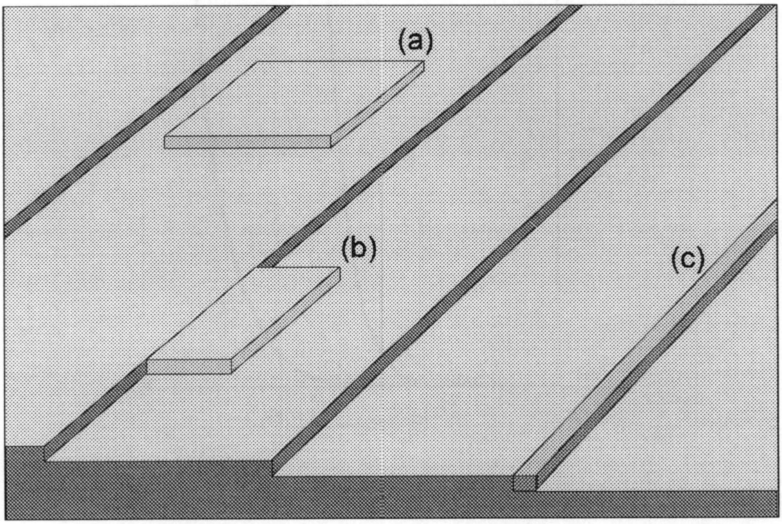

Figure 10: Formation of 2D and 1D Me_{ads} phases on stepped foreign substrates. (a) 2D nucleation on atomically flat terraces; (b) 2D nucleation at monatomic steps; (c) 1D Me_{ads} phase formation along monatomic steps at $\Delta E > \Delta E^*$ for $\Psi_{Me_{ads-Step}} >> \psi_{Me_{ads-ads}}$.

Monoatomic steps at real surfaces do influence significantly not only the nucleation act, but also the spreading and overlapping of 2D islands which determines the shape of potentiostatic current-time transients. Neglecting the adsorptive contribution in eq. [7], i.e. $q_{ep} << q_{cd}$, and assuming a barrierless nucleation and growth of a condensed 2D Me_{ads} phase only at steps with a regular pattern characterized by a constant step spacing, $2\delta_{step}$, the $i_S(t)$-transient is a step function as schematically illustrated in Fig. 12a:

$$i_S(t) = \begin{cases} q_{cd}\,(2\delta_{step})^{-1}\,v = q_{cd}L_S v & \text{for } 0 \leq t < 2\delta_{step}v^{-1} \\ 0 & \text{for } t = 2\delta_{step}v^{-1} \end{cases} \qquad [9]$$

where $(2\delta_{step})^{-1} = L_S$ represents the step density and v is the propagation rate of the growing 2D Me_{ads} overlayer being a function of the supersaturation $zF|E - E_{2D\ Me}|$. Real surfaces exhibit monoatomic steps with irregular pattern, i.e. inconstant $2\delta_{step}$. Thus, $i_S(t)$ becomes dependent on the step distribution on the substrate surface as schematically shown in Fig. 12b for an arbitrary step distribution. In this case, the initial current $i_S(t = 0)$ is related to the step density L_S, however, the shape of the transient depends on the step distribution function. Transients of the type of Fig. 11b were experimentally found in the systems $Ag(111)/Pb^{2+}$ (24-26), and $Au(111)/Bi^{3+}$ (26,27).

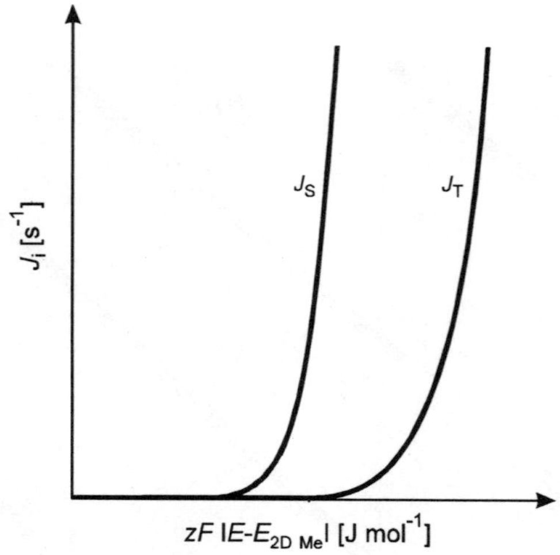

Figure 11: 2D nucleation rates at steps, J_S, and at terraces, J_T, as a function of $zF\,|\Delta E - \Delta E^*|$.

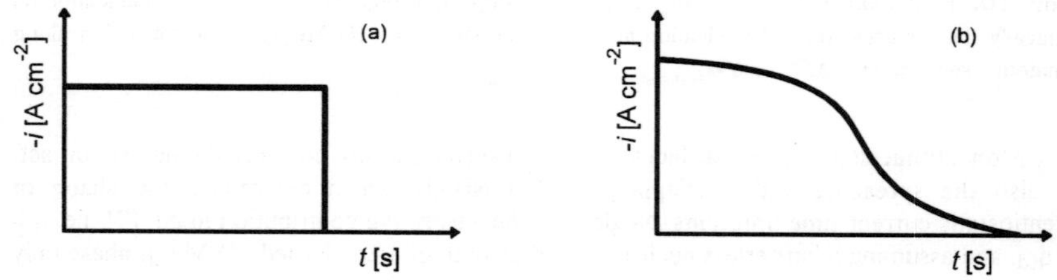

Figure 12: Schematic representation of $i(t)$ transients of condensed 2D Me_{ads} phase formation at monatomic steps. (a): substrate surface with regular step spacing; (b): substrate surface with irregular step spacing (arbitrary step distribution).

A quantitative analysis of transient or impedance data requires an exact knowledge of the substrate surface morphology and the local mechanism of Me UPD depending on the underpotential (1). This information is only available by in situ STM or AFM measurements under electrochemical conditions. For example, experimental results in the system $Ag(111)/Pb^{2+}$, ClO_4^- obtained by Siegenthaler and Ammann (15) show that a first order phase transition, occurs exclusively at monoatomic steps at relatively low supersaturation values, $zF|E - E_{2D\,Me}|$, with respect to the condensed 2D Me_{ads} phase.

CONCLUSIONS

From the structural, thermodynamic and kinetic results follows that the UPD range can be subdivided into three different and limited underpotential ranges concerning 0D, 1D, and 2D Me phase formation for which equivalent terms corresponding to $\Delta E = E - E_{2D\ Me} > 0$ can be introduced.

Now, the experimentally observed limited growth of decorated point defects, monoatomic steps and expanded Me_{ads} overlayers within defined underpotential ranges is quantitatively explained.

ACKNOWLEDGMENT

Financial support for this work given by the Deutsche Forschungsgemeinschaft (DFG), Arbeitsgemeinschaft Industrieller Forschungsvereinigungen (AIF), and Bundesministerium für Bildung und Forschung (BMBF) is gratefully acknowledged.

The authors thank Prof. Dr. E. Budevski, Academy of Sciences, Sofia, Bulgaria, and Prof. Dr. E. Schmidt and Prof. Dr. H. Siegenthaler, University of Bern, Bern, Switzerland for international cooperation and Dr. Obretenov, Dr. U. Schmidt, Dr. S. Vinzelberg, Dr. R. Speck, Dipl.-Ing. A. Froese and Dipl.-Ing. R. T. Pötzschke as coworkers of the Karlsruhe electrochemistry group for their activity.

REFERENCES

(1) E. Budevski, G. Staikov, and W. J. Lorenz, *Electrochemical Phase Formation and Growth - An Introduction in the Initial Stages of Metal Deposition.* VCH, Weinheim, **1996**.

(2) G. Staikov in *Proceedings of NATO Advanced Study Institute on Nanoscale Probes of the Solid/Liquid Interface,* Sophia Antipolis, France, July 10–20, 1993, (Eds.: H. Siegenthaler and A. Gewirth), Kluwer Acad. Publ., Dordrecht, **1995**, p. 193.

(3) G. Staikov, W. J. Lorenz in *Electrochemical Technology: Innovation and New Developments,* (Eds.: N. Masuko, T. Osaka, Y. Ito), Kodansha, Tokyo and Gordon and Breach Publ., Amsterdam, **1996**, p. 241.

(4) E. Bauer, *Z. Krystallogr.* **1958**, *110*, 85.

(5) E. Bauer, J. H. van der Merwe, *Phys. Rev. B* **1986**, *33*, 3657.

(6) I. Markov, S. Stoyanov, *Contemp. Phys.* **1987**, *28*, 267.

(7) I. Markov, R. Kaischew, *Thin Solid Films* **1976**, *32*, 163.

(8) J.W. Matthews, D.C. Jackson, A. Chambers, *Thin Solid Films* **1975**, *26*, 129.

(9) B. C. Schardt, S.-L. Yau, F. Rinaldi, *Science* **1989**, *243*, 1050;

S.-L- Yau, C. M. Vitus, B. C. Schardt, *J. Am. Chem. Soc.* **1990**, *112*, 3677.

(10) S. M. Lindsayin *Meeting Abstracts, Vol. 96-1, Electrochem. Soc., Spring Meeting, Los Angeles, USA,* **1996**, 1319.

(11) E. Bunge, B. Roelfs, H. Baumgärtel, H. Meyer, R. J. Nichols, S. N. Part in *Meeting Abstracts, Vol. 96-1, Electrochem. Soc., Spring Meeting, Los Angeles, USA,* **1996**, 1364; E. Bunge, *Ph.-D. Thesis*, FU Berlin, **1996**.

(12) W. J. Lorenz, F. Mansfeld, *Electrochim. Acta,* **1986**, *31*, 467.

(13) W. J. Lorenz, F. Mansfeld, in *New Classification Concepts of Corrossion Inhibitors, Proceedings of the H. H. Uhlig Memorial Symposium, Vol. 94-26, Electrochem. Soc.,* **1995**, 179.

(14) U. Schmidt, S. Vinzelberg, and G. Staikov, *Surf. Sci.,* **1996**, *348*, 261;

W. J. Lorenz, G. Staikov, *Surf. Sci.* **1995**, *335*, 32.

(15) H. Siegenthaler, E. Ammann, in *Meeting Abstracts, Vol. 96-1, Electrochem. Soc., Spring Meeting, Los Angeles, USA,* **1996**, 1327; E. Ammann, *Master Thesis,* University of Bern, **1995**.

(16) W. J. Lorenz, E. Schmidt, G. Staikov, H. Bort, *Faraday Symposia of the Chem. Soc.* **1977**, *12*, 14.

(17) K. Jüttner, W. J. Lorenz, G. Staikov, E. Budevski, *Electrochim. Acta* **1978**, *23*, 741.

(18) W. Obretenov, U. Schmidt, W. J. Lorenz, G. Staikov, E. Budevski, D. Carnal, U. Müller, H. Siegenthaler, E. Schmidt, *Faraday Disc.* **1992**, *94*, 107.

(19) G. Staikov, E. Budevski, M. Höpfner, W. Obretenov, K. Jüttner, W. J. Lorenz, *Surf. Sci.* **1991**, *248*, 234.

(20) G. Staikov, E. Budevski, W. Obretenov, W. J. Lorenz, *J. Electroanal. Chem.* **1993**, *349*, 355.

(21) W. Obretenov, U. Schmidt, W. J. Lorenz, G. Staikov, E. Budevski, D. Carnal, U. Müller, H. Siegenthaler, E. Schmidt, *J. Electrochem. Soc.* **1993**, *140*, 692.

(22) G. Staikov, W. J. Lorenz in *Proceedings of NATO Advanced Study Institute on Nanoscale Probes of the Solid/lLiquid Interface*, Sophia Antipolis, France, July 10–20, 1993, (Eds.: H. Siegenthaler, A. Gewirth), Kluwer Acad. Publ., Dordrecht, **1995**, p. 215.

(23) G. Staikov, K. Jüttner, W. J. Lorenz, E. Schmidt, *Electrochim. Acta* **1978**, *23*, 305.

(24) A. Bewick and B. Thomas, *J. Electroanal. Chem.* **1977**, *84*, 127.

(25) V. D. Jovic, J. Jovichevic, and A. R. Despic, *Electrochim. Acta* **1985**, *30*, 1455.

(26) M. Hölzle, *PhD Thesis*, University of Ulm, **1995**.

(27) J. W. Schultze and D. Dickertmann, *Faraday Symposia of the Chem. Soc.* **1977**, *12*, 36.

HYDROGEN EVOLUTION AT SILVER ELECTRODES BY HOT ELECTRONS

D. DIESING, G. KRITZLER, A. OTTO
Lehrstuhl für Oberflächenwissenschaft (IPkM), Heinrich-Heine-Universität Düsseldorf, D-40225 Düsseldorf, Germany

Electrons within silver, with energies up to 2 to 3 eV above the Fermi level of silver, are formed in $Al/Al_2O_3/Ag$ tunnel junctions. These hot electrons may induce electrochemical reactions at the silver-electrolyte interface. When the silver surface is smooth, one observes only injection of hot electrons into the conduction band of the electrolyte, whereas hydrogen is developed at rough and activated silver surfaces: at electrochemical potentials where this is thermodynamically possible, but usually kinetically hindered. Intermediate "wet electron" states are not involved in this reaction. There is an evidence for a 60% atomic hydrogen coverage of silver during the hydrogen evolution reaction.

1 Introduction

The role of photochemical processes is also important in electrochemistry. The most emphasis in electrochemical investigations was put on the semiconductor electrolyte interface due to its application in photovoltaics.

The number of investigations of photochemical processes on metal surfaces in UHV has increased since femtosecond lasers became available. Two reaction schemes are in discussion:

1. The incident photon leads to an intramolecular excitation of the adsorbate and for instance to desorption [1].

2. Photogenerated hot electrons of the metal substrate are transfered into empty electronic levels of the adsorbed molecules and may cause desorption, dissociation or a chemical reaction [2,3,4]. We have chosen metal/ insulator/metal junctions with thin film top electrodes as samples. With these samples hot charge carriers can be produced, which reach the thin film surface [5,6]. By this way of producing electronic excitations at metal surfaces with MIM junctions we are able to neglect the process described in [1] and focus our work on the hot charge carrier mechanism [7]. Electrochemical reactions at metal surfaces induced by hot electrons were first performed with redox couples like Fe^{2+}/Fe^{3+} [6].

In this work we focus on the hydrogen evolution reaction, further called HER, since this reaction is of outmost importance in creating a fuel of chemical free enthalpy not based on fossil fuels, but on nuclear or solar energy. HER has been investigated since the time of Tafel [8]. Apart of unravelling the multistep mechanism of the HER an understanding of the different electrocatalytic properties of the various elemental metal electrodes must be achieved by a surface science approach.

Modern textbooks of electrochemistry [9] present the electrocatalytic activity of metals in the HER from acid aquaous electrolytes in diagrams of exchange current versus the adsorption enthalphy of atomic hydrogen $E_{\text{M-H}}$ according to a theory of R. Parson [10] who expects a socalled vulcano curve, if the discharge reaction is rate determining and the adsorbed atomic hydrogen formed from molecular hydrogen obeys a Langmuir isotherm. Parsons traced the exchange current versus the the difference of the free energy ΔG° between H_2 and two adsorbed hydrogen molecules, which is related to $E_{\text{M-H}}$ by

$$E_{\text{M-H}} = \Delta G^\circ + \frac{1}{2}E_{\text{H-H}} \quad ; \quad \frac{1}{2}E_{\text{H-H}} = 216\text{kJ/mol}$$

The data in the vulcano curve in fig. 9.4 of[9] has been adopted from Trasatti [11] which again has as basis of the $E_{\text{M-H}}$ values the Table 1, column "$E^\circ_{\text{M-H}}$, calculated from a)" in[12]. The letter a in the column above means, that only the apparent activation energy was detemined from the temperature dependence of the electrolytic currents, see for instance [13]. Trassati took the values of the exchange current of socalled positively charged metals but extra"-polated the values of the exchange current of the socalled negatively charged metals (Cd, Pb, Zn, Cu, Au).

More recent data from metal-UHV interfaces collected by Christmann [14,15] give values of $E_{\text{M-H}}$ in the range 237-294 kJ/mol for all investigated metals, whereas the values in Fig 9.4 of [9] range between about 105 and 370 kJ/mol. Fig. 1 compares the vulcano curve (fig 9.4 in [9]) with its analogue using the exchange currents from the old figure but the data of $E_{\text{M-H}}$ collected by Christmann.

One would not be inclined to call the pattern of entries in the new figure a vulcano curve. Pt, Rh and Ir electrodes show a socalled underpotential adsorption of hydrogen at potentials anodic from the onset of the HER, and at least in the case of Pt it is believed that these species are not involved in the HER, but more loosely bound hydrogen species [16].

If this holds, data from the metal-vacuum interface would not be a useful entry into Parson's theory.

In the case of silver $E_{\text{M-H}}$= 171 kJ/mol in sulfuric acid and $E_{\text{M-H}}$= 159 kJ/mol in hydrochloric acid was derived by Bystrov and Krishtalik [21] from the measured apparent activation energies of 46 respectively 56 kJ/mol by assuming a barrierless ionization of adsorbed hydrogen [21].

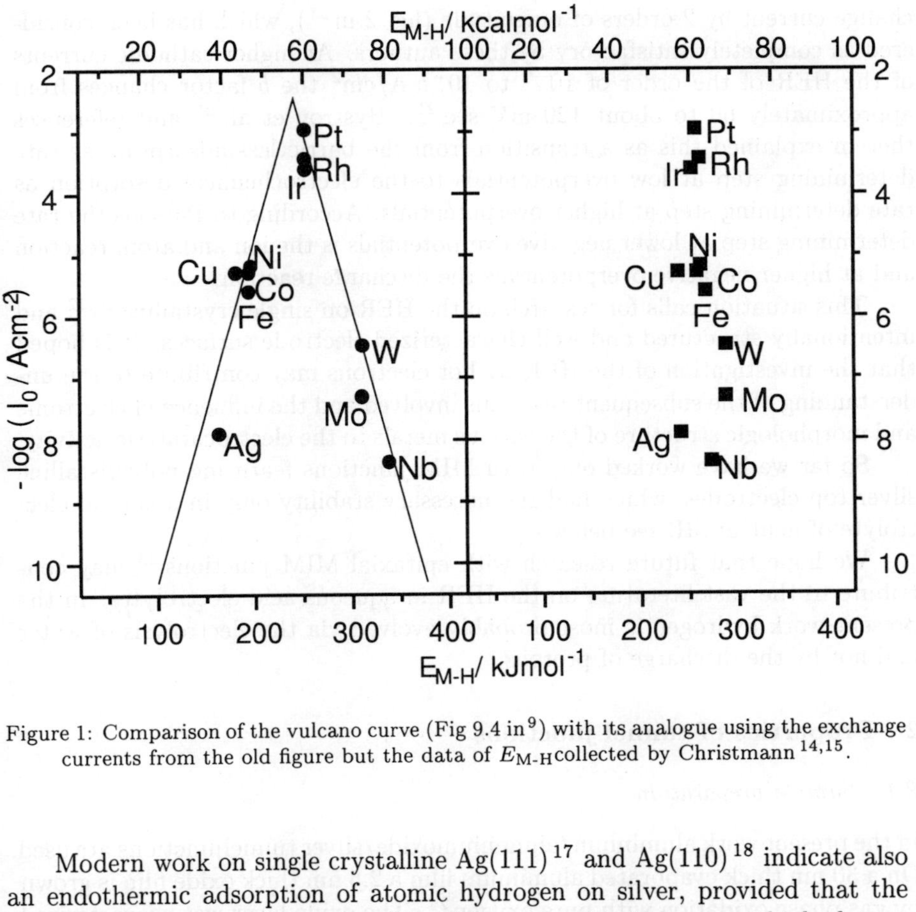

Figure 1: Comparison of the vulcano curve (Fig 9.4 in[9]) with its analogue using the exchange currents from the old figure but the data of $E_{\text{M-H}}$collected by Christmann[14,15].

Modern work on single crystalline Ag(111)[17] and Ag(110)[18] indicate also an endothermic adsorption of atomic hydrogen on silver, provided that the principle of detailed balance is valid. This may not hold because the hydrogen adsorbs at an unreconstructed silver surfaces but desorbs from a reconstructed one[17,18]. Nevertheless there is definitely an activation energy for desorption of the order of 25-42 kJ/mol for H/Ag(111) and a coverage of 0.6 on on the reconstructed Ag(111)[17]. Measured isosteric heats of hydrogen adsorption ΔG° are only known for copper films[19], in this case ΔG°= 40-50 kJ/mol, which seems to be compatible with the activation enthalpies of adsorption and desorption [20]. These facts are not easily reconciled with the assumption of a barrierless ionization of H/Ag at the electrolyte interface and a low hydrogen coverage of the silver electrode. However these points may hold for a hypothetical active hydrogen in analogy to the HER on Pt.

Especially for silver there is a scatter of experimental values of the exchange current by 2 orders of magnitude (fig. 2 in [21]), which has been considered as completely satisfactory by these authors. At higher cathodic currents of the HER of the order of 10^{-6} to 10^{-3} A/cm^2 the b factor changes from approximately 60 to about 120 mV see [21]. Bystrov et al. [21] and references therein explained this as a transition from the barrierless adsorption as rate determining step at low overpotentials to the electrochemical desorption as rate determining step at higher overpotentials. According to Parsons the rate determining step at lower negative overpotentials is the ion and atom reaction and at higher negative overpotentials the discharge reaction.

This situation calls for research on the HER on single crystalline [15,22] and intentionally structured and well characterized electrode surfaces. It is hoped that the investigation of the HER by hot electrons may contribute to the understanding of the subsequent reactions involved and the influence of electronic and morphologic structure of the various metals to the electrocatalytic activity.

So far we have worked only with MIM junctions featuring polycristalline silver top electrodes, which had the necessary stability only in a special electrolyte of neutral pH, see below.

We hope that future research with epitaxial MIM junctions [23] may contribute to the vast literature on the HER in aqueous acid electrolytes. In the present work hydrogen is most probably evolved via the electrolysis of water and not by the discharge of protons.

2 Properties of tunnel junctions

2.1 Sample preparation

In the present work aluminium/aluminiumoxide/silver tunneljunctions are used. On a 30 nm thick evaporated aluminium film a 2.5 nm thick oxide film is grown by gas phase oxidation with pure oxygen [24]. The oxide layer was characterized by impedance measurements, the dielectric permittivity ϵ_R of the oxide film was determined to $\epsilon_R = 11$ [21]. Then a 15 nm thick Ag film (99.999 %) is deposited as shown in [6], see also fig. 2. The details of the evaporation conditions and the sample preparation have been given elsewhere [6,24].

2.2 Characterization of the tunnel junction

Fig. 3 shows the one-electron-energy levels and the thicknesses of the 3 layers of an Al/AlOx/Ag-junction. The silver top electrode is grounded. The applied tunnel voltage is assumed as $U_T = -2.3$ V. The negative value of the voltage means that the electrons tunnel from the aluminium side through the oxide layer to the silver side. This tunnel current i_T is detected with an ampermeter.

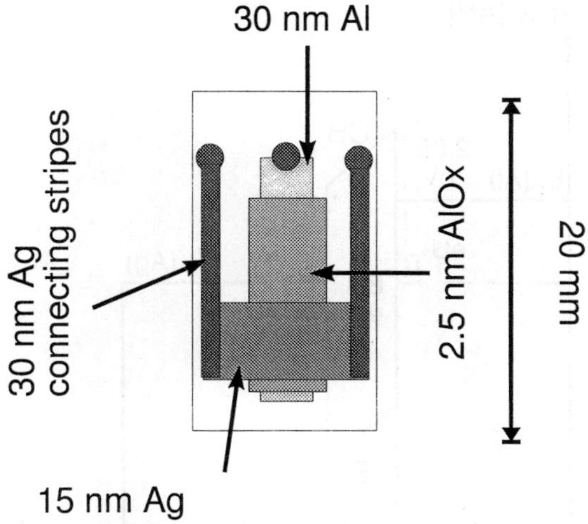

Figure 2: Configuration of Al/AlOx/Ag tunnel junctions and electric connecting stripes, prepared on glass slides.

There are many different values in literature for the barrier height between the metals and the oxide conduction band. For instance, Shepard found a value of 1.1 eV[25] for the contact voltage of aluminium and aluminumoxide, Shu and Ma 2.9 eV[26]. Most often the contact voltages are measured by the photo current method. Here we apply an increasing absolute tunnel voltage $|U_T|$ and measure the tunnel current i_T. If $|U_T|$ becomes greater than the contact voltage the thickness of the tunnel barrier decreases for the hot electrons and socalled Fowler-Nordheim-tunneling sets in[27]. This can be detected with the change of the slope in the current-voltage characteristic.

Fig. 4 shows the $\log(|i_T|)$ versus $|U_T|$ characteristic measured at a temperature T = 77 K. Curve a) shows a change of the slope at a tunnel voltage $U_T = +2.05$ V, indicating the value of the energetic distance between the Fermi level of aluminum and the lower edge of the conduction band of aluminumoxide (we will call it the contact voltage, note this is not necessarily the difference between the work function of Al and the electron affinity of aluminium oxide, because of an unknown dipole layer between both media), see fig. 3. Curve b) shows the change of the slope at $U_T = -2.71$ V. This is the equivalent value at the silver aluminum oxide interface[28], see fig. 3.

The position of the valence band of the aluminum oxide is not exactly known, but we expect electron correlation effects to be small and therefore we position it according to the optical bandgap[29] 8.3 eV below the conduction

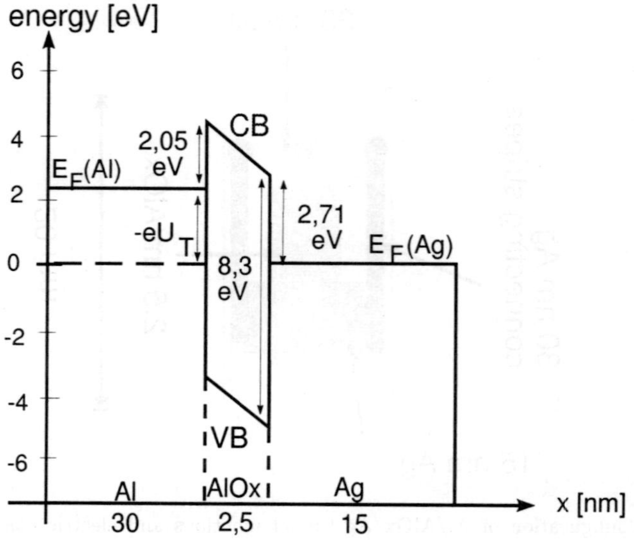

Figure 3: Energy levels and layer thicknesses of Al/AlOx/Ag-junctions. CB is the lower edge of the conduction band and VB the upper edge of the valence band of the aluminumoxide (energy axis in true scale).

band. Since the offset voltages of the valence band with respect to the Fermi levels in the metals is larger than the contact voltages discussed above, tunnelling via the conduction band of the AlOx will prevail in any case over hole tunnelling via the valence band. However, electron tunnelling in the direction from Ag to Al via the conduction band at positive bias is equivalent to hole tunnelling from Al towards Ag. This becomes observable, if a suitable hole acceptor is present in the electrolyte, see [6] and fig. 6 therein. This will not be further discussed in this article.

3 Electrochemical reactions of hot electrons

3.1 Electrochemical set-up

In the following the 15 nm thick silver top electrode MIM junction will be made the working electrode in an electrochemical cell. By changing its electrochemical potential different states of the metal-electrolyte interface, consisting of the metal surface, adsorbates and fully or partly hydrated ions, are achieved. If the electrochemical potential is changed by x V, the levels are shifted by x eV. In this way electron transfer processes are controlled by the setting of the

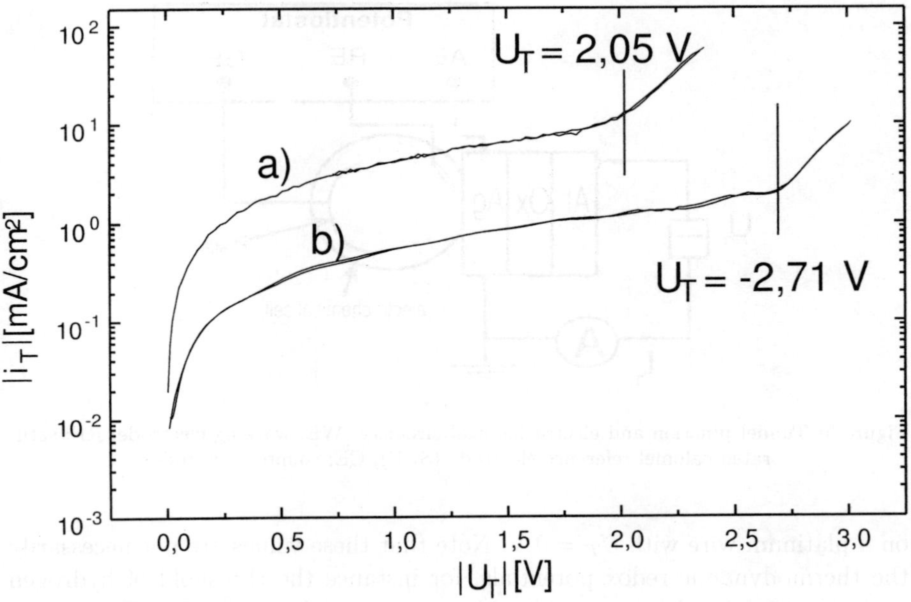

Figure 4: Current-voltage characteristic of a Al-AlOx-Ag-junction at T = 77 K [28] a) at positive bias, b) at negative bias.

electrochemical potential.

To perform these measurements two electrical circuits are necessary (see fig. 5). One circuit contains the voltage source U_T and the ampermeter to control the current i_T through the tunnel junction. For experiments with pulsed tunnel voltages the voltage source U_T is replaced by a pulse generator. The other circuit controls the electrochemical potential E_{SCE} of the silver top electrode of the MIM junction, which is made the working electrode whereas the counter electrode is a platinum wire. The current measurement system contains a three step current voltage converter with a time delay of 300 ns. We use the saturated calomel electrode (SCE) as reference electrode. The silver working electrode is the common ground of both circuits. The active tunnel area of the samples is typically 0.12 cm^2, which is in contact with the electrolyte. Outside this area the sample is covered with a protective lacquer.

Fig. 6 shows the electron energy scheme of the tunnel junctions in the electrolyte, the threshold PE for photoemission into an aqueous electrolyte at 0 V_{SCE}, and the threshold electrochemical potentials (versus SCE) of the hydrogen evolution HER observed with our MIM junctions at $U_T = 0$ V and

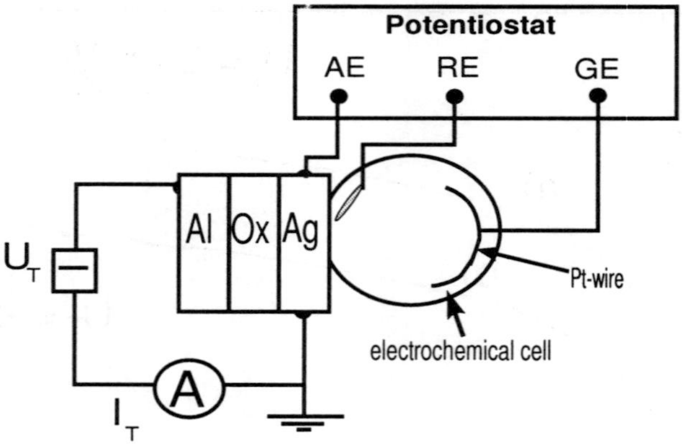

Figure 5: Tunnel junction and electrochemical circuitry. WE: working electrode, RE: saturated calomel reference electrode (SCE), CE: counter electrode.

on a platinum wire with $U_T = 0$ V. Note that these values are not necessarily the thermodynamic redox potentials, for instance the threshold of hydrogen evolution may involves a considerable overpotential, see [22]. The Fermi level of the saturated calomel reference electrode E_{SCE} is given in fig. 6 at the electrochemical potential of 0 V_{SCE}. The threshold potentials are given on the scale of the one electron energy levels in the MIM junction, but they should not uncritically be mistaken as the levels of electronic acceptor as donor states.

3.2 Electron injection into electrolytes on smooth thin silver films

In the following experiments we insert a tunnel junction with a smooth silver film top electrode in a 0.9 M NaAc-buffer (pH=5.9) with 50 % water and 50 % ethylene-glycol after de-airing with pure nitrogen for 30 minutes. This partially organic electrolyte was chosen because the MIM junctions are not as easily failing under hydrogen evolution than when inserted into a pure aqueous electrolyte.

Fig. 7 shows the electrolyte current i_{El} of the silver top electrode of the tunnel junction versus applied tunnel voltage U_T for three electrochemical potentials E_{SCE}. There appears a clear threshold of U_T, shifting with E_{SCE} below which electrons are injected into the electrolyte. These threshold values are well approximated by the relation $E_{SCE} + U_T = -3.3$ V. This value of -3.3 V is in good agreement with the lower boundary of the electron conduction band in aqueous electrolytes [31].

The overall transfer rate of hot electrons, taken as ratio of the extra current

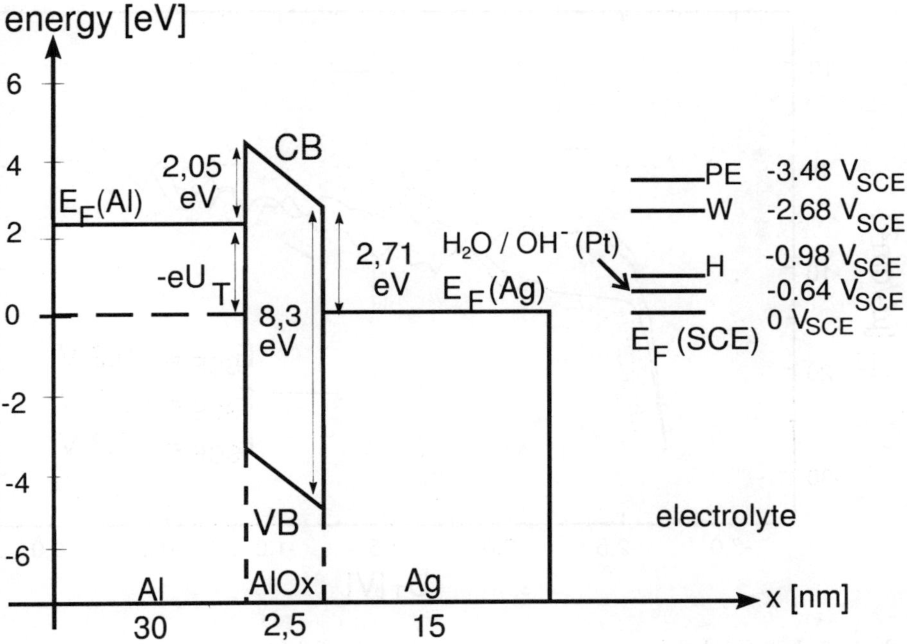

Figure 6: Energy scheme of the tunnel junction (in eV) and of the lower edge PE (= D in [30]) of the electron conduction band and of the onset potential of the hydrogen evolution at a platinum electrode H_2O/OH^-, both measured in 0.9 M NaAc-buffer (pH=5.9) with 50 % water and 50 % ethylene-glycol and the position of the lower edge W of the socalled wet electron state and the center H of the hydrated state of the electron in an aqueous electrolyte [30], all levels at the electrochemical potential of the top silver electrode of 0 V_{SCE}. (In this case the Fermi levels of the silver electrode and of the metallic mercury in the SCE (E_F (SCE)) are at the same level). 1 eV on the energetic scale corresponds to 1 V_{SCE} on the electrochemicalpotential scale.

into the electrolyte divided by the tunnel current is around 10^{-4} til 10^{-3}. In the following this transfer rate is assigned as R.

The clear threshold for the onset of the cathodic current in fig. 7 demonstrates, that primary tunnelling electrons have been injected into the conduction band, but that electrons of lower energy than the position of the electronic conduction band in water are not injected into the electrolyte or are not scavenged in this electrolyte and return to the electrode. This behaviour changes, when the silver electrode is activated, see the following chapter.

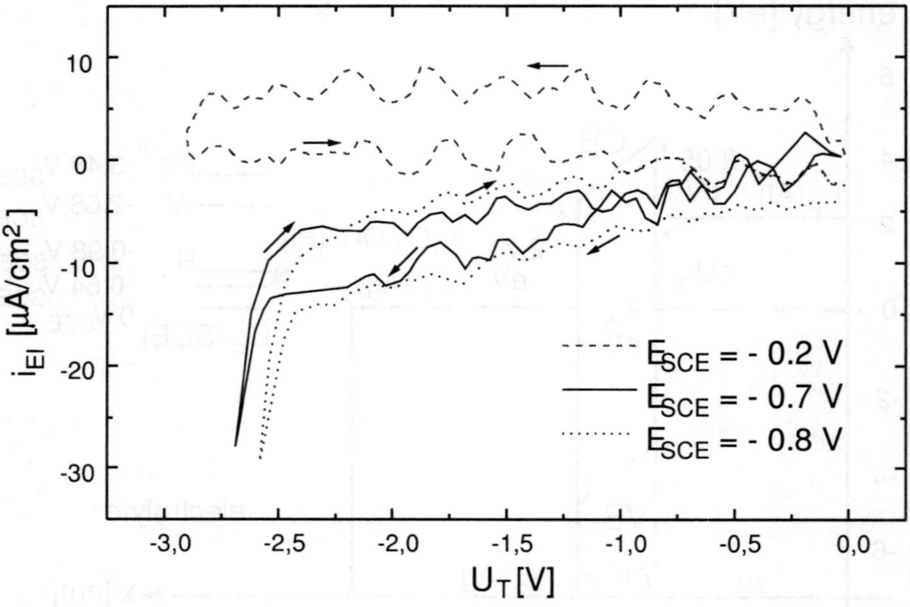

Figure 7: Injected electron current density i_{El} as function of the tunnel voltage U_T at various electrochemical potentials E_{SCE}.

3.3 Reaction with hot electrons on activated silver surfaces

When the MIM junctions are intentionally produced on a rough substrate or activated (see below) hydrogen gas evolves by a hot electron process, at electrochemical potentials, where neither the normal hydrogen evolution reaction by "cold" electrons nor the injection of primary hot electrons into the electronic conduction band of the electrolyte (see section 3.2) is possible.

The roughness of the silver surface is obtained by two methods:

1. Undercoating the tunnel junctions with a CaF_2 layer of 100 nm thickness. Capacity measurements of the metal/electrolyte interface which can be performed with current transients after potential steps yield differential capacities of 19 $\mu F/cm^2$ of the smooth and 21 $\mu F/cm^2$ of the CaF_2 undercoated surface. Accordingly the accessible surface area is increased by about 10 % by the undercoating.

2. Socalled activation of the silver top electrode in the same electrolyte by a potential pulse to 0.5 V_{SCE}, with the oxidation and reduction charge always below 3 mC/cm^2.

Both methods deliver equal experimental results in the hydrogen evolution.

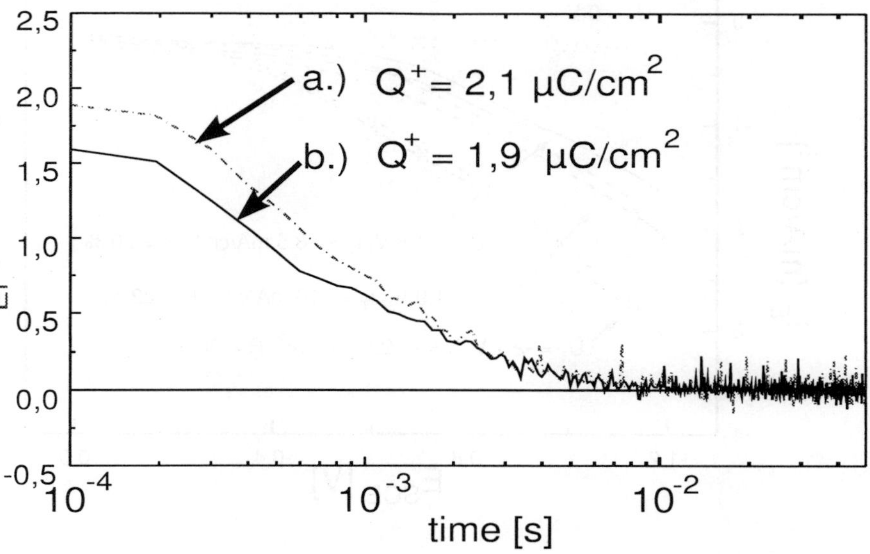

Figure 8: Current transients and corresponding charge Q^+ after a potential step from $E_{SCE} = -0.9$ V to -0.8 V in a 0.9 M NaAc-buffer with 50 % water and 50 % ethylene-glycol of a smooth MIM junction b.) and a MIM junction on 100 nm CaF_2 a.).

Fig. 9 shows voltammograms of a tunnel junction prepared on CaF_2 at tunnel voltages $U_T = -1.8$ V, -1.9 V and -2.1 V [6].

The voltammograms depend on the applied tunnel voltage even though the hot electrons are certainly below the threshold of injection into the electronic conduction band of the electrolyte as discussed above. The onset of the hot electron induced cathodic current is shifted 300 mV to positive potentials when the tunnel voltage is raised by 300 mV. The ratio R, defined in section 3.2, increases at $E_{SCE} = -1.2$ V from $R = 20\%$ at $U_T = -1.8$ V to $R = 26\%$ at $U_T = -2.1$ V. These R values are about a factor of 100 larger in comparison to the measurements with "smooth" tunnel junction surfaces.

Nevertheless we cannot exclude from these experiments that the roughness of the silver electrode opens direct pathways to the wet electron states, which are composed of the localized nondegenerate p states of the electron, broadend by the fluctuations of the electrolyte, which in turn may be scavenged, leading by some unknown subsequent reactions to hydrogen evolution. However we can exclude bielectronic reactions, for instance:

$$e^-_{aq} + e^-_{aq} + 2H_2O \rightarrow H_2 + 2OH^-$$

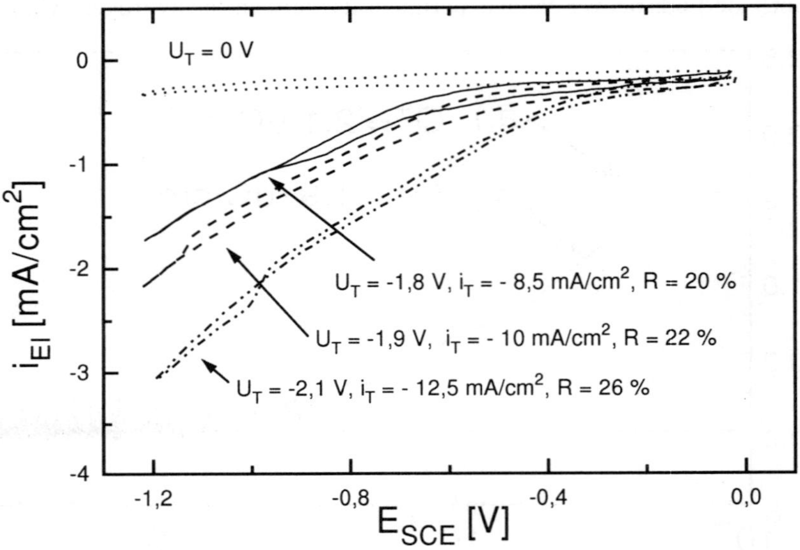

Figure 9: Cyclic voltammograms of a silver top electrode of a MIM junction on CaF_2 in a 0.9 M NaAc-buffer with 50 % water and 50 % ethylene-glycol at tunnelling voltages 0 V, -1.8 V, -1.9 V and -2.1 V. The corresponding tunnelling currents at $E_{SCE} = -1.2$ V are indicated. For R see text.

since the initial rate of hydrogen evolution is proportional to the cathodic charge, see section 3.4 and fig. 14 below.

In order to exclude pathways via the wet electron states we performed analogeous experiments at tunneling voltages U_T between 0 and -0.82 V, see fig. 10 into which the voltammograms of fig. 9 at more negative U_T have been included.

Clearly the energy of the hot electrons at the low tunneling voltages are considerably below the wet electron states, nevertheless about the same fraction R of the tunneling current is continuing as cathodic current into the electrolyte, as is evident by comparing the ratios R given in fig. 10. The hot electrons overcome the kinetic hindrance of the HER, the hot electron reaction takes place in the electrochemical potential range between the potentials of the HER at platinum at about -0.74 V_{SCE} and at silver at about -1.2 V_{SCE}. However weak reactions are observed even at $E_{SCE} > -0.74$ V. This may be

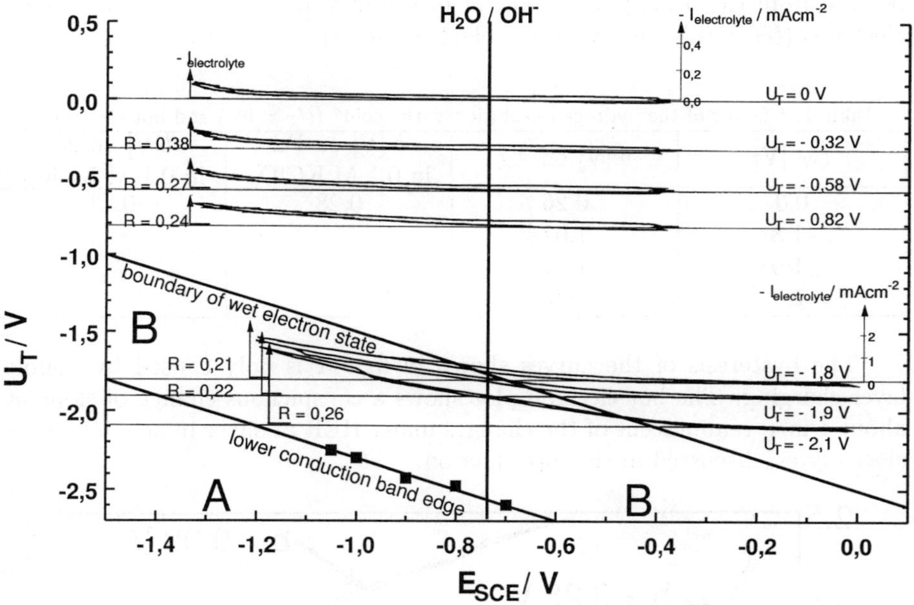

Figure 10: Region A: Range of tunneling voltage U_T and of electrochemical potential E_{SCE} in which electron emission into the electronic conduction band of the electrolyte is possible. Range B: Range of U_T and E_{SCE} in which electronemission into the wet state (p type state) of the elctron in the electrolyte must be considered. Seven voltammograms (Cathodic current as function of E_{SCE}) at different values of U_T. The base lines ($i_{\text{electrolyte}} = 0$) of these are positioned at the proper values of U_T. R is the ratio of $i_{\text{electrolyte}}$ induced by the tunneling current i_T and i_T at the most negative values of U_T. The vertical line H_2O/OH^- is the onset potential of the hydrogen evolution for a Pt electrode in the same electrolyte.

caused by the overpotential of H_2 oxidation if H_2 is created by a hot electron mechanism. The overpotential for H_2 oxidation is for instance confirmed by the absence of an according anodic charge after a long HER and a potential step to $E_{SCE} = -0.2$ V. This overpotential for H_2 adsorption at the silver electreode corresponds to the experimentally well confirmed activation barrier for H_2 absorption at noble metal-vacuum interfaces, for instance for copper [32,33].

The fully hydrated s-like state of the electron (level H in fig. 6) plays no role in the HER because it is a stable state not known to split water according to reaction (1) below.

The hydrogen evolution may be characterised by the b factor

$$b = -\frac{dE_{SCE}}{d\log(i_{El})}$$

derived from the Tafel-plots in fig. 11 of the hydrogen evolution with "cold" electrons ($U_T = 0$ V) and with hot electrons at $U_T = -2.1$ V.

Table 1: b factor of the hydrogen evolution with "cold" ($U_T = 0$V) and hot electrons.

U_T [V]	b [V] on Ag	b [V] on Pd in 0.1 M $KClO_4$	b[V] on Ni in 0.1 M $KClO_4$
0.0	0.26	0.28	0.23
-1.8	1.07		
-1.9	1.01		
-2.1	0.9		

The hysteresis of the curves shown in fig. 11 is only caused by double layer charging. The hot electron plot shows a conspicuous change of slope at about 1 mA, reminescent of the changes under HER at silver in acid aqueous electrolytes, discussed in the introduction.

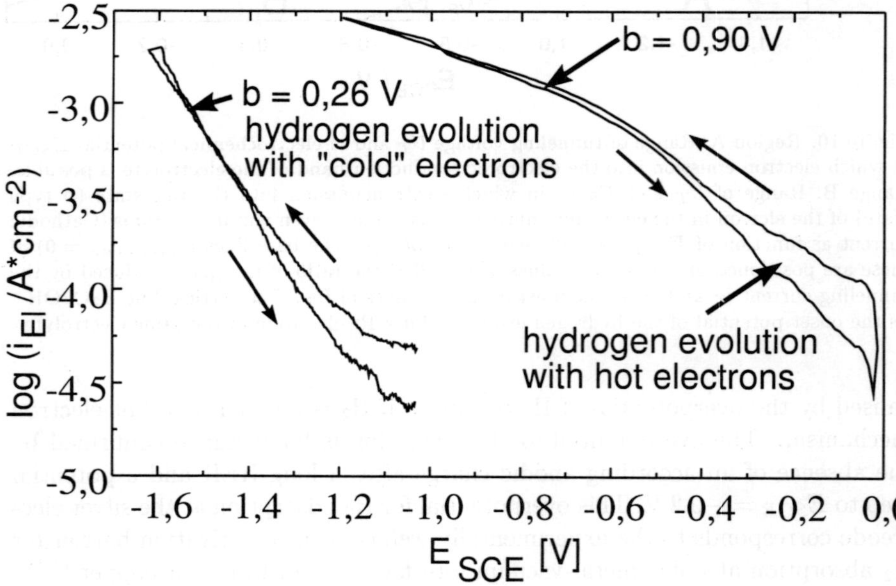

Figure 11: Tafel plots and b-factors of the hydrogen evolution by "cold" electrons and hot electrons at $U_T = -2.1$ V, in a 0.9 M NaAc-buffer with 50 % water and 50 % ethylene-glycol. Scan velocity $dE/dt = 50$ mV/s.

The relation between the b factor and the tunnel voltage shows clearly that the dynamic of the hydrogen evolution reaction is determined by the hot electrons. It should be noted that at a tunnel voltage $U_T = -2.1$ V hot electrons reach the threshold for injection into the electrolyte only at $E_{SCE} < -1.2$ V (see fig. 10). Therefore the hydrogen evolution by hot electrons in fig. 9 is not a scavenger process.

The b factor of the hydrogen evolution with hot electrons above a current of about 1 mA/cm^2 is about four times greater than in the case of HER by "cold" electrons. The values of b for the MIM junctions and different tunnel voltages U_T and for "cold" electron HER on a Pd and Ni wire in a neutral $KClO_4$ electrolyte are given in tab.1.

We assign the values for the b factors of around 200 mV for cold electron HER to mass transfer (diffusion limited reaction) but not to the reactions at the metal electrolyte interface. On the other hand the hydrogen evolution with hot electrons shows much greater values for the b factors. Therefore this process is not limited by the diffusion.

3.4 Experiments with pulsed tunnel voltages

Fig. 12 shows a voltammogram with $U_T = -2.1$ V and a second voltammogram with $U_T = 0$ V in which at the times A and B a tunnel voltage $U_T = -2.1$ V is switched on, and is switched off at times C and D. One can observe the reversibility between reactions of the hot electrons and the "cold" electrons.

After switching off at C and D there are transients of the anodic current. The following experiments study these phenomena with pulsed tunnel voltages and a time resolution of 1 μs.

Fig. 13 shows i_{El} versus the logarithm of time after the application of three pulses of the tunnelling voltage U_T from 0 V to -3 V and back to $U_T = 0$ V after 10^{-2} s, 10^{-1} s and 1 s. The electrochemical potential is $E_{SCE} = -0.8$ V. The measurements demonstrate that the electrolyte current reaches a constant value after 10^{-3} s, corresponding to R = 0.33. This time is given by the charging of the MIM capacity and thus by the transient of i_T, as measured separately by impedance spectroscopy and current transient spectroscopy of the MIM junction [28]. This delay is 10 to 100 times smaller than the charging time of the double layer capacity [28,34]. The electrolyte current driven by the hot electrons is constant after a charge transfer of less than 1.5 μC, which corresponds to less than 10^{-2} of a monolayer of ions. Both the short time and the small charge exclude ionic effects within the oxide barrier driven by the electrochemical potential. Apparently the silver film is tight enough to not allow potential modulated migration of ions or atoms from the silver-electrolyte interface into the tunnelling barrier, influencing the tunnelling current.

After switching off the tunnel voltage there is an anodic current transient

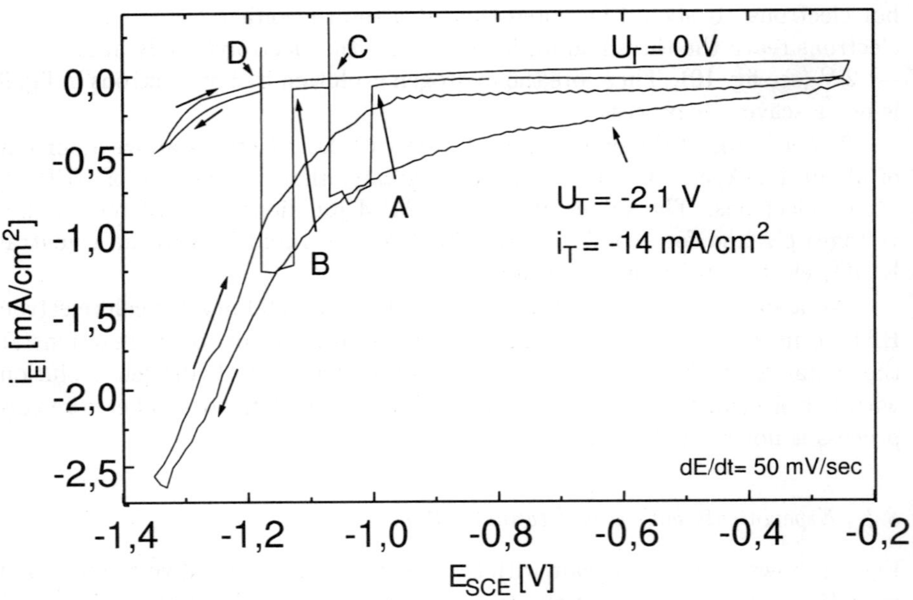

Figure 12: Switching between reactions of hot electrons and "cold" electrons, see text.

within about 10^{-1}s. The related charge density is $120\mu C/cm^2$ in all the three cases in fig. 13, though the cathodic charges injected by the pulses is different: by pulse a) of 1 s duration 15 mC/cm^2 but by pulse c) of 0.01 s duration only 150 $\mu C/cm^2$. Further experiments with shorter pulse durations (5 ms up to 100 ms) show that the anodic charge density saturates as function of the pulse duration at a value of 115 $\mu C/cm^2$. This is shown in fig. 14 for the example of $U_T = -2.2$ V. Up to the cathodic charge of 115 $\mu C/cm^2$ all the charge is recovered in the anodic transient.

Similar results with an a similar saturation density of the anodic charge could be delivered after cathodic potential pulses on a slightly activated Ag(111) surface in a pure aqueous electrolyte of $KClO_4$ [22]. From both studies follows, that the state of the silver top electrode of the MIM junction after switching off the negative tunnel voltage is equivalent to the state of a normal metal electrode after ending the cathodic polarization.

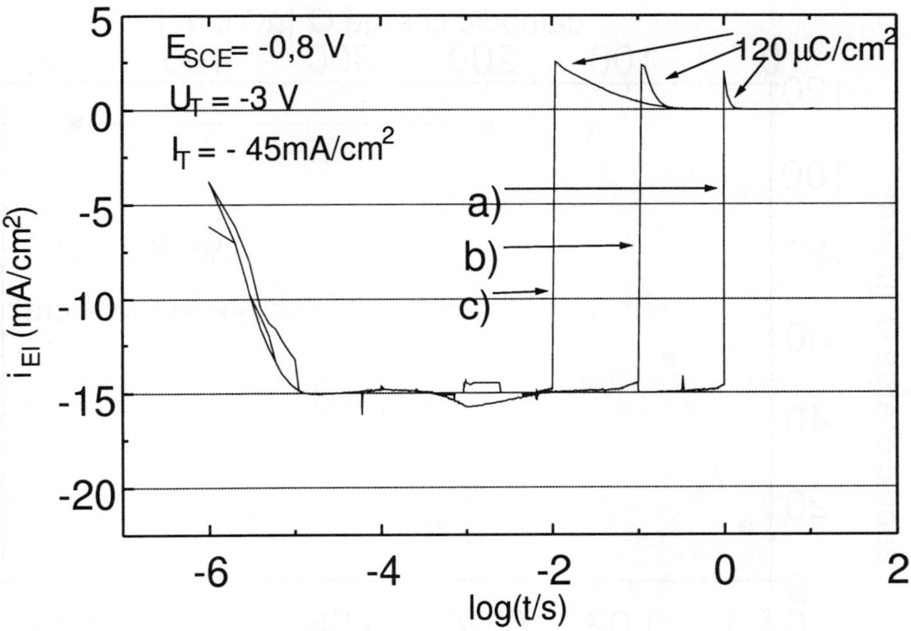

Figure 13: Electrolyte current during and after pulses of U_T from 0 V to -3.0 V and back at constant E_{SCE}. Time t starts at the onset of the pulses. The electrolyte is 0.9 M NaAc in 50 % water, 50 % ethylene-glycol.

4 Discussion

In the present work we have shown, that the overpotential of the HER in neutral solutions can be reduced by electronically excited electrodes.

The first step in the HER at a silver electrode in neutral electrolytes is the Volmer reaction, which is given in neutral electrolytes by [22]

$$H_2O + e^- \rightarrow OH^- + H_{ad} \tag{1}$$

because the proton discharge cannot prevail due to the low proton concentration. For the same reason the socalled electrochemical desorption or Heyrovski reaction (in surface sciene called Eley Rideal reaction)

$$H_{ad} + H^+ + e^- \rightarrow H_2(\text{dissolved}) \tag{2}$$

is unlikely.

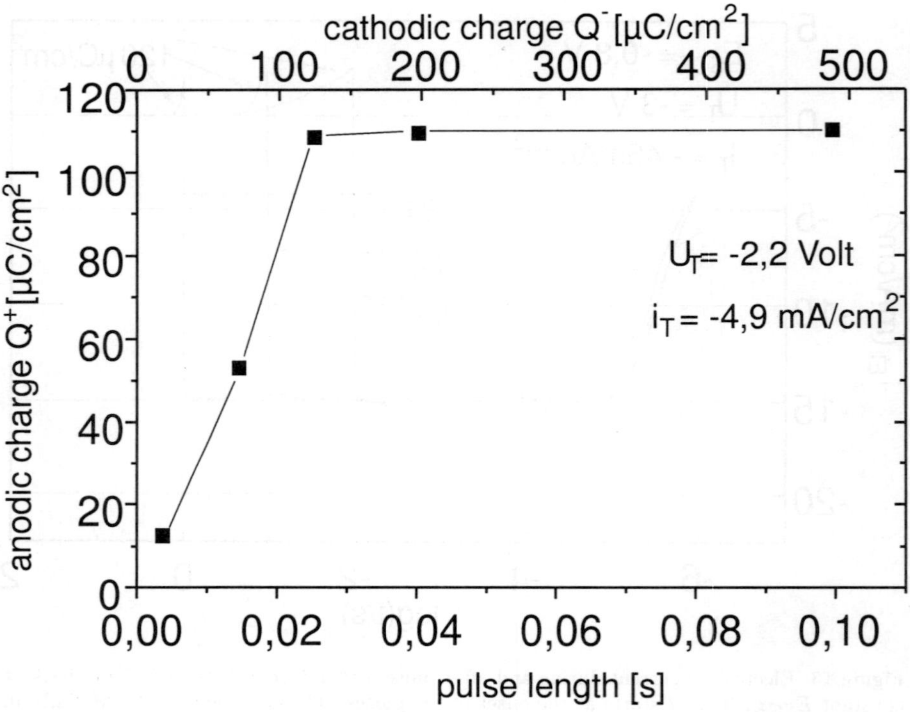

Figure 14: Charge density Q^+ in the anodic transient versus hot electron pulse length (below) and corresponding cathodic charge density Q^- (above), calculated from the indicated values of U_T and i_T.

Rather the so-called catalytic reaction or Tafel reaction (in surface science called a Langmuir Hinshelwood reaction)

$$H_{ad} + H_{ad} \rightarrow H_2 \tag{3}$$

is prevailing.

(We cannot exclude that the layer of H_{ad} formed in reaction (1) is not involved in the further reactions as it is believed for the hydrogen adsorbed at underpotential on Pt. In this case reaction (3) should be replaced by

$$H_{active} + H_{active} \rightarrow H_2 \tag{4}$$

where H_{active} is assumed to be the same species involved in HER as on Pt).

We believe that the sequence of the steps of HER induced by hot electrons and by cold electrons are the same. Apparently, the mass transport of the OH^-

into the electrolyte is not rate determining during the HER by hot electrons, but rather the pathway of the hot electrons to H_2O in the inner Helmholtz layer and the subsequent reaction (1) and/or (4). From the fast response of the HER to a pulse of hot electrons follows, that no anions or cations other than OH^- are transported into or out of the double layer during this process.

The rather large *b* factor for the HER with "cold" electrons might be caused by diffusion or coadsorption effects of ethylene-glycole or acetate anions or hydroxyl on the silver surface. The oxidation reaction at the end of the hot electron pulses, which also has been observed after pulses into the potential range of "cold" electron HER with "activated" MIM junctions at zero tunneling voltage[35] and at slightly roughened Ag(111)[22] may be explained in two ways:

1. The layer of H_{ad} is oxidized and protons are released into the electrolyte and form H_2O with the OH^- ions produced during the HER (1):

$$H_{ad} + OH^- \rightarrow H_2O + e^-(Ag) \quad (5)$$

2. After the end of the HER by hot or "cold" electrons the surface coverage of H_{ad} desorbs by reaction (3), the new empty sites are covered within 0.1s by hydroxyl groups formed by the reaction:

$$Ag + OH^- \rightarrow Ag - OH + e^-(Ag) \quad (6)$$

This process implies the assumption that silver is covered by hydroxyl groups at neutral and basic pH.

Since during and after the HER by hot electrons the electrochemical potential is constant this implies in any case that reaction (6) runs also during the HER by hot electrons, thus competing with the cathodic reaction (1).

Though we have no means as yet to differentiate between these mechanisms we maintain, that under the HER at activated silver electrodes in neutral and basic aqueous electrolytes the (111) facets are covered by atomic hydrogen with a coverage of about 60% correponding to that found at Ag(111) in ultra high vacuum at low temperatures[17].

We have no firm explanation for the missing HER by hot electrons at smooth MIM junctions (see fig. 6 above) nor for the missing anodic transient after steping out of the HER at smooth MIMs and at smooth Ag(111) films[22]. One cannot argue that the anodic transient is H_{ad} at "defect sites", because the rather high coverage of about 60 % excludes this explanation. Rather we are looking for explanations based on new pathways for electrons opened by the introduction of atomic scale roughness.

Hydrogen adsorption reconstructs the Ag(111) surface[17]. This raises the question whether a Ag(111) electrode is reconstructed during the HER. The missing change of the the electrical surface resistance of Ag(111) films in 0.1 M $KClO_4$ aqeous electrolyte [22] during the HER may be compatible with an ordered hydrogen adsorption and an ordered reconstruction of the film, because elastic scattering of electrons at the Fermi level by a transfer of momentum parallel to the surface corresponding to the reconstruction is possible only for a small fraction of the electrons (no "nesting"). We think the search for extra diffraction spots would be a worthwile experiment for in situ surface X-ray diffraction during HER.

In a former work it was shown, that the overpotential of the HER on monocrystalline silver surfaces in neutral aqueous solutions depends on the crystallograhic orientation [22]. Therefore our aim is to investigate the change of HER in pure aqeous electrolytes with respect to the energetic level of the electrons of the metal (hot electron injection) at electrodes of various single crystallographic orientations.

Acknowledgment

We thank K. Christmann and W. Schmickler for discussions.

5 References

1. P.Feniner, D.Menzel, in *Laser Spectroscopy and Photochemistry on Metal Surfaces*, edited by H.-L. Dai and W. Ho (World Scientific, Singapore, 1995), p.625
2. W.Ho, J. Phys. Chem. **100**, (1996) 13050-13060
3. D. G. Busch and W. Ho, Phys.Rev.Lett. **77**, (1996) 1338-1341
4. J. A. Misewich, T. F. Heinz, P. Weigand, and A. Kalamarides, in *Laser Spectroscopy and Photochemistry on Metal Surfaces*, edited by H.-L. Dai and W. Ho (World Scientific, Singapore, 1995), p.764
5. H. Janssen, D.Diesing and A. Otto, Surf.Sci. **331-333**, (1995) 1267
6. D. Diesing, H. Janssen, A. Otto, Surf.Sci. **331-333**, (1995) 289
7. A. Otto, D. Diesing, H. Janssen, M. Hänisch, A. Schaak, S. Schatteburg, D. Körwer, G. Kritzler, H. Winkes. S. Rüße, M.M. Lohrengel, Surface reactions of hot electrons at Ag/UHV and Ag/electrolyte interfaces. IUPAC *Chemistry for the 21st Century* Monograph *Interfacial Science* editor M.W. Roberts, in press.
8. J. Tafel, Z.Phys.Chem. **50**, (1905) 641
9. W. Schmickler, Interfacial Electrochemistry, Oxford University Press 1996
10. R. Parson, Transactions Faraday Soc. **54**, (1958) 1053

11. S. Trassatti, J.Electroanal.Chem. **39**, (1972) 163
12. L. I. Krishtalik in *Adv.Electrochem. Electrochem.Engin.* , ed.P.Delahay, Vol.7 (1970) 283
13. U. Frese, W. Schmickler, Ber.Bunsenges.Phys.Chem. **92**, (1988) 1412
14. K. Christmann, Surface Science Reports **9**, (1988)
15. K.Christmann in *Electrocatalysis*, ed.P.N.Ross, J.Lipkowski, in press
16. A. Peremans, A. Tadjeddine, Phys.Rev.Letters **73**, (1994) 3010
17. G. Lee,E. W. Plummer, Phys.Rev. **B51**, (1995) 7250
18. K. Christmann, private communication
19. C. S. Alexander, J.Pritchard, Faraday Transactions I **68**, (1972) 202
20. U. Bischler, P.Sandl, E. Bertel, T.Brunner, W.Brenig, Phys.Rev.Lett. **70**, (1993) 3603
21. I. Bystrov, L. I. Krishtalik V Elektrokhimiya **3**, (1967) 1345
22. D. Diesing, H. Winkes, A. Otto. Phys. stat. sol.(b), in press
23. A. Schaak, Diploma Thesis, Heinrich Heine University Duesseldorf 1995.
24. D. Diesing, S. Rüße and M. M. Lohrengel, Proc. Electroceramics IV, Vol II (Verlag der Augustinus Buchhandlung Aachen, 1994) p. 1295.
25. W. Shepard, J. Appl. Phys. **36**, (1965) 796
26. Q. Q. Shu, W. G. Ma, Appl.Phys.Lett. **61**, (1992) 2542
27. R. H. Fowler, L. Nordheim, Proc.Roy.Soc. **A119**, (1928) 173
28. D. Diesing, Dissertation, Düsseldorf 1996
29. American Institute of Physics Handbook, 3.ed., chap 9, p 20, McGraw-Hill 1972
30. A. Staib, D. Borgis, J.Chem.Phys. **103(7)**, (1995) 2642
31. T. E. Furtak, K.L. Kliewer, Comments on Solid State Physics **4**, (1982) 103
32. K. D. Rendulic, A. Winkler, Surf.Sci. **299-300**, (1994) 261
33. A. Gross, Surf.Sci. **363**, (1996) 1
34. G. Kritzler, Diploma Thesis, Heinrich Heine University Duesseldorf, 1996
35. D. Diesing, S. Rüße, A. Otto and M.M. Lohrengel, Ber.Bunsenges Phys.Chem. **99**, (1995) 1402

DIRECT, COHERENT AND INCOHERENT INTERMEDIATE STATE TUNNELING AND SCANNING TUNNEL MICROSCOPY (STM)

J. Halbritter
Forschungszentrum Karlsruhe, Institut für Materialforschung I, Postfach 3640, 76021 Karlsruhe, Germany

Theory and experiment in tunneling are still qualitative in nature, which hold true also for the latest developments in direct-, resonant-, coherent- and incoherent-tunneling. Those tunnel processes have recently branched out of the field of 'solid state tunnel junctions' into the fields of scanning tunnel microscopy (STM), single electron tunneling (SET) and semiconducting resonant tunnel structures (RTS). All these fields have promoted the unterstanding of tunneling in different ways reaching from the effect of coherence, of incoherence and of charging in tunneling, to spin flip or inelastic effects. STM allows not only the accurate measurements of the tunnel current and its voltage dependence but, more importantly, the easy quantification via the (quantum) tunnel channel conductance and the distance dependence. This new degree of freedom entering exponentially the tunnel current allows an unique identification of individual tunnel channels and their quantification.

In STM measurements large tunnel currents are observed for large distances d > 1 nm explainable by intermediate state tunneling. Direct tunneling with its reduced tunnel time and reduced off-site Coulomb charging bridges distances below 1 nm, only. The effective charge transfer process with its larger off-site and on-site charging at intermediate states dominates tunnel transfer in STM, biology and chemistry over distances in the nm-range. Intermediates state tunneling becomes variable range hopping conduction for distances larger than d$>$ 2 nm, for larger densities of intermediate states $n_I(\varepsilon)$ and for larger temperatures T or voltages U, still allowing high resolution imaging.

1. INTRODUCTION

Tunnel barriers are of growing importance in research and applications. Reviews exist mainly on voltage and temperature spectroscopy of direct tunnel channels of solid, extended junctions,[1,2] resulting in barrier shapes $\phi(x,y)$ and density of states $n_m(E)$ in the metallic banks. Corresponding reviews on scanning tunnel microscopy (STM), single electron tunneling (SET) or resonant tunnel structures (RTS) devices are not known to the author.

Barriers are never 'perfect' and intermediate states make them transparent at specific energies even for barrier widths $d_o = 2d$ above 2 nm.[2-7] This became obvious by the new STM distance tunnel spectroscopy (DTS).[7] There, the slow decrease of the tunnel current I(d) with d at low voltages ($eU \ll \phi_o$).

$$I(d) \propto \exp(-2\,\kappa_{eff}\, d_o)\ ;\ \kappa_{eff} = \sqrt{2m\phi_{eff}}\ /\ \hbar$$

can be related to intermediate state tunneling:[5-7] n intermediate states in equal separation $d_o/(n+1)$ bridge the tunnel gap d_o, so that $\kappa_{eff}d_o = \kappa_o d_o/(n+1)$ has to be used. This is translated misleadingly as barrier reduction $\phi_{eff} = \phi_o/(n+1)^2$. For $n > 1$ the dependencies of Table 1 relate to incoherent tunneling, made obvious by the activated temperature and voltage dependencies.

Table 1: Approximate dependencies of intermediate state tunnel conductance with n intermediate states. The reduced effective tunnel distance $d_{eff} = d_o/(n+1)$ is usually translated in an effective tunnel barrier $\phi_{eff} = \phi_o/(n+1)2$. For several intermediate states of non identical energy thermal or voltage energy has to be supplied. This yields the indicated activation, often named inelastic.

n_I	distance d	voltage U	temperature	
0	$\propto \exp(-2\,\kappa d)$	$\propto U$	const.	elastic
1	$\propto \exp(-\,\kappa d)$	$\propto U$	const.	elastic
2	$\propto \exp(-2\,\kappa d/3)$	$\propto U^{7/3}$	$\propto T^{4/3}$	activated
3	$\propto \exp(-\,\kappa d/2)$	$\propto U^{7/2}$	$\propto T^{5/2}$	activated
n	$\propto \exp(-2\,\kappa d/(n+1))$	$\propto U^{n+1-2/(n+1)}$	$\propto T^{n-2/(n+1)}$	activated

The relations of incoherent to coherent intermediate state tunneling are summarized in Sect. 2. In this transition tunnel times τ, on-site and off-site charging energies play a crucial role, omitted in most treatments. In Sect. 3.1, the nature of the intermediate states is analyzed, reaching from atomic-like, localized states to more extended resonances, like, electron attachement states. In Sects. 3.2 - 3.4, the theory is compared to experimental observations. There it is shown that tunnel charging makes most intermediate state tunneling incoherent. Despite this incoherence and charging, extended states with low charging energies are very effective in charge transfer and imaging at low voltages and temperatures and may bridge tunnel distances d_0 in excess of 2 - 10 nm, still allowing high spatial resolution. In STM individual, quantized tunnel channels are encountered, which, by taking the obstruction by charging into account, allow quantitative distance estimates in STM tunnel studies worked out for the first time in Sect. 3. The enforced charge transfer in other fields of research, like 'super exchange in biology' is discussed in its relation to intermediate state tunneling in Sect. 3.4.

2. TUNNEL CHANNELS

Charge transfer through a complex barrier, sketched in Fig. 1, in various thicknesses and with various amounts of intermediate states needs a unified treatment of metal and insulator properties. Following Refs. 2b and 3 in Sects. 2.1 - 2.3, direct and resonant tunneling are summarized with a discussion of charging in tunneling in Sect 2.4.

2.1 Transfer Hamiltonian

In order to discuss the different tunnel channels modified by coupling to internal degrees of freedom of the tunnel barrier, the treatment based on the stationary Schrödinger equation is not well suited. A somewhat more flexible technique, the method of the transfer Hamiltonian $\hat{H}_T$, was first introduced by Bardeen [8]:

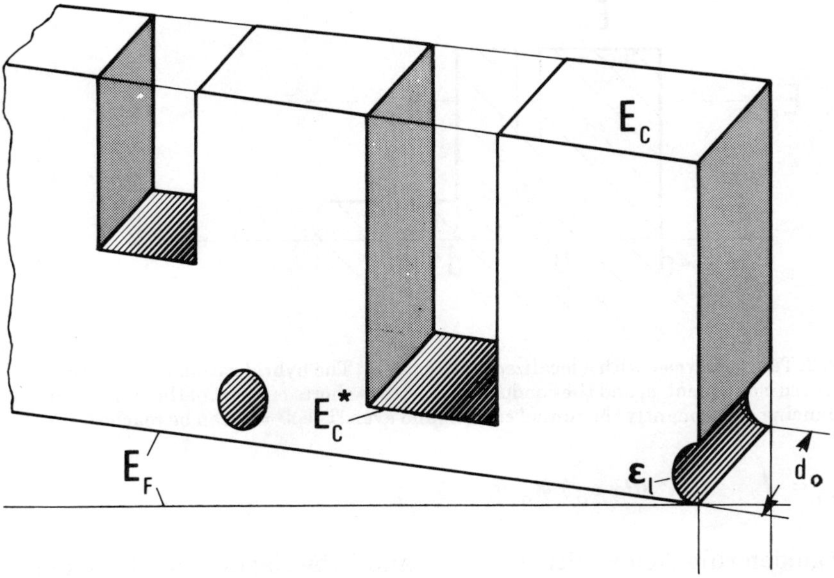

Fig. 1: Sketch of tunnel barrriers with $E_c(r)$ as lower edge of the conduction band. Until now barriers $E_c(r) = E_c(x)$ being uniform along the electrodes have been treated, which cannot describe microcrystalline or inhomogeneous barriers, like Nb_2O_5 or adsorbates. The model depicted corresponds to $(E_c - E_F) = \Phi = 1eV$ for crystallites, $(E_c^* - E_F) = \Phi \simeq 0.1 - 0.2$ eV for channels, and $(E_F - \varepsilon_l) \geq 5$ meV for resonant tunneling via localized states symbolized as holes in the barrier. Any actual barrier is not step-like and cannot be depicted so simply.

$$\hat{H}_T = \sum_{i,f} T_{fi}\, \alpha_f^* \alpha_i \,, \tag{1}$$

where 'i' and 'f' denote the initial and final states. Consider a barrier containing a localized state sketched in Fig. 2. Following Refs. 1 to 3, a wave function $\psi_i(r)$ is introduced, which is the solution of the Schrödinger equation in the region $\kappa < -d$ and which is matched to an exponentially decaying function $\propto \exp | -\kappa\delta x |$ with $\kappa = \sqrt{2}\, m\, (E_c - E_F)/\hbar$ in the barriere region $x > -d$ of Fig. 2.

The function ψ_i is thus a correct solution for $x < -d$ and almost everywhere in the barrier ψ_i is not a correct solution for $x > d$ and in the region V_l region housing the localized state. The function ψ_f, being a solution for $x > d$, can be written in a way similar to that for ψ_i. Then with the matrix element

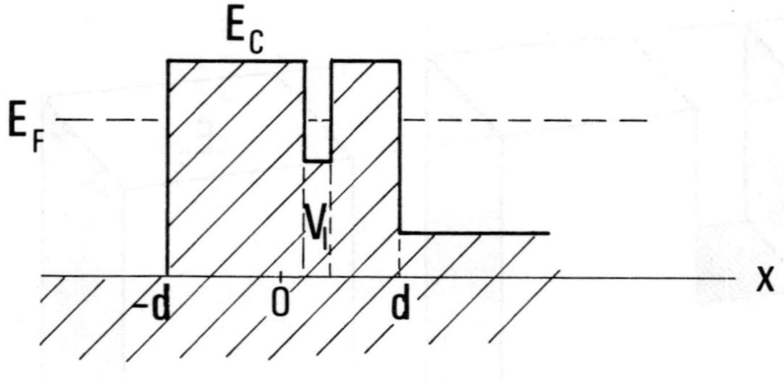

Fig. 2: Tunnel barrier with a localized state ψ_l at V_l. The hybridization between the localized component ψ_l and the conduction electrons shorts out part of the tunnel barrier enhancing so resonantly the tunnel current and even $|T(\varepsilon_l)|^2 = 1$ can be reached.

$$A_{kn} = \int \psi_k^* (\hat{H} - \varepsilon_n)\, \psi_n \, dr, \tag{2}$$

the golden rule yields with Refs. 1 - 3 and 8 the elastic tunnel current

$$I = \frac{4\pi e}{h} \sum_i \sum_f |T_{if}|^2 \left[f(\varepsilon_i) - f(\varepsilon_f + eV) \right] \delta(\varepsilon_i - \varepsilon_f), \tag{3}$$

with $T_{if} = T_{if}^D + T_{if}^R$, where $T_{if}^D = A_{if}$ is the direct tunnel matrix element and

$$T_{if}^R = A_{il} \, A_{lf} \, / \left(\varepsilon_i - \bar{\varepsilon}_l + i\Delta_l \right) \tag{4}$$

is the resonant tunnel matrix element, with $\bar{\varepsilon}_l = \varepsilon_l + \delta\varepsilon_l$. Here, $\delta\varepsilon_l$ = Re A_{lf} describes the shift of ε_l due to hybridization. By symmetry, $\delta\varepsilon_l$ is negligible for $\varepsilon_l \approx E_F$.[2] Δ_l describes the broadening due to hybridization[2]. The T^D term in Eq. 3 describes the direct tunneling as obtained by Bardeen[8]. The T^R term describes tunneling via the localized state being resonant at $\varepsilon_i \approx \bar{\varepsilon}_l$[1-3], i.e. making then a noticable contribution. In resonant tunneling[2], states in the middle of the barrier dominate, yielding a thickness dependence $T^{R2} \propto \exp(-\kappa d_o)$. This thickness dependence is slower than in direct tunneling $T^{D2} \propto \exp(-\kappa 2d_o)$. Hence T^R dominates with increasing d over T^D, as discussed in Refs. 2 or 7 , where the integrated, averaged tunnel current has been given. Beyond the scope of this paper is a detailed analysis of I(U,r) around a localized state revealing interference patterns for $T^R \approx T^D$, energy- and space-wise.[2] In summary, the total tunnel Hamiltonian

$$\hat{H} = \sum_{i,f} T_{fi}\ \alpha_f^{\dagger}\ \alpha_i \quad \textit{is given by} \quad \hat{H} = \hat{H}^{D} + \hat{H}^{R} \quad \textit{with}$$

$$\hat{H}^{D} = \sum_{i,f} T_{fi}^{D} \alpha_f^{\dagger} \alpha_i \ , \quad \hat{H}^{R} = \sum_{i,f} T_{fi}^{R} \alpha_f^{\dagger} \alpha_i \ .$$

Both tunnel channels can excite states - vibrations, phonons or electronic states - of the barriers. The charging in resonant tunneling $\hat{H}^R$, discussed in Sect. 2.4, enforces inelastic processes.[2c]

2.2 *Direct tunneling*

Tunneling through a potential barrier $\phi(r) = E_c(r) - E_F > 0$ is given by the tunnel matrix element T^D and by the density of initial and final states $n_m(E)$. As indicated in the introduction, up till now mostly 1-dimensional barriers, i.e. $\phi(r) = \phi(x)$, have been treated, which were mostly trapezoidal in shape. For a rectangular barrier of thickness $2d = d_o$, sketched in Fig. 2 for electrons with the momentum $\hbar k$ approaching the barrier from the left, neglecting effects of other valence or conduction bands, the transmission coefficient T^D is given by: [2b,3]

$$T^D\,(U=0)^2 = \left[\cosh^2 2\kappa_x\, d + \left(\frac{k_x^2 - \kappa_x^2}{2k_x\, \kappa_x} \right)^2 \right]^{-1} \propto \exp\,(-\ 2\kappa_x\ (d), \tag{5}$$

with $\kappa^2_x = 2m\,\{E_c - [E - (\hbar_\| k_\|)^2 \cdot 2m]\}/\hbar^2$. For a finite voltage U across the barrier, κ_x declines and thus I(U) increases. For a rectangular barrier with $\phi_1 = \phi_2 = 0.3$ eV, I(U) computed by the WKB approximation[2,7] is shown in Fig. 3. The roughly exponential increase of I(U), is in line with the decrease of

$$\kappa_{eff} = \left[\ 2m\ (\ \phi - |\,eUx/d\,|\ )\ \right]^{1/2} / \hbar.$$

with U. But as indicated in Fig. 3, for voltages $eU \geq \phi$ the current increases more rapidly, like $T^{D2} \propto \exp\,(-\,c\phi^{3/2} / E_\perp)$.

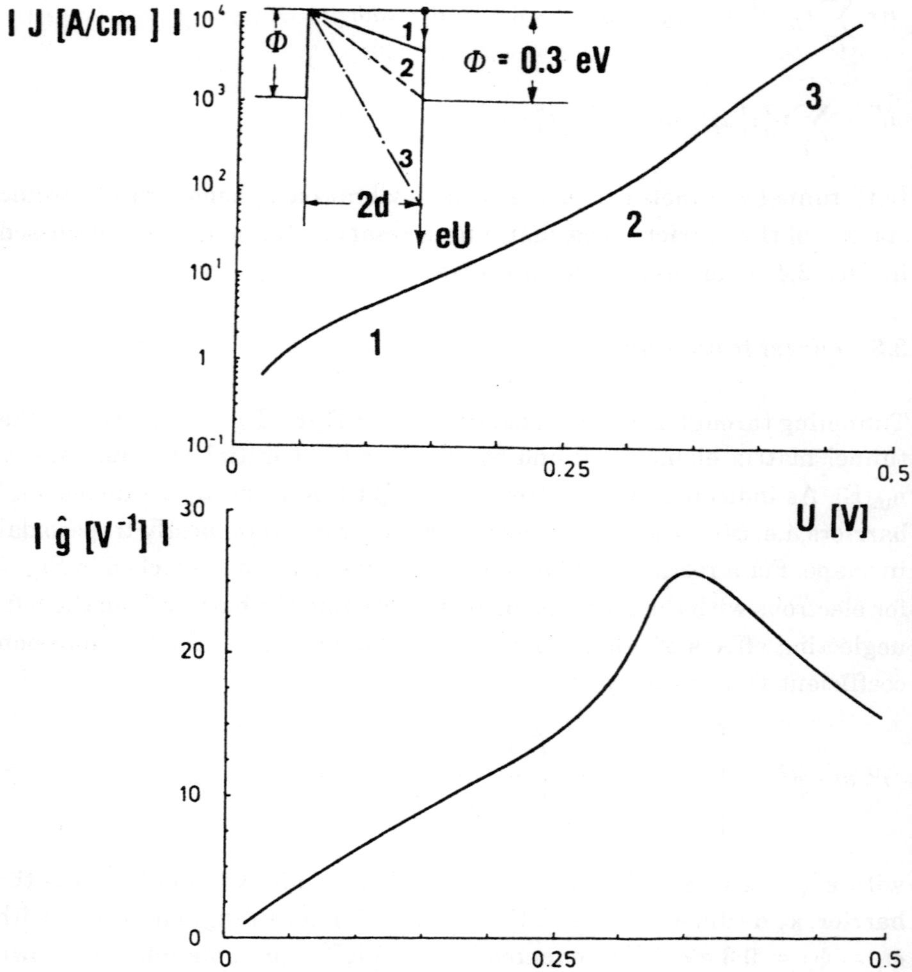

Fig. 3: I(U, 4.2 K) dependence for a symmetric ($\phi_L = \phi_R$) tunnel barrier $\phi(x) = \phi = 0.3$ eV of width $2d = 4$ nm. The logarithmic derivative $\hat{g} = d \ln (I/U)\, dU$ at 4.2 K emphasizes the current increase at $eU \geq \phi$, where the barrier $\phi - |eU|/2 + eUx/2d$ in addition becomes narrower. It should be mentioned that $\hat{g}_{max}$-values depend very much on the rounding of $\phi(x)$ at $x = \pm d$ and increase with d and ϕ.

This stronger I(U) increase is due to the fact - see insert in Fig. 3 - that for $eU > \Phi$ the barrier becomes with U not only lower but, in addition, shorter, like in field emission. The Fowler-Nordheim increase of I(U) for $eU \geq \Phi$ is shown most clearly by the logarithmic derivative $\hat{g} = d \ln (I/U) / dU$ depicted in Fig. 3. As shown there for 4.2 K, the maxima occur at $\pm eU_{max}$

$\simeq 1.2\ \phi$. The height of the maxima increases with barrier thickness 2d, because the angular cone in tunneling shrinks with 2d. By the same reason, the $\hat{g}(U)$ maximum increases with barrier height ϕ. Both dependencies allow a unified estimation of 2d and ϕ - aside of $1.2\phi = e\ U_{max}$ by g_{max} in Fig. 3. $\phi(x)$ does not rise abruptly to ϕ at $x = \pm\ d$, but is graded by interdiffusion or chemical reactions. Such graded interfaces yield lowered $\hat{g}(U)$ values worked out in Ref. 2 for Nb-Nb_2O_5-PbO-Pb junctions. Such junctions in Fig. 4 show a graded $\phi(x)$-dependence at the Nb-Nb_2O_5 interface, but not at the PbO-Pb-interface. Thus the grading is not due to mirror charge effects.[9,10] It is due to interface chemistry because otherwise Nb_2O_5 with $\varepsilon_r \approx 20$ would yield the steepest grading. The mirror charge correction to the tunnel barrier are discussed in Sect 2.4.
For a symmetric barrier and symmetric metallic banks $\phi(x) = \phi\ (-x)$ an antimetric $I(U) = -I(-U)$ holds. More typical for tunnel barriers is $\phi(x) \neq \phi(-x)$, e.g., due to $\phi_L \neq \phi_R$. Because of the 3-dimensional metallic banks, this yields $I(U) \neq -I(U)$ with maxima at $1.2\ \phi_I$ and $1.2\ \phi_R$. As worked out first in Ref. 2 also for $\phi_R = \phi_L$ but $\phi(d) \neq \phi(-d)$ $I(U) \neq -I(-U)$ is possible, e.g., by asymmetries in the transition layers. Examples for $\hat{g}(U)$ of such asymmetric transition layers are shown in Fig. 4.[2] It is obvious that $\phi(r) \equiv \phi\ (x)$ is an approximation not appropriate for microcrystalline tunnel barriers. To approximate the inhomogenities extending deep into the barrier, as a first step, $\phi(x)$ consisting of a barrier ϕ and channels $\phi^* < \phi$ (Fig. 1) is introduced in Ref. 2. This explains tunneling through Nb_2O_5-barriers consistently, if resonant tunneling is added - see Section 2.3 and Fig. 4.

2.3. *Resonant tunneling*

Besides ϕ^*-channels in a $\phi(>\phi^*)$ barrier, localized confined regions with $\phi(r_I) < \phi$ or ϕ^*, may yield localized states $n_l(r_l, \varepsilon_l)$, modelled in Fig. 2. Whereas states with $\varepsilon_l > \phi^*$ are difficult to separate from the integrated tunnel current, states with $\varepsilon_l < \phi^*$ have been separated in extended tunnel barriers in Ref. 2. To evaluate the resonant tunnel current, the change of the barrier with U is neglected in the following. Assuming plane waves k incident from the left on the barrier and averaging over the {y,z} plane

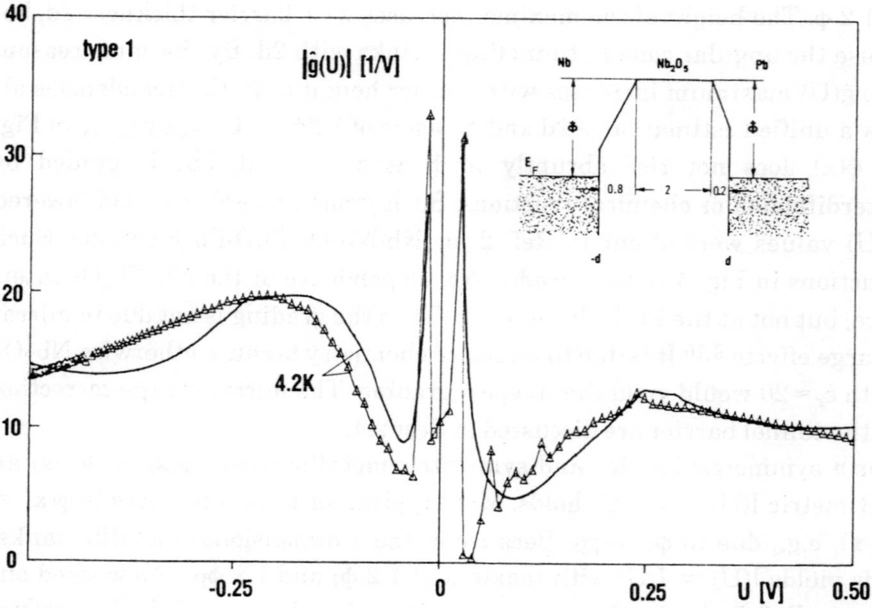

Fig. 4: Absolute value (Δ) of the logarithmic derivative $\hat{g}(U) = d \ln (I/U)/dU$ as a function of voltage for the $Nb\text{-}Nb_2O_5\text{-}Pb$ junction, see also Fig. 1. The oxide has a thickness of about 2 d ≈ 3 nm. In accordance with Sect. 2, the narrow symmetric peaks at $\pm U \approx 20$ mV are explained by resonant tuneling via states ≥ 10 meV below E_F. The broad peaks at $U \approx \pm 0.2$ V in $\hat{g}(U)$ fit to barrier tunneling with ϕ (Nb) $\approx \phi$ (Pb) ≈ 0.2 eV. Besides this tunneling through channels, the flat portions in $\hat{g}(\approx \pm 0.5$ eV) are compatible with the crystallite 1 eV barrier. These three tunnel channels contribute to the tunnel current I (50 mV) according to the (-) fit shown by: $\Phi \approx$ 1eV barrier (35 %); $\phi^* \approx$ 0.2 eV channels (45 %); RT at $\Delta U^* \approx$ 10 meV (20%). The asymmetry of $\hat{g}(|U| > 50$ mV) is explained by ϕ(Nb) = ϕ (Pb) with a transition layer at the $Nb\text{-}Nb_2O_5$ interface depicted in the insert. Despite the low and long ϕ^*-barriers no evidence of mirror charge effects are obvious.

with $n_l(x)$ as the averaged density of localized states, one obtains as transmission coefficient T^R near resonance $E \approx \varepsilon_l(x)$ with Eq. (4)[1-3]:

$$T^R(k,x)^2 = \frac{m}{k_x}\frac{2n_l(x)}{h^2}\frac{|S(k,x)|^2\ \Delta_l^-(x)}{\left[E-\varepsilon_l(x)\right]^2+\Delta_l^2(x)} \tag{6}$$

with

as resonantly enhanced transmission coefficient via the localized state ψ_l. In Eq. 7 $A = \int dr\ V_l(r)\ \psi_l(r)$ is the 'cross section' and T^D the matrix element for direct tunneling of Eq. 5. The decay widths

$$S(k,x_l) = |A|^2 \; T^D \left[\cosh^2 \kappa_x (d-x_l) + \frac{k_x^2}{k_x^2} \sinh^2 \kappa_x (d-x_l) \right] \times \left[(E-\varepsilon_l)^2 + \Delta_l^2 \right]^{-1} \tag{7}$$

$$\Delta_l^{\pm} = \frac{h}{\tau_l \pm} = \frac{k\kappa}{4\pi(E_c - E)(d \pm x)} |A|^2 \exp[2\kappa(d \pm x)]\, g(d, \Delta U^*) \tag{8}$$

give the decay probability of the state (ψ_l) to the left (-) or right (+). The factor $g(d, \Delta U^*) < 1$ in Eq. (8) describes the fact, that the mirror charge build-up obstructs tunneling by factors of the order of 100,[4] which as many body effect discussed in Sect. 2.4 cannot be treated consistently, yet.
In summary, T^{R2} in Eq. 2 describes the resonant capturing ($\propto$ S) of an electron in a localized state and the decay of this state ($\propto \Delta_l$) to the right side. The resonant enhancement is largest for the smallest decay probability $\Delta_l = \Delta_l^+ + \Delta_l^-$, i.e. for $\Delta_l^+ = \Delta_l^-$ in Eq. 8, occuring the middle of the barrier. For $E \simeq \varepsilon_l$ and $\Delta_l^+ = \Delta_l^-$ $T^{R2} \approx 1$ is obtained,[11] i.e., the barrier becomes transparent independent of d. These states dominate[2] j^R and thus will be assumed in the following for simplicity. For these states, after integrating over E, one obtains as ratio of resonant tunnel current j^R to direct tunnel current j^D (U= 0) with $E_l = \varepsilon_l - E_F$.

$$\frac{\overline{j^R}}{j^D\,(U=0)} = g n_l(\varepsilon_l)\; |A|^2 \exp(2\kappa d)\, \frac{\sinh(eU/2kT)}{\cosh(eU/2kT) + \cosh(E_l/2kT)} \tag{9}$$

In Eq. 9 the fact has been omitted that j^R (U,r) and j^D (U, r) interfere (Fano effect) becoming observable for $j^R \approx j^D$.

2.4 *Charge transfer and hybridization of states adjacent to surfaces*

The above sections dealt with direct and resonant tunneling through barriers sketched in Figs. 1 and 2. The effect of charging is discussed in this section. In bulk, classic metals the electron charge is compensated by the positive ionic background and the nearly free electron motion yielding a high relative dielectric constant $\varepsilon_r \rightarrow \infty$. In the 'dielectric tunnel barrier' a

charge is no longer effectively screened and creates electric fields described mainly by mirror charges, i.e. surface plasmons with their characteristic energy $\hbar\omega_S \approx 1$ eV and build-up time of the order of $1/\omega_S$.
The lifetime τ_l in an intermediate state, separated by a distance δx from the metal, is given by

$$\hbar/\tau_l\ (\delta x) \approx \Delta_l\ (\delta x) \propto e^{-2\kappa\delta x} \tag{10}$$

- see Fig. 2 and Eq. 8. After residence times τ_l longer than ω_S^{-1} the charging of such a state depicted in Fig. 2 yields an energy change

$$\Delta U^*(\delta x) = \frac{0.28\,d}{\varepsilon_r\ (d/2 - x)\ (d/2 + x)} [eV] \propto 1/d \tag{11}$$

This corresponds to the creation of surface plasmons[10] with their characteristic time constant ω_S^{-1}. For typical intermediate states $\tau_l \gg 1/\omega_S$ holds and so the charging in tunnel exchange is governed by the relation of Δ_l to ΔU^*. Eq. (11) assumes atomic like localization. By delocalization over an area $A \geq (\delta x)^2$ $\Delta U^* = e^2/2C$ decreases by smearing out the charge over A. Quantitatively for charges in the middle of the tunnel gap this corresponds to the reduction of $C = 4\pi\varepsilon_r\varepsilon_o d/2$ in Eq. (11) to $C = \varepsilon_r\varepsilon_o A/d/2$ i.e.,

$$\Delta U^* = e^2/2\varepsilon_r\varepsilon_o\Delta/d/2 \propto d \tag{11'}$$

with the opposite distance dependence of Eq. (11).
For $\Delta_l < \Delta U^*$, i.e. slow tunnel exchange, charging described by the energy shift ΔU^* blocks the tunnel exchange, i.e, by the many particle effect, the creation of surface plasmons. The latter reduces the tunnel probability by about 2 orders of magnitude, i.e., $g \approx 10^{-2}$ in Eq. 8 in the transition from $eU < \Delta U^*$ to $eU > \Delta U^*$.[2,4] $\Delta_l < \Delta U^*$ yields incoherent exchange across a tunnel barrier. If n intermediate states equally spaced are involved in this tunnel process, they reduce the effective tunnel distance d_o to:

$$d_{eff} \approx d_o\ /(n+1) \quad or \quad \phi_{eff} = \phi_o\ /(n+1)^2 \tag{12}$$

For kT or eU larger than ΔU^* and $\delta\varepsilon_l$ as energy difference of adjacent states, the incoherent tunneling is summarized in Table 1 and discussed in Refs. 6 and 7 in detail. There the formula for the U- or T-dependencies are given showing, e.g., a conductivity for $kT \ll eU$

$$G_n \propto S\,(d_o U n_l^2/\kappa^2 n) U^{-2/(n+1)} \exp(-2\kappa d_o/(n+1)) \quad (12')$$

due to the activation energy needed to bridge the energy differences $\delta\varepsilon_l \geq \Delta U^*$. With $\Delta_l \approx \Delta U^*$ or $\Delta_l \gg \Delta U^*$ coherent tunnel processes become important,[12] no longer hindered by ΔU^* and, therewith, the excitation of surface plasmons. The coherent exchange is given by the band picture for the overlap

$$W(\delta x) = W(0)\, exp\,(-\,\delta x\, \kappa) \quad (13)$$

with the conduction band width W(0). For n = 1 intermediate state coherent resonant tunneling follows with $g \approx 1$ in Eq. 8. The case of n = 2 intermediate states has also been evaluated yielding optimum coherent charge transfer of the states separated by δx_{L1} and δx_{R2} from the left and right electrode with[13]

$$\delta x_{L1} + \delta x_{R2} = d = d_o/2 \quad (14)$$

$$\delta x_{L1} \approx \left(1+\sqrt{3}/2\right) d \text{ and } \delta x_{R2} \approx \left(1-\sqrt{3}/2\right) d$$

The coherent tunnel conductivity writes as

$$G_2^{coh} = \frac{e^2}{\pi\hbar} N_2 \text{ and } N_2 = n^3\, n_l^2/\kappa\, d_o^3\, H_{12}\cdot S \quad (15)$$

Here S is the tunnel area and

$$H_{12} \propto exp\,(-2\kappa d) \quad (15')$$

the overlap between the two states. Thus, n = 2 corresponds to $d_{eff} \simeq d_o/2$ or $\Phi_{eff} \simeq \Phi_{eff} \simeq \Phi_o/4$ over a wide δd range because of the asymmetry obvious by Eq. (14). In analogy, for n = 3 the distance d/2 is bridged by two states yielding $d_{eff} \simeq d_o/4$ and $\Phi_{eff} \simeq \Phi_o/16$. But $n \geq 3$ states cause incoherent tunneling, already, because several intermediate states and channels in parallel make coherence unlikely,[12] see also comparison with experiments in Sect. 3.

The above analysis of coherent and incoherent intermediate state tunneling yields directly a discussion of mirror charge correction in direct tunneling.[10] In this case the tunnel time τ_T substitutes τ_l in the above analysis. τ_T is still in debate but as crude estimate[9,10]

$$\tau_T = d_o / \sqrt{2\Phi / m} \tag{16}$$

is useful and yields for a typical tunnel barrier $\tau_T \leq 10^{-15}$ sec, i.e. $\omega_S \approx 1/\tau_T$. Thus, in direct tunneling only a part of the mirror potential is active in lowering the tunnel barrier.[10]

3. COMPARISON WITH EXPERIMENT

The analysis of 'solid' tunnel junctions is based on voltage tunnel spectroscopy (VTS) which averages over various states $n_l(x)$ and $\phi(x,y)$-dependencies.[1,2,7,12] With the advent of STM, besides morphological studies, a distance tunnel spectoscopy (DTS) becomes possible, in addition, elaborated in Ref. 7, which allows, e.g., the identification of intermediates states energy- and space-wise. The quantization of the channel conductance[4,13] in units $e^2/\hbar$ of yields quantitative information on distances, if charging is taken into account. Because in STM the tip approaches the substrate from large distances, the slowest decaying tunnel current I(d) is picked up, being most likely an intermediate state tunnel channel with its exponentially enhanced transmission probability of Eqs. (12).
Direct tunneling and coherent tunneling via one to two intermediate states conserves energy. Then, voltage tunnel spectroscopy (VTS) and distance tunnel spectroscopy (DTS) are straight forward.[7] In the case of direct tunneling, VTS yields the barrier shape[2] $\phi(x,y)$ and the density of states n_m in the metallic banks.[1] In the case of coherent intermediate state tunneling VTS and DTS give mainly information on energy- and spatial-distribution $n_l(\varepsilon,x)$ of the intermediate states. Incoherent tunneling is growing in importance with increasing density of intermediate states n_l and growing tunnel distance d_o. Such channels smear out energy information due to the activation energies kT or eU, being larger than ΔU^* of Eq. (11). DTS has been used successfully to measure the spatial distibution $n_l(\varepsilon \approx E_F,x)$ of those states.[7] The most detailed analysis covering the transition of tunneling to variable range hopping is given in Ref. 6. STM imaging, i.e. the lateral tunnel resolution depends crucially on type of tunnel channel and the number of intermediate states. For n ≥ 4 and {kT, eU}

$\geq$0.05 eV variable range hopping dominates, which smears information energy- and space-wise, but still allow imaging.

3.1 Intermediate states in tunneling

Atomic-like localized states, like, oxygen vacancies in transition metal oxides, are well known as cause for intermediate state tunneling.[1,2] Their decay width Δ_l has been estimated in Ref. 2. Such localized states at Si-SiO_2 interfaces are well known as centers causing 'inelastic' surface scattering or noise.[2,14] Mesoscopic devices have recently become feasable allowing detailed theoretical and experimental analysis of transition probabilities of more extended states.[12] In adsorbates on metals or oxides atomic-like intermediate states are rarly observed. Instead, extended resonances known from electron scattering at molecules in vacuum[15] act often as intermediate states.[7] Such resonances are either 'shape' or 'Feshbach' resonances.[14] The shape resonance corresponds to the optimum tunnel coupling to 'center of mass motion' in quantum dots.[12] In a shape resonance an electron ($E_{kin} \geq 0$) is captured either in the field of the ground state or in the field of an excited state through a number of mechanisms,[15,16] being metastable with a lifetime ranging from $\approx 10^{-15}$ to 10^{-2} sec. As capture mechanisms one considers the combined effect of the long-range polarization potential and the short-range repulsion due to the centrifugal motion. For example, the attachement states are delocalized and avoid high concentration of bonding electrons in molecules as to reduce the on-site Coulomb charging ΔU_c, discussed in Sect. 3.2. In Refs. 15, cross section and energies of e^--attachement states of hydro-, chloro- and fluoro-carbons are summarized - see also Fig. 5a.

Their cross section increase to values above $\geq 10^{-15}$ cm at $E_{kin} \geq 0.4$ eV is worth mentioning. In line with the above description their energies $E_{kin} \geq 0$ are lower for longer chained molecules and for Cl or F substitution enhancing the dipole moments. In molecules with a dipole moment above 1.6 Debye a resonance at $E_{kin} \approx 0$ is found due to bonding of the electron in the static dipole potential[15,16]. The cross section reaches 10^{-15} cm^2, the half width of this resonance is around 0.15 eV with a lifetime $\tau \approx 10^{-13}$ sec.

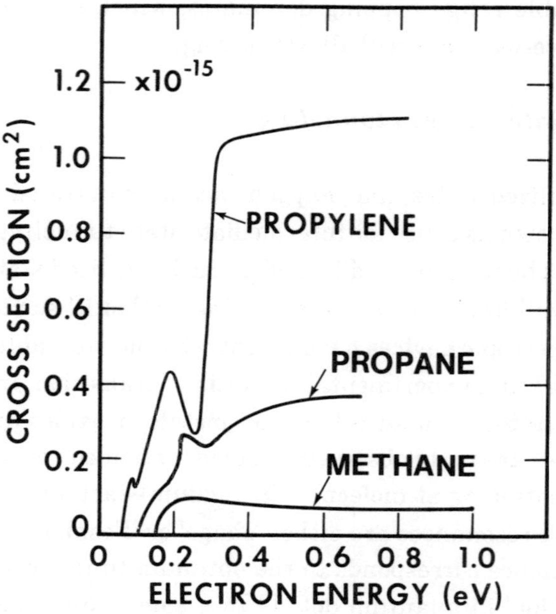

Fig. 5a: The cross section at low kinetic electron energy grow with chain lenghts of the hycrocarbons and with their dipole moment, e.g., due to Cl or Fl. The onset at $E_{kin} \approx 0.4$ eV with $\sigma \geq 10^{-15}cm^2$ may give rise to intermediate state tunneling.

In Fig. 5b the cross section of the H_2O-dipole resonance[16] is depicted.

The resonance curves in Figs. 5a or 5b are obtained by swarm experiments or electron scattering at individual molecules in vacuum.[15,16] In adsorbates the interaction with adjacent molecules or metallic banks may shift the resonance energy ε_l and causes off-site Coulomb charging ΔU^* analyzed in Sect. 3.2. Such energies decrease by delocalization or separation (Eq. 11'), e.g., the extended dipole resonances of a oriented water-hydroxide layer on Nb_2O_5 have yielded a resonance at $E_{kin} \approx 0$, observed by SEEDC shown in Fig. 5c. As compact part of the double layer the oriented water - hydroxide dipole layer has a lateral spatial extension and a life time increasing with bonding strength, e.g. to oxides. In incoherent tunneling such states show $E_{kin} \approx 0$ and $\sigma \approx 10^{-15}$ cm² (Fig. 5c) needed for the application of the above theory. For incoherent tunneling via hydrocarbons states with $E_{kin} \geq 0.4$ eV and $\sigma \approx 10^{-15}cm^2$ are adventageous, also.

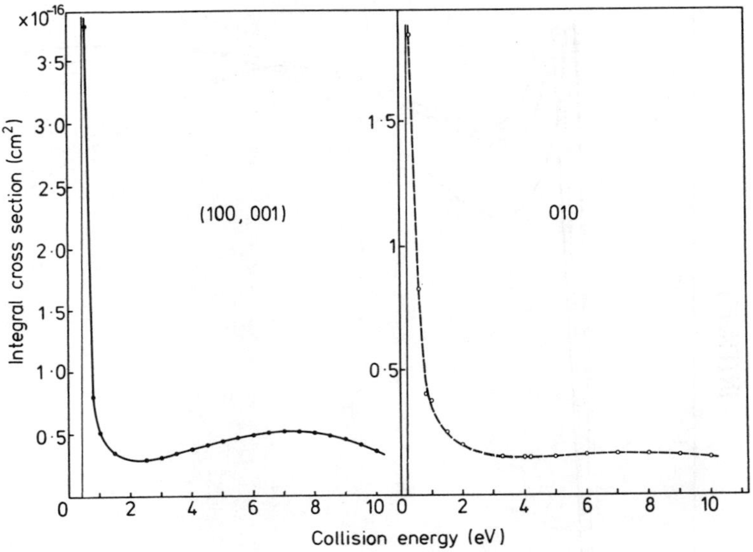

Fig. 5b: Integral excitation functions for (100, 001) and (010) excitation of H_2O. Two distince resonances, at the threshold (dipole) resonance and a broad resonance region around 6 - 8 eV, are superimposed on a background from direct excitations. [15,16] The dipole resonance of $E_{kin} \approx 0$eV with $\sigma \geq 10^{-15}$ cm² gives rise to incoherent intermediate state tunneling.

3.2 Tunneling and image charges

Atomic contacts start with the one channel, quantum conductance

$$G_o = 2e^2/\hbar = 12.9\ k\Omega \tag{17}$$

observed experimentally.[17,18] In direct tunneling barriers with $\phi \simeq \phi_{vac}$, and in dielectrics $\phi = E_c - E_F$ govern the decay in Eq. 5 with, e.g., $\kappa \approx 7.7$/nm for $\phi \simeq 1$ eV and an effective electron mass 1 m_e. To obtain the barrier height ϕ or the barrier width d, the logarithmic derivative g(U) shown in Fig. 3, is most useful. As elaborated in Ref. 2 and in Fig. 4, the shape $\phi(x,y)$ and various tunnel channels can be separated by g(U)-analysis. For tunneling through a barrier $\phi = \phi(x)$ an off-site image charge correction ΔU^* has often been proposed[5] assuming that the electron in tunneling has time to excite surface plasmons.

For direct tunneling the image charge corrections introduced in Sect. 2.4 yield a rounding of $\phi(x)$. Fig. 4 shows no effect of a mirror potential, which

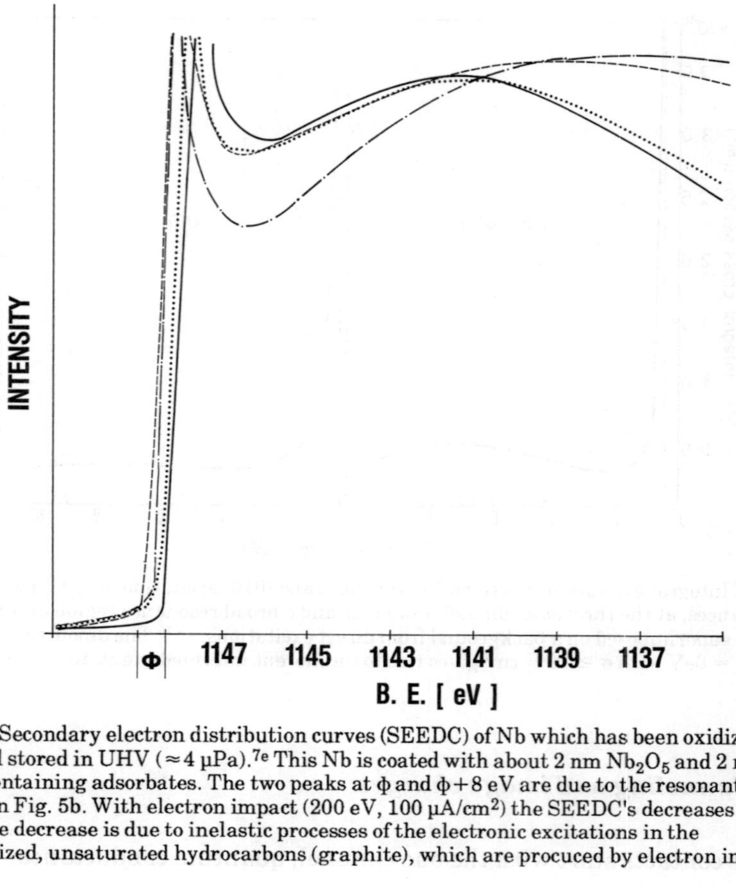

Fig. 5c: Secondary electron distribution curves (SEEDC) of Nb which has been oxidized in H_2O and stored in UHV ($\approx 4\ \mu Pa$).[7e] This Nb is coated with about 2 nm Nb_2O_5 and 2 nm C and O containing adsorbates. The two peaks at ϕ and $\phi + 8$ eV are due to the resonant states of H_2O in Fig. 5b. With electron impact (200 eV, 100 $\mu A/cm^2$) the SEEDC's decreases with dose. The decrease is due to inelastic processes of the electronic excitations in the polymerized, unsaturated hydrocarbons (graphite), which are procuced by electron impact.

would yield at the Nb_2O_5-Nb interface a steeper grading than of the PbO side. This is due to the fact, that even for $\phi \approx 1$ eV and $\phi^* = 0.1$ eV and d ≈ 3 nm the tunnel time τ_T of Eq. 16 does not allow a fully developped mirror potential rounding $\phi(x)$.[10] Instead, the interface chemistry Nb-Nb_2O_5 is the reason for the grading. Similarely to the build-up time of the mirror potential $\tau_S \sim \omega_S^{-1}$ is the time τ_P for the interaction of tunneling electrons with the barrier polarization. Especially for water only a small part of its large polarization will be active in reducing the tunnel barrier on the time scale $\tau_T \approx 10^{-15}$ sec of Eq. (16). Thus the computed barrier reduction by actual, image charge corrected electron trajectories in water by 1.3 eV seems an overestimate for d ≤ 2 nm.[21]

Below, the conditions are discussed where image charge corrections may occur in intermediate state tunneling. Tunneling via intermediate states is given by their cross section $\sigma(\varepsilon)$, their resonance energy ε_l and their overlap Δ_l to neighboring states, either localized at or in the electrodes. In relating ε_l of the intermediate state in distance δx from the electrodes to the energies observed by voltage tunnel spectroscopy (VTS) several corrections have to be applied. The states showing up in VTS in I(U) at ε^*_l

$$\varepsilon_l^* = \varepsilon_l - e(d-x)U/2d + \Delta U_\infty + \Delta U^* + \Delta U_c \tag{18}$$

are related[2] to ε_l via the energy shift $\delta x e U/d_o$ in the barrier,[14] the correlation and screening effects ΔU^* by the metallic electrodes, namely the so-called image charge corrections[14] and the energy change by polarizing adjacent molecules, ΔU_∞ by charging of neighboring intermediate state tunnel channels and ΔU_c the on-site Coulomb charging energy for an additional electron (n=1) on the site.[7a] For several adjacent eletrodes or charges $\Delta U = 1/2\ \Sigma e_i e_j / C_{ij}$ is a complex quadratic form with C_{ii} as standard capacitances.[19] For example

$$\Delta U^* = e^2/2C^* \quad \text{and} \quad \Delta U_c = e^2/4\pi\,\varepsilon_r\varepsilon_o t^* \quad \text{with} \quad C^* = 4\pi\,\varepsilon_r\varepsilon_o\delta x \tag{11'}$$

holds for charges in a distance δx in front of a metal surface where the charge is delocalized over t^* as maximal linear dimension. For states extended over an area $A > (\delta x)^2$

$$C^* = \varepsilon_r\varepsilon_o A/\delta x \quad \text{and} \quad \Delta U_c \simeq e^2/\,4\pi\varepsilon_r\varepsilon_o \sqrt{A} \tag{11'}$$

hold, showing decreasing ΔU^*- and ΔU_c- values with increasing delocalization area A.

In layers with a high density of dipole resonances as intermediate states, strong tunnel charge transfer can occur along the layer.[7d,20] If biased against tip and substrate this yields a floating, current biased electrode, i.e. $\Delta U^*(d)$ and $\Delta U_c(d)$ describe a Coulomb barrier with charging energies above ≥ 0.01eV.[7] For atomicly localized states ε_l^* is a step function named devils stair case with $\Delta U^* \gg 0.1$ eV and $n\Delta U_c$ with $\Delta U_c \approx 1$ eV.[7] In such slowed down intermediate state tunnel processes[1], the electron is localized long enough, so that delocalization and screening by surface plasmons excitation, i.e. ΔU^* can develop.

3.3 Coherent intermediate state tunneling

For states adjacent ($\delta x < 0.5$ nm) to classical metals $\Delta_l \geq \Delta U^*$ hold and thus coherent tunnel exchange with $W(x) \propto \exp(-\kappa\delta x)$ of Eq. 13 may occur. As distance dependence Eqs. 14 and 15 yield $\phi_{eff} \approx \phi_o/4$ for two states about $\delta x \approx d_o/4$ separated form the tip and substrate. Those individual states in front of the metal should show specific electron energy distribution curve (EEDC) centered at ε_l. In contrast, in field emission already individual metal atoms on a nanotip show EEDC's shifting with applied voltage, which is a proof for $\Delta U^* > \Delta_l > W(\delta x \approx 0.2$ nm). Thus, the tip states yield incoherent, intermediate state tunneling by a reduced overlap with the nantotip. This is made obvious by heating at the localized state in Fig. 6a.[22] Seemingly this result indicates that only 'two dimensionally extended states' adjacent (≤ 0.5 nm) to a planar metal surface may have sufficient overlap for coherent intermediate state tunneling.

3.4 Incoherent intermediate state tunneling

Incoherent intermediate state tunneling suffers from the slowed down tunnel process by the off-site Coulomb charging $\Delta U^* > \Delta_l$. This yields a reduction of the transmission probabilities by factors of about $1/100 = g(U)$ in Eq. 8, i.e. the channel conductance (Eq. 17) is reduced to $(1M\Omega)^{-1}$ observable for $\{kT, eU\} \geq \Delta U^*$

$$G^* (d_o \leq 0.3\, nm) \approx 1/1\, M\Omega. \tag{17'}$$

Incoherent intermediate state tunneling has been resolved in Ref. 6 as fuction of temperature and voltage for α-Si junctions and in Refs. 7 by distance tunnel spectroscopy (DTS) and voltage tunnel spectroscopy (VTS). In Sect. 3.4.2, Eq. (17') is used as starting point for distance estimations and classifications of the tunnel channels involved.

The most detailed tunnel information are available by field emission (FE) via the measurement of electron energy distribution curves (EEDC) shown in Fig. 6 a. This is superior to VTS discussed in Sect. 4.3.1, which averages over lateral dependencies.[22,23] With the calibration by the Fermi energy E_F by FE the intermediate states in energy ε_l and position δx are obtained.

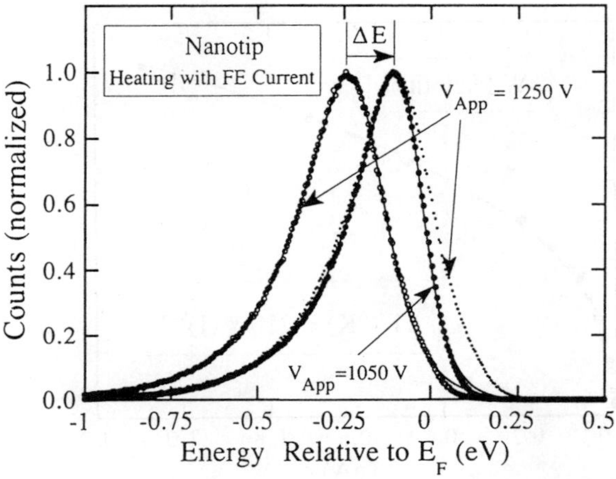

Fig. 6a: Total electron energy distribution curves (TEEDC) from a protusion tip with one localized state: room temperature and 590 K. The higher temperatures is created by a loop heating current. Spectra from the same nanotip for different applied voltages and different emission currents. The spectrum at higher voltage shifts to lower energy. It has been numerically shifted by ΔE to the position of the lower voltage peak (small dots) to show the broadening of the high-energy side of the spectrum that is related to the temperature increase.

States with $\varepsilon_l^* \approx 0.2$ eV below E_F by FEEDC[22] have been observed which fit nicely to ΔU^*,[7a] i.e. to the off-site Coulomb charging energy of atomic-like intermediate states with $\varepsilon_l \approx 0$ according to Eq. (11). Despite the small distance $\delta x \approx 0.2$ nm the overlap of such a state with the adjacent tip states shows $\Delta_l < \Delta U^* \simeq \varepsilon_l^*$. By the strong Coulomb charging inelastic effects at the localized site occur which give rise to the drastic broadening shown in Fig. 6b.[22]

Similar heating has been found also for intermediate state tunneling via thiol.[23] For larger distances and larger delocalization ΔU^* decreases and so does the image force and therewith inelastic processes producing heat. But how much the heating may be reduced, is not known.

3.4.1 Incoherent voltage tunnel spectroscopy (VTS)

In standard STM analysis EEDC's cannot be obtained. So VTS serves as alternative. For small temperatures or voltages a Coulomb barrier at $\{kT, eU\} \approx \Delta U^*$ is encountered usually. This has been observed α-Si-junctions,

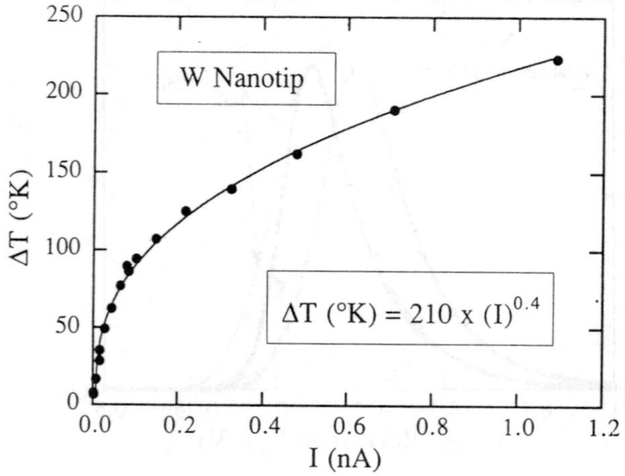

Fig. 6b: Variation in temperature versus FE-current as determined by fitting the TEEDC's from nanotips.

despite uniform distribution of localized states,[4] or for floating electrodes on tips.[7c,7d] In the latter case $\Delta U^* \gg \Delta U_c$ holds, yielding $I \propto U^2$ for $\{kT, eU\} < \Delta U^*$, i.e. a current biased electrode. For separated atomic-like, intermediate states $\Delta U_c \geq \Delta U^*$ holds yielding a devils stair case.[7d,24] In α-Si-junctions with n_l $(\varepsilon \approx E_F) \simeq 8 \cdot 10^{18}/cm^3$, the off-site charging energy ΔU^* acts as Coulomb barrier i.e. an increase of $I(|eU| \leq \Delta U^*)$ till it saturates at $\{kT, eU\} \approx \Delta U^*$.[4] For $\{kT, eU\} \geq \Delta U^*$ the incoherent intermediate state tunnel theory - see Table 1 - has been worked out and compared to experiments successfully,[4-7] even in its transition to variable range hopping (VRH).[6] Applied to molecules, VTS yields ε_l of attachement state[23,25] modified by ΔU^*, where ε_l^* is broadened seemingly by heating.[23] To quantify $\{\varepsilon_l, \Delta_l, \Delta U^*\}$ by VTS is difficult, because with increasing $|eU|$ often inelastic effects and additional intermediate states show up in tunneling and also the dipole moments may change modifying the barrier shape. So, the application of VTS is limited to solid junctions or inert interfaces.

In the above analysis, the mirror charge yields $\Delta U^* > 0.01$ eV. In the case of a dipole resonances of a water dipole layer directly on a metal the situation is more complex. A dipole resonance is extended and creates in the metal a mirror dipole resonance, by definition. But dipoles have small stray fields only, in addition the resonance is extended reducing

ΔU^*(dipole) and ΔU_c further. This is confirmed by room temperature measurements showing no sign of a Coulomb barrier for $HOHOH^-$ dipole layers.[7] The large extension of shape resonances of organic molecules reduces ΔU^* drastically also. This seems to be a prerequisit for the imaging strenght by those resonances.[7d,20]

3.4.2 Incoherent distance tunnel spectroscopy (DTS)

As guide line for tunnel channel and distance estimates one can use Eq. (17') with

$$G_n(d) \simeq \exp(-2\,\kappa d/(n+1))/10^6\Omega \qquad (17'')$$

where n describes the number of equally spaced intermediate states or resonances at $E_{kin} \approx 0$. One intermediate state supporting incoherently tunneling yields $R_1 \simeq 10^6\Omega\exp(+\kappa d) \simeq 10^7\Omega$ for $d \simeq 0.2$nm, i.e. one water layer on tip or substrate. In this range G. Nagy[26] obtained STM results on tunneling via one water layer, confirming the estimate. One intermediate state and $R \approx 10^8\Omega$ of Berthe is in line with $d \approx 0.5$ nm.[7c] In the range of $R_1 \simeq 10^8\Omega$ to $R_{3\text{-}4} \simeq 10^{10}\Omega$ Refs. 7 results in $n = 1$ - 4 intermediate states caused by the dipole layers of tip and substrate for $d \leq 1$ nm. This confirms the use of $R_n \approx 10^6\ \Omega \exp(2\kappa d/(n+1))$. It is obvious, that n depends crucially on cleanliness. For example, the peroxy insulated tips of G. Repphun yielded, loosely said, (n+1) intermediate states as compared to n-states observed by St. Vinzelberg[27] using apezon wax insulation. Those differences may be related to lifetime effects of the water dipole layers despite the continuous exchange of water with the electrolyte. Especially on oxides or by impurities those layers will be more stable and yield so more intermediate states. For 'clean' Au- or Ag-surfaces $\Phi_{eff} \approx 0.8$ - 1.3 eV have been observed for $I \simeq 40$ - 5 nA and $U = 0.2$ V,[27] i.e., for about 5 to 50 MΩ. For $I \leq 5$ nA, i.e. $\delta d > 0.4$ nm often Φ_{eff}-increases are observed. Despite the observed scatter of Φ_{eff}, a fresh Pb coating always yielded 10 % higher Φ_{eff}-values which is in contradistinction to direct tunneling. It should be mentioned that $\Phi_{eff} \approx 1$ eV fits to $d \approx 0.5$ nm and $n = 1$ or $n = 2$ 'coherent tunneling', where the scatter hints to 'dipole moment' effects. Computations using mirror charge derived electronic potential yielded[21] $\phi_{eff} \approx 2.5$ eV, in clear disagreement with of the observations.[7,27]

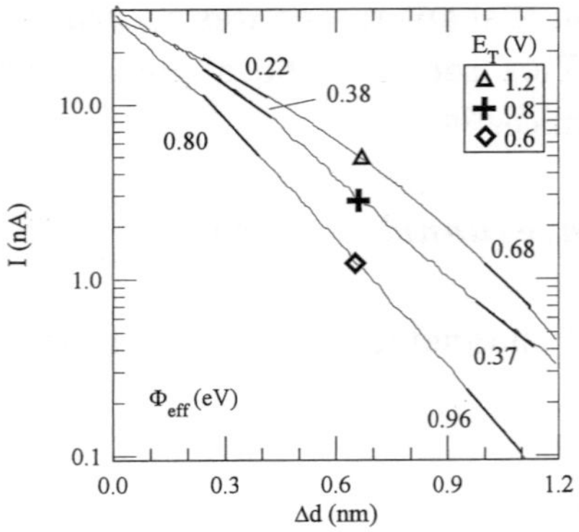

Fig. 7: Ln (I_T) vs. Δd depencence for three different tip potentials of PtIr-tip corresponding to different interface states.

As shown in Fig. 7,[7] lnI vs d is curved downward below 5 - 1 nA and varies with interface composition given by the tip-electrolyte potential E_T.[7] Whereas the E_T-dependence around 30nA$\approx$I can be explained by equally separated dipole resonances, the curvature, i.e. the ϕ_{eff} increase by a factor 2 in Fig. 7 above d ≈ 0.5 nm, i.e., I $<$ 1nA needs a separate explanation. As worked out to Refs. 7, the curvature is due to the confinement of the dipole resonances in the compact part of the electrolytic double layer, i.e. to a narrow interface region $\delta x_l \leq 0.4$ nm. Such a $\phi_{eff}(d)$ curvature has been found by several researchers even reaching $\phi_{eff} \approx 4$eV for I ≤ 0.1 nA, i.e. above 10^{10} Ω or d $>$ 1 nm.[28,29] Similar results have been obtained with in-situ and emersion methods.[30] The water dipole layers confined to anions or cation as 'crystal water' yield dipole resonances as found in distance dependencies of Ag-surfaces with MgF_2-H_2O coverage.[31] Such solvated ions exist also in electrolytes acting as intermediate state in tunneling. This has been observed in buffered solution containing 10 mM Tris, ph 7.4, 100 mM NaCl, 1 m M Mg SO_4.[32] There, for R $\geq 10^{10}$ Ω and kT $\approx$ 50 meV variable range hopping is encountered showing up as $\phi_{eff} < 0.1$eV and telegraph noise.[7,32] $\phi_{eff} \leq 0.1$ eV is equivalent roughly to a linearly increasing resistance, i.e. ρ (300 K) $> 10^3$ Ωcm.[6]

In all cases, the STM-tip approaching from large distances will first pick up a channel with maximal n-values, depending on minimal STM current.

So imaging for $d \geq 30$ nm and for $I < 20$ pA is reported.[24,32] In this case, the high concentration of solvated ions in distances of about 5 nm yield hopping conduction, so that electron transport dominates over electrolytic ion diffusion.

3.4.3 Incoherent tunneling along proteins

Above, water and other dipole layers with $E_{kin} \approx 0$ and $\sigma \approx 10^{-15}$ cm^2 act as intermediate states. Intermediate states at $E_{kin} \approx 10$ meV have been observed in Nb_2O_5.[2] In proteins the charge transfer is channeled along the hydrocarbon chains with $2\ \kappa_{eff} \approx 10$/nm i.e. $\phi_{eff} \approx 0.7$ eV for 6 to 14 methylen units independent of side-groups.[33,34] This so-called 'superexchange' is proposed to be due to incoherent tunneling via the resonances obvious in Fig. 5a at $E_{kin} \approx 0.4$ eV. The incoherence is due to the 1-dimensional nature and due to the high kT- and eU-values involved. This tunnel coupling along the chain corresponds to the coupling along water dipole layers as shown by similar large cross section, charging energies and hence, similar tunnel dynamics.

3.4.4 Lateral resolution in STM

Incoherent, intermediate state tunneling does not obstruct lateral resolution or even atomic resolution and even brighter images by water interfaces have been reported.[20] The lateral resolution is limited by the density of intermediate states or there localization at the substrate. As discussed above, a distance $L^* < 10^{-8}$ cm can be assumed with specific polarization and dipole moments at specific sites. Thus oriented OH^-HOH layers will not limit the resolution but their magnifying action may change contrast and details on a scale below L^*. In addition to those dipole resonances, intermediate states farer away (>0.2 nm) from the dipole layers will allow a resolution of the order of their mean distance L^*. For example, $n = 3$ did allow atomic resolution in electrolytic surrounding.[7] For $n \geq 4$ a smearing out is likely, but even variable range hopping ($n \geq 8$) at distances $d \geq 30$ nm allow a lateral resolution of about 1 nm because of $L^* < 5$ nm.[32]

3.4.5 Imaging of adsorbed molecules

In Section 3 shape resonances, especially dipole resonances, have been presented as intermediate states at E_F with small ΔU^*- and ΔU_c-values because of the delocalization. For example, the above discussed solid like OH^-HOH-layer with their dipoles can be considered as resonance extending laterally about 2 - 10 nm. Lateral, i.e. charge transfer along this layer with $\rho^{\|} \approx 10\Omega cm$ has been discussed in Refs. 7 and 20 and related to the 'superexchange' in Sect. 3.4.3.. The internal charge distribution of adlayers is sampled by shape resonances as to minimize ΔU^*- and ΔU_c-values, being also discussed in Refs. 7d, 23 and 25. ΔU^*- and ΔU_c-values can be measured by VTS combined with DTS.

3.4.6 Interaction with other barrier degrees of freedom

The author summarized in Ref. 2b various effects in barrier tunneling. Whereas on-site Coulomb correlations have been nicely analyzed in Ref. 35 the corrresponding spin flip scattering in resonant tunneling is mentioned in Ref. 2c. Inelastic processes yield in direct tunneling an up to 3 % tunnel current enhancement.[1,8] In intermediate state tunneling ΔU^* yield strong electron phonon coupling as made obvious by the activated tunnel conduction and the heating shown in Fig. 6.[22]

Acknowledgement

Fruitful discussion with R. Ahlrichs, K. Dransfeld, H. Hörber, D. M. Kolb, C. Miller, G. Repphun, U. Schmidt, S. Vinzelberg, and the participants of the Adriatic Research Conference at Trieste on "Electron and Ion Tranfer in Condesed Media" (July 1996) are thankfully acknowledged. Thanks are going to G. Nagy for critical reading of the manuscript.

References

1. e.g., E. L. Wolf, Principles of Electron Tunneling Spectroscopy (Oxford Science, London, 1989)
2. a. J. Halbritter, Surf. Sci. 122, 80 (1982)
   b. J. Halbritter, Surf. Sci. 159, 509 (1985)
   c. J. Halbritter, J. Appl. Phys. 58, 1320 (1985)
3. H. Knauer, J. Richter, P. Seidel, Phys. Status Solids a44, 303 (1977)
4. S. J. Bending, M. R. Beasley, Phys. Rev. Lett. 55, 324 (1985); M. Maito, M. R. Beasley, Phys. Rev. B 35, 2548 (1987)
5. Y. Xu, A. Matsuka, M. R. Beasley, Phys. Rev. B 42, 1492 (1992)
6. Y. Xu, PhD-thesis (Appl. Phys. Stanford University, 1992)
   Y. Xu., D. Ephron, M. R. Beasley, Phys. Rev. B 52, 2843 (1995)
7. a. J. Halbritter, G. Repphun, S. Vinzelberg, G. Staikov, W. J. Lorenz, Electochim. Acta 40, 1385 (1995)
   b.G. Repphun, J. Halbritter; J. Vac. Sci. Technol. A13, 1693 (1995)
   c. B. Berthe, J. Halbritter, Phys. Rev. B 43, 6880 (1991)
   d. J. Halbritter, J. Low Temp. Phys. 135, 1249 (1996) and submitted to Phys. Rev. B
   e. J. Halbritter, M. Grundner, J. Appl. Phys. 51, 5396 (1980) and J. Halbritter, IEEE Trans EI-20, 671 (1985)
8. J. Bardeen, Phys. Rev. Lett. 6, 57 (1961); M. Cohen, L. Falicoy, J. Philips, Phys. Rev. Lett. 8, 316 (1962)
9. E. H. Hauge, J. A. Stovneng, Rev. Mod. Phys. 61, 917 (1969) and references therein
10. B. N. J. Persson, A. Baratoff, Phys. Rev. B 38, 9616 (1988)
11. V. Kalmeyer, R. B. Laughlin, Phys. Rev. B 35, 9805 (1987)
12. R. H. Blick. R. J. Hang, J. Weis, D. Pfannkuche, K. v. Klitzing, K. Eberl, Phys. Rev. B 53, 7899 (1996); D. Pfannkuche, S. E. Ullov, Phys. Rev. Lett. 74, 1194 (1995)
13. A. I. Larkin, K. A. Matveev, Sov. Phys. JETP 66, 580 (1988); T. V. Shahbazyan, M. E. Raikh, Phys. Rev. B 49, 17123 (1994)
14. H. H. Müller, M. Schulz, J. Material Science: Mat. in Electronics 5, 329 (1994)
15. G. J. Schulz: in Principles of Laser Plasmas (Ed. G. Bekafi, Wiley, N. Y. 1976), p. 33;
16. G. Seng, F. Linder, J. Phys. B 9, 2539 (1976)
17. J. M. Kraus, C. J. Müller et al., Phys. Rev. Lett. 48, 14721 (1993)
18. J. K. Gimzewski, R. Möller, Phys. Rev. B 36, 1284 (1987); L. Olesen et al., Phys. Rev. Lett. 6, 1485 and 1276 (1996)
19. L. D. Landau, E. M. Lifshitz, Electrodynamics of Continous Medial (Pergamon Press N.Y., 1960), Chapt. I
20. M. Heim, G. Cevc, R. Guckenberger, H. F. Knapp, W. Wiegäbe, Biophysical J. 69, 489 (1995); M. Heim, R. Eschrich, A. Hillebrand, H. F. Knapp, G. Cevc, R. Guckenberger, J. Vac. Sci, Technol. B
21. W. Schmickler et. al., J. Electroanal. Chem. 290, 283 (1990); W. Schmickler, Sur. Sci. 335, 416 (1995)
22. Vu Thien Binh et al., Adv. in Imaging and Electron Physics 95, 63 (1996)
23. S. T. Purcell, N. Garcia, Vu. Thien Binh, L. Jones, J. M. Tour, J. Am. Chem. Soc. 116, 11987 (1994)
24. e.g., A. E. Hama, M. Tinkham, Phys. Rev. B 44, 5919 (1991)
25. N. J. Tao, Phys. Rev. Lett. 76, 4066 (1996)
26. Gabor Nagy, J. Electroanalyt. Chem. 409, 19 (1996) and
27. St. Vinzelberg, PhD-thesis (Uni Karlsruhe, 1995); St. Vinzelberg and U. Schmidt, private communication.
28. J. Pan et al., J. Phys. Chem. 98, 4205 (1994)

29. M. Strobel et al.,Microsystem Technologies 3, 10 (1996) and in Chemistry of the 21. Century (Eds. W. Lorenz, W. Plieth, IUPAC, Oxford, 1996)
30. D. M. Kolb, Z. Physikalische Chemie, Neue Folge 154, 179 (1987) and references therein
31. C. Bechinger, private communication
32. H. Hörber et al., to be published
33. J. Winkler, this volume , C. Liang, M. D. Newton, J. Phys. Chem. 97, 3199 (1993)
34. C. J. Miller, Physical Electrochemistry (Ed. I. Rubinstein, Dekker, N. Y., 1995), p. 27
35. D. Ephron et al., Phys. Rev. Lett. 69, 3112 (1992)

ION DIFFUSION RELATED TO STRUCTURE IN MOLTEN SALTS

M. P. TOSI

Istituto Nazionale di Fisica della Materia and Classe di Scienze, Scuola Normale Superiore, Piazza dei Cavalieri 7, I-56126 Pisa, Italy
and
International Centre for Theoretical Physics, Strada Costiera 11, I-34014 Trieste, Italy

Abstract

A model first proposed by Zwanzig to evaluate transport coefficients in cold dense fluids directly from the Green-Kubo time correlation formulae allows one to relate macroscopic diffusion coefficients to the local fluid structure. Applications to ionic diffusion processes in bulk molten salts and in the interface between a molten salt and a charged electrode are reviewed.

1. Introduction

As is well known, the self-diffusion coefficient of an atom in a fluid is a macroscopic manifestation of the autocorrelations of the particle velocity in time. The main focus of the present short review is on a model invoking a combination of vibrational and jumping motions of the ions in a molten salt for the purpose of evaluating the ionic diffusion coefficients. The model was originally proposed by Zwanzig [1] starting from an idea of Stillinger and Weber [2] and was used by him to derive a Stokes-Einstein-type relation between diffusion coefficient and shear viscosity in a cold dense liquid directly from the Green-Kubo time correlation formulae for transport coefficients. Further analysis and applications to monatomic fluids have been given by Mohanty [3] and by Tankeshwar *et al.* [4-6].

In essence, the model assumes that the configuration space of the many-body system is divided into "cells", each cell being associated with a local minimum on the free energy hypersurface. The configuration of the melt remains near one of these minima, performing approximate harmonic vibrations about it until a saddle point is found. Diffusion occurs at that point as a jump process out of the local free energy minimum into another cell, the effects of the jump being to rearrange the local configuration of the particles in some subvolume V* and to interrupt the oscillations within it so that the motions in V* before and after the jump are uncorrelated.

As extended by Tankeshwar and Tosi [7] to ion diffusion in bulk molten salts, the model involves a characteristic Einstein vibrational frequency and a characteristic residence time for each ionic species. With the help of sum rules these parameters can be evaluated in terms of pair potentials and partial pair distribution functions for the ionic species in the melt. Through this connection with local liquid structure the model accounts for (i) the large difference in mobilities of the two ionic species in superionic-conductor melts such as molten AgI or CuCl [7], and (ii) the so-called Chemla effect, i.e. the reversal of the mobilities of the two cations with varying concentration in liquid (Li,K)Cl mixtures [8]. These results, as well as their extension to evaluate the anisotropic diffusion coefficients of ions in a model electrical double layer [9], are briefly reviewed below.

2. Ionic diffusion in bulk 1:1 melts

The Green-Kubo formula for the diffusion coefficients $D_{\pm}$ of cations and anions is

$$D_{\pm} = \frac{k_B T}{N_{\pm} m_{\pm}} \int_0^{\infty} dt \sum_{i=1}^{N} < v_i^{\pm}(o) v_i^{\pm}(t) > / < (v_i^{\pm})^2 > \quad (1)$$

where $m_{\pm}$ and $v_i^{\pm}(t)$ are the masses and the velocities of cations and anions. The realization of Zwanzig's model replaces the sum over the ions in Eq. (1) by a sum over a frequency spectrum for normal modes localized in subvolumes V* and describing optic-like rattling motions of the ions in the structural cages provided by their neighbours. Approximating the spectrum by two Einstein frequencies $\omega_{\pm}$ and introducing a waiting time distribution of the form $\mathrm{sech}(t/\tau_{\pm})$ for cell jumps destroying coherence in any V*, we have a crude model for the velocity autocorrelation functions in Eq. (1), which reproduces their gross behaviours as obtained from computer simulation work on molten salts (for a review see [10]). This model yields

$$D_{\pm} = \frac{\pi k_B T \tau_{\pm}}{2 m_{\pm}} \mathrm{sech}(\pi \omega_{\pm} \tau_{\pm} / 2) \quad . \quad (2)$$

Evidently, many of the details of the ionic motions become irrelevant in performing the time integral shown in Eq. (1).

The Einstein frequencies and the relaxation times can now be evaluated from the short-time expansion of the velocity autocorrelation functions, with the results [7]

$$\omega_{\alpha}^2 = [5 A_{\alpha}{}^2 - B_{\alpha}] / (4 A_{\alpha}) \quad (3)$$

and

$$1/\tau_{\alpha}^2 = [B_{\alpha} - A_{\alpha}^2] / (4 A_{\alpha}) \quad (4)$$

where

$$A_\alpha = \frac{1}{m_\alpha}\sum_{\alpha'} n_{\alpha'} \int d\mathbf{r} g_{\alpha\alpha'}(r) \frac{\partial^2 U_{\alpha\alpha'}(r)}{\partial r_x^2}$$

and

$$B_\alpha \approx \frac{1}{2}A_\alpha^2 + \frac{1}{m_\alpha}\sum_{\alpha'} n_{\alpha'}\left(\frac{1}{m_\alpha}+\frac{1}{m_{\alpha'}}\right)\sum_{x'}\int d\mathbf{r} g_{\alpha\alpha'}(r)\left(\frac{\partial^2 U_{\alpha\alpha'}}{\partial r_x \partial r_{x'}}\right)^2 . \quad (6)$$

In these equations $g_{\alpha\beta}(r)$ are the pair distribution functions and $U_{\alpha\beta}(r)$ are the interionic pair potentials. From this formulae it is evident that the main contributions to the model parameters for each species, and hence to its diffusion coefficient come from the shells of first (unlike) neighbours and of second (like) neighbours. From Eq. (3) and (4) we also see that for $B_\alpha \rightarrow 5(A_\alpha)^2$ the Einstein frequency vanishes yielding gas-like behaviour, while for $B_\alpha \rightarrow (A_\alpha)^2$ the waiting time diverges yielding solid-like behaviour.

For an illustration of the quantitative usefulness of the theory one may take a simple molten salt such as NaCl: at T = 1340.5 K one finds [7] $D_+ = 16$ and $D_- = 13$ against values of 14.1 and 12.2 from molecular dynamics and values of 17.3 and 12.9 from experiment (here and in the following, diffusion coefficients are in units of 10^{-5} cm^2 s^{-1}). Evidently, neither the theory nor the simulation are fully quantitative. The calculation shows that (i) the contributions from the Coulomb part of the interionic potentials largely cancel out as a consequence of charge neutrality, thus emphasizing the role of short-range interactions and hence, in essence, of ionic sizes; and (ii) the observed similarity of $g_{NaNa}(r)$ and $g_{ClCl}(r)$ in molten NaCl implies similar back-scattering for each ionic species from its shell of second neighbours and hence a reduction of the ratio D_+/D_- to a value of order 1.3, appreciably smaller than one may surmise from ionic sizes and ionic masses.

2.1 Diffusion in superionic-conductor melts

In contrast to the observation made just above for molten NaCl, a strong asymmetry in the diffusion coefficients for the ionic species in the melts of superionic conducting solids was emphasized in molecular dynamics work by Trullas *et al.* [11]. As can be seen from Table 1, they found that the ratio D_+/D_- is of order 4 in molten CuCl and CuBr and reaches values of about 7 in molten CuI and of order 10 in molten AgI.

For CuBr, CuI and AgI the asymmetry of diffusion coefficients in the melt is a liquid-state remnant of the superionic behaviour that they attain in the solid state at high temperature. In CuCl no structural phase transition to a fast-ion conducting state occurs before melting, but the mobility of the Cu ions and the ionic conductivity reach rather high values in the hot crystal. It appears that melting preempts in CuCl a superionic phase transition in the solid state [10].

The theory connects the strong asymmetry of diffusion in these melts to the strong asymmetry in the pair distribution functions for like neighbours. This is well known for molten CuCl from neutron diffraction experiments carried out by the isotopic enrichment technique [12, 13]. The diffraction data show typical molten-salt ordering in the Cu-Cl and Cl-Cl correlations combined with a broad, featureless and heavily damped Cu-Cu structure factor. The Cu-Cu distribution function shows little structure except for excluded volume. Structural back-scattering of a Cu ion by its second-neighbour shell of Cu ions is therefore very ineffective, yielding from Eq. (3) - (6) a relatively low vibrational frequency and a relatively high jump frequency for the Cu ion.

A quantitative comparison between the model and the data from molecular dynamics is reported in Table 1 [7]. The calculated values of the ratio D_+/D_- are in good agreement with the data, whereas the absolute values of the diffusion coefficients are not consistently reasonable. In summary, the theory relates the residual traces of solid-state superionic behaviour in the melt to the poor short-range order in the fast ion component. Of course, both the structural and the transport properties ultimately derive from special features in the pair potentials, which are grossly summarized in an unexpectedly small effective size for the Cu ion.

Table 1. Calculated diffusion coefficients in superionic-conductor melts (in units of 10^{-5} cm^2 s^{-1}), compared with the results of molecular dynamics (MD). From ref. [7].

Salt	T(K)	D_+	$(D_+)_{MD}$	D_-	$(D_-)_{MD}$	D_+/D_-	$(D_+/D_-)_{MD}$
AgI	873	3.6	3.8	0.34	0.3	11.	12.7
CuCl	773	20.	10.	4.3	2.5	4.8	4.0
CuBr	880	13.	10.	2.3	2.7	5.6	3.9
CuI	923	18.	8.8	2.2	1.3	8.1	6.8

2.2 The Chemla effect in molten (Li,K)Cl mixtures

In measurements of the electrical mobilities of Li^+ and K^+ in molten (Li,K)Br mixtures Pèriè and Chemla [14] observed that the mobility of K^+ overtakes that of Li^+ with increasing content of KBr. The Chemla effect has since been reported for many other melts with two monovalent cations and a common anion (see e.g. [15]).

It has been proposed [8] that this behaviour arises from a structural liquid-state analogue of the so-called "off-centre" ions in the crystalline state (see e.g. [16]). A well-known instance of the latter is provided by KCl:Li, where the Li^+ impurity can occupy sites displaced along the eight equivalent <111> directions from a cation lattice site and tunnel between them. The locking of Li^+ in this configuration arises from the combined effect of its small radius, allowing close approach to Cl^- ions, and of the polarization energy gain ensuing from the off-centre displacement.

The transport behaviour of the two cations in the molten mixture is related to the structural features of the cation-anion distribution function, combined with the volume dilation which accompanies the increase of KCl content. Computer simulation studies [17, 18] as well as x-ray diffraction experiments [19] show that the first-neighbour peak in the Li-Cl distribution function is considerably sharper and lies at a considerably shorter distance than that in the K-Cl distribution function and that the coordination of Li^+ by chlorines is approximately fourfold rather than six-fold. From the simulation study of Lantelme and Turq [18] it is also seen that the velocity autocorrelation function for Li^+ shows marked remnants of an oscillatory behaviour at short times.

Table 2 reports a comparison between the values of the ionic diffusion coefficients at different values of the mole fraction and of the temperature in the liquid (Li,K)Cl mixture, as calculated form the pair distribution functions in the mixture [8] and as obtained directly from the simulated velocity autocorrelation functions [18]. It is seen from the Table that at 1096 K the diffusion coefficient of Li^+ becomes smaller than that of K^+ as the KCl content increases above 50-60 % and that lowering the temperature shifts this crossover towards lower KCl content. The behaviour of the diffusion coefficients with composition and temperature is a consequence of the special sharpness of the first-neighbour Li-Cl structural correlations, which increases with decreasing mole fraction of LiCl and with decreasing temperature. This yields from Eq. (3) - (6) a relatively high vibrational frequency and a relatively long waiting time between jumps for the Li^+ ion. It should also be noticed from the Table that the magnitude of the calculated diffusion coefficients tends to exceed that from the simulation study.

Table 2. Calculated self-diffusion coefficients in liquid (Li,K)Cl mixtures, compared with the results of molecular dynamics (in parentheses). The units are 10^{-5} cm^2 s^{-1}. From ref. [8].

MF of LiCl/T(K)	D_{Li}	D_K	D_{Cl}	D_K/D_{Li}
0.898/1096	14.7	13.4	19.4	0.91
	(11.1)	(10.1)	(8.95)	(0.91)
0.583/1096	11.2	13.0	18.6	1.16
	(7.78)	(7.57)	(7.06)	(0.97)
0.370/1096	8.2	12.8	19.7	1.56
	(5.95)	(7.02)	(6.68)	(1.18)
0.102/1096	4.7	12.5	14.1	2.65
	(2.91)	(7.05)	(6.85)	(2.42)
0.583/646	3.0	3.8	6.6	1.27
	(2.02)	(2.25)	(1.75)	(1.11)

3. Diffusion in the double layer at a model electrode/molten-salt interface

Because of the complexity of real electrochemical systems, computer simulations of greatly simplified models of electrified metal-electrolyte interfaces have had a very useful role in providing controlled data to test advances in statistical mechanics (see e.g. [20]). Esnouf *et al.* [21] (see also Williams [22]) have extended such studies to the molten-salt regime of the electrolyte. In their model a slab of ionic liquid representing molten KCl is enclosed between two oppositely charged hard walls. The single-particle density profiles show layering of the charge density in the melt along the z direction perpendicular to the walls and the diffusion coefficients of the two ionic species, which are obtained from their mean square displacements as functions of time, show appreciable anisotropy.

In extending the model reviewed in § 2 to diffusion in a slab one needs to introduce local vibrational and jump frequencies for each species, which are different for diffusion parallel or perpendicular to the walls [9]. For instance, for the parallel motion of a K^+ ion at position z_1 the quantity in Eq. (5) is first replaced by

$$A_{K\parallel}(z_1)=\frac{1}{m_K}\int_{-\infty}^{+\infty}dx_{12}\int_{-\infty}^{+\infty}dy_{12}\int_{-d/2}^{d/2}dz_2\Big[\frac{\rho_{KCl}(z_1,\mathbf{r}_{12})}{\rho_K(z_1)}\frac{\partial^2 U_{KCl}(r_{12})}{\partial x_{12}^2}+\frac{\rho_{KK}(z_1,\mathbf{r}_{12})}{\rho_K(z_1)}\frac{\partial^2 U_{KK}(r_{12})}{\partial x_{12}^2}\Big] \quad (7)$$

involving the inhomogeneous one-body and two-body densities, with d the separation distance between the walls. An average over the one-particle distribution yields

$$A_{K\parallel}=\Big[\int_{-d/2}^{d/2}dz_1\rho_K(z_1)A_{K\parallel}(z_1)\Big]\Big/\Big[\int_{-d/2}^{d/2}dz_1\rho_K(z_1)\Big] \quad .$$

(8)

The approximation

$$\rho_{\alpha\beta}(z_1,\mathbf{r}_{12})\cong\rho_\alpha(z_1)\rho_\beta(z_2)g_{\alpha\beta}(r_{12}) \quad (9)$$

for the two-body densities in terms of bulk distribution functions then allows an evaluation of the anisotropic diffusion coefficients in the slab.

The results are shown in Table 3 in comparison with the simulation data, at various values of the wall charge qe/d^2 per unit area and of the temperature. The trends shown in the Table arise from the layering of the charge density in the molten salt slab parallel to the walls. For any given ion in a layer, the most effective back-scattering comes from its unlike neighbours lying in the same layer for parallel diffusion and in the two adjacent layers for perpendicular diffusion. The back-scattering becomes relatively more important for perpendicular diffusion with increasing charge density on the electrodes, so that the ratios $D_{\parallel}/D_{\pi}$ are larger than unity and increase with the electrode charge density. Allowing for spill out of electrons at the electrodes in a jellium model reinforces these trends, by favouring increased ionic layering in the liquid slab through increased adsorption of counterions at the electrodes [9].

Tankeshwar and Tosi [9] have also taken advantage of the fact that in some of the simulation runs of Esnouf *et al.* [19] the liquid slab is quite thick to construct local diffusion coefficients for counter-ions and co-ions in a single double layer

Table 3. Anisotropic self-diffusion coefficients in model electrochemical cells in the molten salt regime, compared with the results of molecular dynamics (averages over the two ionic species, in parentheses). The units are 10^{-5} cm^2 s^{-1}. From ref. [9].

q/T(K)	$D_{K\|\|}$	$D_{K\pi}$	$D_{K\|\|}/D_{K\pi}$	$D_{Cl\|\|}$	$D_{Cl\pi}$	$D_{Cl\|\|}/D_{Cl\pi}$
0/1309	9.0	8.9	1.01	10.6	10.5	1.01
1/1326	9.1	8.7	1.04	10.8	10.5	1.03
2/1320	8.9	8.4	1.06	11.3	10.5	1.07
	(9.3)	(8.6)	(1.08)			
9/1406	11.6	7.4	1.58	14.7	10.3	1.43
	(12.9)	(8.9)	(1.45)			

single-particle densities. While parallel diffusion is found to change relatively little with distance from the charged wall across the layered charge density profile, strong correlations are found between the perpendicular diffusion processes and the local structure in the double layer. The correlation is similar for counter-ions and co-ions, except for their relative phase difference, and is simply summarized by the observation that perpendicular diffusion of each ionic species is slowed down in the regions where its density piles up.

4. Concluding remarks

In this brief review I have stressed in a number of examples the connection between structure and ionic transport in disordered ionic systems. The simple formulae obtained for the diffusion coefficient in terms of pair potentials and radial distribution functions have allowed us to highlight (i) the role of back-scattering by second neighbours in superionic-conductor melts, (ii) the role of structural trapping of small ions by their first neighbours in (Li,K)Cl mixtures, and (iii) the role of anisotropy and inhomogeneity in the double layer at a charged-wall/molten-salt interface.

It may be appropriate to explicitly recall at this point that from a basic viewpoint structure in a fluid is a consequence of its dynamical behaviour. Models leading from structure to transport are justified by the fact that structure is so much more easily accessible. Very recent work on a variety of liquids and disordered systems [23] has given evidence for an explicit relationship between the diffusion coefficient, after reduction to dimensionless form by scaling by the atomic collision frequency and the atomic diameter, and the pair distribution function through the excess entropy calculated in the two-particle approximation.

Finally, recent developments in the statistical mechanics of disordered systems [24-26] allow the solution of non-equilibrium structural problems relating to the structure taken by a fluid which permeates the micropores of a quenched matrix. It turns out that such a system in partial equilibrium can be mapped into a special multicomponent system at full equilibrium. In particular, a binary ionic melt may thereby be contrasted with a model superionic glass. From very different models of the interaction potentials [27, 28] a dynamical consequence of partial quenching is predicted to be the decreased localization of the annealed component.

References

[1] R. Zwanzig, *J. Chem. Phys.* **79** (1983) 4507.
[2] F. H. Stillinger and T. A. Weber, *Phys. Rev.* A **28** (1983) 2408.
[3] U. Mohanty, *Phys. Rev.* A **32** (1985) 3054.
[4] K. Tankeshwar, K. N. Pathak and S. Ranganathan, *J. Phys.* C **20** (1987) 5748.
[5] K. Tankeshwar, B. Sigla and K. N. Pathak, *J. Phys.: Condens. Matter* **3** (1991) 3173.
[6] R. K. Sharma, K. N. Pathak, K. Tankeshwar and S. Ranganathan, *Phys. Chem. Liquids* **29** (1995) 59.
[7] K. Tankeshwar and M. P. Tosi, *J. Phys.: Condens. Matter* **3** (1991) 7511.
[8] K. Tankeshwar and M. P. Tosi, *Solid State Commun.* **84** (1992) 245.
[9] K. Tankeshwar and M. P. Tosi, *J. Phys.: Condens. Matter* **3** (1991) 9817.
[10] M. Rovere and M. P. Tosi, *Rep. Progr. Phys.* **49** (1986) 1001.
[11] J. Trullas, A. Girò and M. Silbert, *J. Phys.: Condens. Matter* **2** (1990) 6643.
[12] I. D. Page and K. Mika, *J. Phys.* C **4** (1971) 3034.
[13] S. Eisenberg, S. F. Jal, J. Dupuy, P. Chieux and W. Knoll, *Phil. Mag.* A **46** (1982) 195.
[14] J. Pèriè and M. Chemla, *C. R. Acad. Sci.* **250** (1960) 3986.
[15] M. Chemla and I. Okada, *Electrochim. Acta* **35** (1990) 1761.
[16] A. M. Stoneham, *Theory of Defects in Solids*, Chap. 21 (Clarendon Press, Oxford 1965).
[17] C. Caccamo and M. Dixon, *J Phys.* C **13** (1980) 1887.
[18] F. Lantelme and P. Turq, *J. Chem. Phys.* **77** (1982) 3177.
[19] I. Okada, H. Okano, H. Ohtaki and R. Takagi, *Chem. Phys. Lett.* **100** (1983) 436.
[20] S. L. Carnie and G. M. Torrie, *Adv. Phys. Chem.* **56** (1984) 141.
[21] R. M. Esnouf, A. C. D. Smith and P. J. Grout, *Phil. Mag.* A **58** (1988) 27.
[22] D. I. T. Williams, *Ph. D. Thesis* (Oxford University, 1990).
[23] M. Dzugutov, *Nature* **381** (1996) 137.
[24] W. G. Madden and E. D. Glandt, *J. Stat. Phys.* **51** (1988) 537.
[25] J. A. Given and G. Stell, in *Condensed Matter Theories*, ed. L. Blum and F. Bary Malik (Plenum, New York 1993), p. 395,
[26] M. P. Tosi, *N. Cimento* D **16** (1994) 169.
[27] H. Tatlipinar, G. Pastore and M. P. Tosi, *Phil. Mag. Lett.* **68** (1993) 357.
[28] M. Cardenas and M. P. Tosi, *Physica* B **222** (1996) 143.

RELAXATION BEHAVIOR OF ION CONDUCTING GLASSES

A. BUNDE[1], W. DIETERICH[2], P. MAASS[2], AND M. MEYER[1,3]
[1]*Institut für Theoretische Physik, J.-L.-Universität*
D-35392 Giessen, Germany
[2]*Fakultät für Physik, Universität Konstanz*
D-78434 Konstanz, Germany
[3]*Sonderforschungsbereich 173, Universität Hannover*
D-30176 Hannover, Germany

We investigate by Monte Carlo simulations the diffusion of ions in an energetically disordered lattice, where the Coulomb interaction between the mobile ions is explicitely taken into account. We show that the combined effect of Coulomb interaction and disorder can account for the ionic ac-conductivity in glasses and the recently discovered non-Arrhenius behavior of the dc-conductivity in glassy fast ionic conductors. Our results suggest that glassy ionic conductors can be optimized by lowering the strength of the energetic disorder but that the ionic interaction effects set an upper bound for the conductivity at high temperatures.

1 Introduction

The low frequency dynamic response of many glassy materials is governed by the transport of mobile ions or other charged mobile defects, and a large number of experimental techniques has been applied to get an understanding of the underlying microscopic mechanism. Among them are tracer experiments [1], conductivity measurements (modulus spectroscopy) [2–4], nuclear spin lattice relaxation [5–7], quasielastic neutron scattering [8,9], internal friction, ultrasonic absorption, and Brillouin scattering measurements [10–12]. In all these experiments the measured quantities show pronounced deviations from the standard behavior one would expect for a purely random motion of the mobile ions.

The most common quantity studied is the dynamic conductivity $\hat{\sigma}(\omega)$. For fixed temperature T, $\hat{\sigma}(\omega)$ exhibits a dc-plateau at low frequencies (below some crossover frequency $1/\tau_\sigma$), and follows an approximate power law behavior up to far-infrared frequencies [13],

$$\hat{\sigma}(\omega) \sim \begin{cases} \sigma_{\mathrm{dc}}, & \omega\tau_\sigma \ll 1, \\ (i\omega)^{n_\sigma}, & \omega\tau_\sigma \gg 1. \end{cases} \tag{1}$$

The dc-conductivity σ_{dc} usually shows an Arrhenius behavior below the glass transition temperature,

$$\sigma_{\mathrm{dc}} T = A_\sigma \exp(-E_a/k_B T)\,. \tag{2}$$

The crossover frequency τ_σ^{-1} is also thermally activated with an activation energy E_σ approximately equal to E_a. The exponent $n_\sigma > 0$ tends to increase, if the temperature is lowered or if the frequency is increased by several orders of magnitude. From standard random walk theory one would expect no dispersion to occur, i.e. $n_\sigma = 0$. The widespread occurrence of an anomalous low frequency dielectric behavior in disordered solids characterized by power laws was first pointed out by Jonscher [14] and is known as the "universal dielectric response".

Recently it was found that in some glassy fast ion conductors the dc-conductivity shows strong deviations from a simple Arrhenius law even below the glass transition temperature [15,16]: The values of σ_{dc} at ambient temperatures are significantly smaller than expected when extrapolating the Arrhenius law valid at low temperatures T. Formally, the deviation from the Arrhenius law can be described by a temperature dependent (apparent) activation energy $E(T)$, which decreases with increasing T. We like to note that a non-Arrhenius behavior was found earlier in ion conducting glasses but disappeared after annealing [17].

From a theoretical point of view, the ionic transport in solids is a very complex phenomenon and rigorous solutions are not available. For an ordered host lattice a mode-coupling theory has been developed to study the effect of long range Coulomb interactions between the mobile ions [18], but it has turned out recently that also the disorder plays an essential role to describe the experimental situation properly [19,20]. A rigorous solution of this complex problem is impossible to obtain, and various phenomenological and semi-microscopic approaches have been sucessfully applied. Prominent examples are the coupling scheme proposed by K.L. Ngai [21], the jump relaxation model proposed by K. Funke [8], and the diffusion-controlled relaxation model elaborated by S.R. Elliott and A.P. Owens [22]. Attempts have been made to map the dynamics of the many body problem onto the dynamics of a single particle moving in a complex energy landscape [23–25].

For a more microscopic description of the ionic transport one is dependent upon numerical investigations. In this paper we summarize the results of a detailed Monte Carlo study of a simple lattice gas model by which the effects of both long range Coulomb interaction between the mobile ions and energetic disorder in the host lattice is investigated in a systematic way. In contrast to other methods (e.g. molecular-dynamic simulations), the Monte-Carlo method is not restricted to the short time regime of the ion dynamics and can explore the behavior of the basic quantities for the relevant time and frequency scales.

2 Model and Numerical Technique

To describe the ionic transport one has to take into account at least (a) the Coulomb interaction between the mobile charge carriers and (b) the structural disorder of the host system. Clearly, the implementation of the Monte Carlo procedure requires to specify the kind of structural disorder, which dominates the transport process in a given material. In the context of network glasses at low doping with ions the influence of randomly placed immobile counterions has been studied recently [26]. Here we focus on general effects of disorder, pertaining, for example, to strongly doped materials or superionic glasses.

As a model for the energetically disordered substrate, we choose a simple cubic lattice of size $L = 40\xi$ with periodic boundary conditions and lattice spacing ξ, where ξ is the typical distance over which the energies along the ionic diffusion pathways are correlated. To each lattice site i we assign an energy ϵ_i drawn from a Gaussian distribution $P(\epsilon)$ with zero mean and variance σ_ϵ^2. The number of mobile ions per lattice site is $c \ll 1$ and $\rho = c/\xi^3$ denotes the ionic number density. We assume that sites cannot be occupied by more than one ion. The strength of the Coulomb interaction relative to the thermal energy k_BT is characterized by V_c/k_BT, where $V_c \equiv q^2/R$ and $R \equiv (3/4\pi\rho)^{1/3}$ is the half mean distance between the mobile ions.

To model the diffusion process we use a standard Monte Carlo algorithm: In each elementary step of the simulation, an ion is chosen randomly, and a nearest neighbor site is also chosen, to which the ion attempts to jump. If the neighboring site is occupied by another ion, the jump is rejected. If the neighboring site is vacant, the ion jumps to it with probability $w = \min\{1, \exp(-\Delta E/k_BT)\}$, where ΔE is the change of the total energy caused by the jump and is given by $\Delta E = \Delta E_s + \Delta E_c$. Here $\Delta E_s = \epsilon_j - \epsilon_i$ is the site energy difference and ΔE_c is the energy difference due to the Coulomb interaction of the jumping particle with all other particles. For the calculation of ΔE_c one has to take care of the image charges caused by the periodic boundary conditions, which is done by using Ewald's method [27]. After each elementary step, the time t is incremented by τ_0/N, where τ_0^{-1} is an attempt frequency [a].

Initially the particles are randomly distributed over the system. In order to reach a thermalized state at the final simulation temperature, the system is cooled down from a high temperature, $k_BT = 10\max\{\sigma_\epsilon, V_c\}$, to the simulation temperature by a linear increase of $1/T$. After the cooling process, the

[a] The spatial coarse graining on a length scale ξ represents an approximate procedure, which allows one to study the low-frequency dynamics. The attempt frequency is chosen as $\tau_0^{-1} \equiv v_{th}^2/\gamma\xi^2$, where $v_{th} = (k_BT/m)^{1/2}$ is the thermal velocity and γ the friction constant.

temperature is held constant for a time equal to the cooling time; then the quantities of interest are determined. The time for the thermalization process is chosen such that the mean total energy of the ion system does not change significiantly during the constant temperature phase, and is typically 3 to 5 times larger than the simulation time itself.

At time $t = 0$ after thermalization, all particle positions $\mathbf{r}_i(0)$ are stored. The mean square displacement of a tracer particle $\langle r^2(t)\rangle$, at time $t > 0$ is calculated from

$$\langle r^2(t)\rangle = \frac{1}{N}\left\langle \sum_{i=1}^{N} \mathbf{r}_i^2(t)\right\rangle , \tag{3}$$

where $\mathbf{r}_i(t) \equiv \mathbf{x}_i(t) - \mathbf{x}_i(0)$ is the displacement of particle i, and $\langle\ldots\rangle$ denotes an average over many configurations. It is convenient to define a time dependent diffusion constant $D(t) \equiv \langle r^2(t)\rangle/6t$. In the limit of large times, $t \to \infty$, $D(t)$ approaches the tracer diffusion constant D_{tr}.

The mean square collective displacement of the particles at time t is

$$\langle r_c^2(t)\rangle \equiv \frac{1}{N}\left\langle \left(\sum_{i=1}^{N} \mathbf{r}_i(t)\right)^2\right\rangle = \langle r^2(t)\rangle + \frac{1}{N}\sum_{i\neq j}\langle \mathbf{r}_i(t)\cdot\mathbf{r}_j(t)\rangle \,. \tag{4}$$

Since the second derivative of $\langle r_c^2(t)\rangle$ is proportional to the current autocorrelation function, $\langle r_c^2(t)\rangle$ is related to the conductivity via the Kubo formula [28],

$$\hat{\sigma}(\omega) = -\frac{\rho q^2}{k_B T}\frac{\omega^2}{6}\lim_{\epsilon\to+0}\int_0^{\infty}\langle r_c^2(t)\rangle e^{i\omega t-\epsilon t}dt\,, \tag{5}$$

where q denotes the charge of the mobile ions. For non-interacting particles, the cross-terms for $i \neq j$ in (4) vanish, and $\langle r_c^2(t)\rangle$ can be replaced by $\langle r^2(t)\rangle$ in (5). It was shown elsewhere [29] that for disordered systems studied here even with strong Coulomb interactions between the mobile ions, the cross terms are not relevant for the overall behavior of $\hat{\sigma}(\omega)$, and the conductivity can be calculated in good approximation from the mean squared displacement of a tracer particle.

3 Results

3.1 Non-Interacting Particles

First we neglect the long range Coulomb interaction. Similar models of non-interacting particles in a disordered energy landscape have been studied earlier [23,30]. Figure 1a shows the time dependent diffusion coefficient $D(t) \equiv$

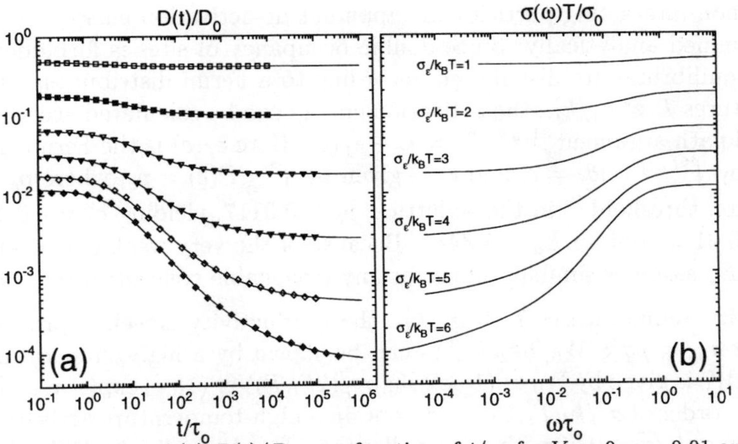

Figure 1: Plot of (a) $D(t)/D_0$ as a function of t/τ_0 for $V_c = 0$, $c = 0.01$ and $\sigma_\epsilon/k_BT = 1$ (□), 2 (■), 3 (▽), 4 (▼), 5 (◇), and 6 (◆), and (b) $\sigma'(\omega)T/\sigma_0$ as a function of $\omega\tau_0$ for the same paramters as in (a).

$\langle r^2(t)\rangle/6t$ in units of $D_0 \equiv \xi^2/6\tau_0$ as a function of t/τ_0 for $V_c = 0$, $c = 0.01$, and different values of σ_ϵ/k_BT. For $t/\tau_0 \ll 1$, $\langle r^2(t)\rangle$ is proportional to the total number of hops, which increases linearly with time and therefore $D(t)$ is constant, $D(t) = D_{\rm st} = W_0\xi^2/6$, where W_0 denotes the mean hopping rate. For $t/\tau_0 > 1$, $D(t)$ decreases with time t and finally approaches $D_{\rm tr}$. At low temperatures, $\sigma_\epsilon/k_BT \gtrsim 3$, $D_{\rm tr}$ is much smaller than $D_{\rm st}$, and one can identify an intermediate time regime $t_1 < t < t_2$, where $D(t)$ shows approximate power law behavior,

$$D(t) \sim t^{-n_{\rm D}}, \quad t_1 < t < t_2. \tag{6}$$

In the intermediate time regime, the data points roughly follow a straight line in the log-log plot. The upper crossover time t_2 and the exponent $n_{\rm D}$ increase with decreasing temperature, while the lower crossover time t_1 is approximately independent of temperature and of the order of τ_0.

The conductivity can be calculated from $\langle r^2(t)\rangle$ by using (5) and neglecting the cross-correlation terms in $\langle r_c^2(t)\rangle$. Figure 1b shows the real part of the conductivity times temperature, $\sigma'(\omega)T$, in units of $\sigma_0 \equiv \rho q^2 D_0/k_B$ as a function of $\omega\tau_0$. Evidently, the behavior of $\sigma'(\omega)T$ is strictly analogous to the behavior of $D(t)$. At low frequencies, $\omega \ll t_2^{-1}$, the conductivity is constant, $\sigma'(\omega)T = \sigma_{\rm dc}T = \sigma_0 D_{\rm tr}/D_0$. In an intermediate frequency regime, $t_2^{-1} \ll \omega \ll t_1^{-1}$, $\sigma'(\omega)$ shows approximate power law behavior $\sigma'(\omega) \sim \omega^{n_{\rm D}}$. At high frequencies $\omega \gg t_1^{-1}$ the conductivity becomes constant again, $\sigma'(\omega)T = \sigma_{\rm hf}T = \sigma_0 D_{\rm st}/D_0$.

For non-interacting particles the apparent dc-activation energy $E(T)$ can be determined analytically. Since double occupancy of sites is forbidden, the ions in equilibrium are distributed according to a Fermi distribution. At low temperatures $T \ll \sigma_\epsilon/k_B$, the activation energy can be calculated according to a critical path argument [31,32], $E_a = \epsilon_c - \epsilon_f(c)$. Here $\epsilon_f(c)$ is the Fermi energy, defined by $\int_{-\infty}^{\epsilon_f} P(\epsilon)d\epsilon = c$, and ϵ_c is given by $\int_{-\infty}^{\epsilon_c} P(\epsilon) = p_c$, where p_c is the percolation threshold [33] in the sc-lattice, $p_c \simeq 0.3117$, yielding $\epsilon_c = -0.491\sigma_\epsilon$. For $c = 0.01$ we obtain $E_a = 1.84\sigma_\epsilon$. Because of the very weak dependence of ϵ_f on c, E_a assumes similiar values for any reasonable concentration $c \ll 1$.

At high temperatures $T \gg \sigma_\epsilon/k_B$, the conductivity is well approximated by $\sigma_{\rm dc} \simeq \sigma_{\rm hf} = \rho q^2\xi^2 W_0/6k_BT$. As can be shown by a high-temperature expansion, W_0 is given by $W_0 \simeq (1-c)\text{erfc}(\sigma_\epsilon/(2k_BT)) \simeq (1-c)\exp(-\sigma_\epsilon/\sqrt{\pi}k_BT)$, in leading order of σ_ϵ/k_BT. Hence we obtain a high-temperature activation energy $E_0 = \sigma_\epsilon/\sqrt{\pi} \cong 0.56\sigma_\epsilon$ that is smaller than E_a. Accordingly, $E(T)$ changes from E_a for low temperatures to E_0 at a crossover temperature $T_x \simeq \sigma_\epsilon/k_B$. Figure 2a shows the simulation results for $\sigma_{\rm dc}$ (data points) in comparision with the analytical results (lines). Except for the crossover regime $T \approx T_x$, the calculated activation energy agrees perfectly with the simulation data.

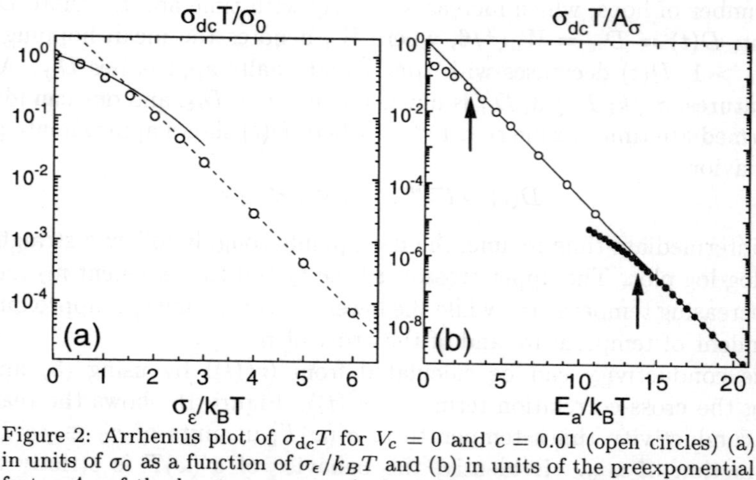

Figure 2: Arrhenius plot of $\sigma_{\rm dc}T$ for $V_c = 0$ and $c = 0.01$ (open circles) (a) in units of σ_0 as a function of σ_ϵ/k_BT and (b) in units of the preexponential factor A_σ of the low-temperature Arrhenius law as a function of E_a/k_BT. In (a) the full line shows the high temperature approximation (see text), the dashed line corresponds to $A_\sigma \exp(-E_a/k_BT)$, where $E_a = 1.84\sigma_\epsilon$ and A_σ was taken to get the best fit to the data. For comparision, the experimental data for $0.4AgI+(0.6)[0.525Ag_2S+0.475(B_2S_3{:}SiS_2)$ are shown in (b) (filled symbols, redrawn from [15]). The solid line is drawn as guide for the eye.

At first glance, the crossover at high temperatures seems to be very similar to recent experimental results for fast ion glasses [15]. Notice however, that both E_a and $k_B T_x$ are of the same order of magnitude, determined by σ_ϵ. This is a serious disagreement to the experiments, where the non-Arrhenius behavior sets in at temperatures more than one order of magnitude smaller than E_a/k_B. The disagreement can be seen clearly in Fig. 2b, where we have plotted both the simulation results in the absence of Coulomb interactions (open circles) and the experimental data [15] (filled circles) as a function of E_a/k_BT. The arrows indicate the crossover temperatures and show that both disagree by about an order of magnitude.

3.2 Interacting Particles

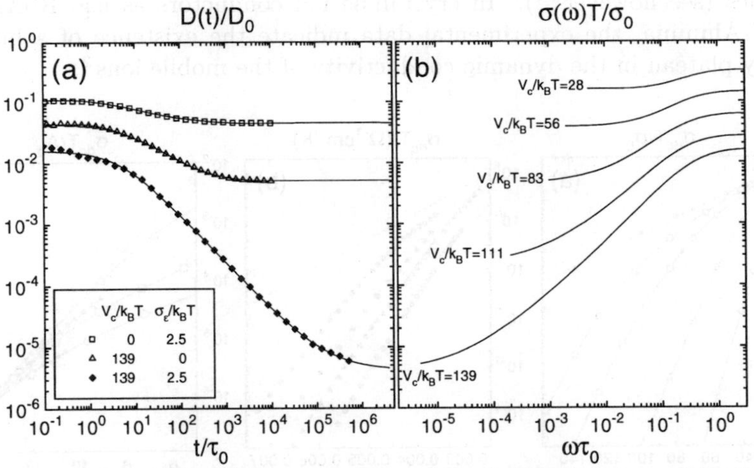

Figure 3: Plot of (a) $D(t)/D_0$ as a function of t/τ_0 for $V_c = 0$ and $\sigma_\epsilon = 2.5k_BT$ ($\square$), $V_c = 139k_BT$ and $\sigma_\epsilon = 0$ ($\triangle$), and $V_c = 139k_BT$ and $\sigma_\epsilon = 2.5k_BT$ ($\blacklozenge$), and (b) $\sigma'(\omega)T/\sigma_0$ as a function of $\omega\tau_0$ for $\sigma_\epsilon/V_c = 2.5/139 = 0.018$ constant, and V_c/k_BT =28, 56, 83, 111, and 139.

Next we include the Coulomb interaction. The model now is characterized by the interaction energy V_c and the disorder energy σ_ϵ. Figure 3a shows $D(t)$ in units of D_0 as a function of t/τ_0 (*i*) for uncharged particles in a disordered environment, $V_c = 0$, $\sigma_\epsilon = 2.5k_BT$, (*ii*) for ions with a strong Coulomb repulsion in an ordered environment $V_c = 139k_BT$, $\sigma_\epsilon = 0$, and (*iii*) for interacting particles in an disordered environment $V_c = 139k_BT$, $\sigma_\epsilon = 2.5k_BT$. For the cases (*i*) and (*ii*), the difference between the short and long time diffusion constant is less than one order of mangitude. For case (*iii*) in contrast, the difference is more than three orders of magnitude. This shows clearly that the combined effect of both energetic disorder and interaction is *not* just a simple superposition of the single ingredients. Because of this highly nonlinear effect, *neither* the disorder *nor* the interaction can be neglected [19].

The conductivity was obtained by using (5) and neglecting the cross correlations. Figure 3b shows $\sigma'(\omega)T$ in units of σ_0 as a function of $\omega\tau_0$ for $\sigma_\epsilon/V_c = 0.018$ and several values of V_c/k_BT. Again the behavior is strictly analogous to the time dependent diffusion constant. At low temperatures, the curves show power law behavior, $\sigma'(\omega) \sim \omega^{n_D}$, over several orders of magitude with an exponent $n_D \simeq 0.8$. The overall behavior is quite similar to the experimental situation. The plateau at high frequencies, however, is usually not found in glasses, because dynamical processes not included in the lattice gas model, e.g. vibrations of the glassy matrix, become dominant at high frequencies (see however [34]). In crystalline ion conductors as e.g. $RbAg_4I_5$ or Na-β''-Alumina, the experimental data indicate the existence of a high frequency plateau in the dynamic conductivity of the mobile ions [2].

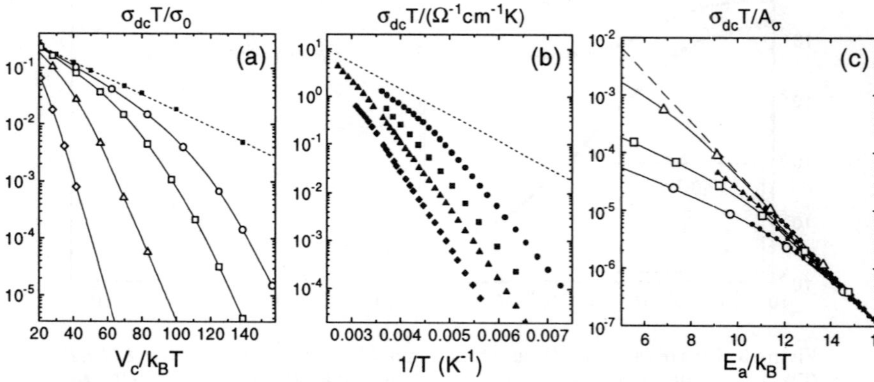

Figure 4: Arrhenius plots of the dc-conductivity $\sigma_{dc}T$ (a) in the model in units of σ_0 for $\sigma_\epsilon/V_c = 0$ (■), 0.0115 (○), 0.018 (□), 0.036 (△), and 0.072 (◇), and (b) for zAgI+$(1-z)[0.525Ag_2S+0.475(B_2S_3{:}SiS_2)]$ in units of $\Omega^{-1}cm^{-1}K$ for $z = 0$ (♦), 0.2 (▲), 0.3 (■), and 0.4 (●) (redrawn from [15]). The dashed lines indicate the upper mobility limit predicted by the model. In (c) the data from (a) and (b) are shown together as functions of E_a/k_BT and are normalized with respect to the preexponential factors A_σ in the corresponding Arrhenius laws. The solid lines in (a) and (c) are drawn as guide for the eye.

In the following we will concentrate on the apparent dc activation energy $E(T)$ as a function of temperature T and disorder strength σ_ϵ/V_c. Figure 4a shows $\sigma_{dc}T$ in units of σ_0 as a function of V_c/k_BT for $c = 0.01$ and $\sigma_\epsilon/V_c = 0.0115$, 0.018, 0.036, and 0.072. At low temperatures, each curve follows a straight line corresponding to an Arrhenius law with constant activation energy E_a, and E_a decreases with decreasing σ_ϵ. The Arrhenius law is valid up to a crossover temperature T_x, where the curves bend toward lower diffusivities. In all cases, the crossover temperature T_x is of the order of σ_ϵ/k_B. For comparison we have redrawn in Fig. 4b the experimental conductivity data [15] for $z\text{AgI}+(1-z)[0.525\text{Ag}_2\text{S}+0.475(\text{B}_2\text{S}_3{:}\text{SiS}_2)]$ with mole fractions z between zero and 0.4. Evidently, when increasing z, the experimental behavior is analogous to the model behavior when decreasing σ_ϵ: E_a becomes smaller and the non-Arrhenius behavior starts to occur at lower T.

The similarity between the results found in the model and in the experiment becomes even more evident in Fig. 4c, where the data from Fig. 4a and Fig. 4b are plotted in the same way as in Fig. 2b. The experimental curve for $z = 0.4$ is almost perfectly reproduced by the model when $\sigma_\epsilon = 0.0115V_c$ (see the filled and open circles in Fig. 4c). The experimental curves for $z = 0$, 0.2 and 0.3 correspond to disorder strengths within a range $0.015V_c < \sigma_\epsilon < 0.036V_c$. It is remarkable that the model not only gives a good fit to the overall shape of the conductivity curves but also reproduces the small values of k_BT_x/E_a.

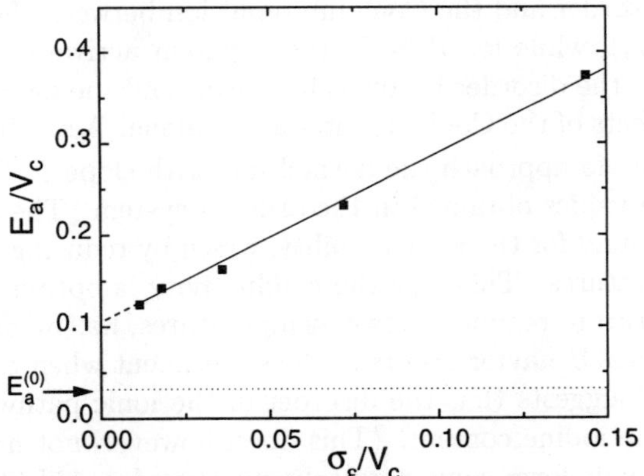

Figure 5: Simulated activation energy E_a/V_c as a function of the strength of the disorder σ_ϵ/V_c. The dotted line indicates the activation energy $E_a^{(0)}$ in the ordered system ($\sigma_\epsilon = 0$), which is about 3 times smaller than the limiting value for $\sigma_\epsilon \to 0$.

In order to understand why k_BT_x/E_a becomes much smaller than one for small σ_ϵ, we need to know how E_a depends on σ_ϵ quantitatively. We therefore plotted in Fig. 5 E_a/V_c (resulting from our computer simulations) as a function of σ_ϵ/V_c and find that E_a/V_c increases linearily with σ_ϵ/V_c,

$$E_a = a_1 V_c + a_2 \sigma_\epsilon \,, \tag{7}$$

where $a_1 = 0.10$ and $a_2 = 1.84$ for $c = 0.01$. Since $T_x \simeq \sigma_\epsilon/k_B$, we obtain $k_B T_x/E_a \simeq \sigma_\epsilon/(a_1 V_c + a_2\sigma_\epsilon) \sim \sigma_\epsilon/a_1 V_c$ for $\sigma_\epsilon \to 0$, and accordingly $k_B T_x/E_a$ becomes much smaller than one for small σ_ϵ. We note that for large $\sigma_\epsilon \gg V_c$, $E_a \simeq a_2\sigma_\epsilon$, which agrees with the result $E_a = 1.84\sigma_\epsilon$ for $\sigma_\epsilon/V_c \to \infty$, i.e. in the disordered system without Coulomb interaction (see the critical path analysis given above). However, when extrapolating Eq. (7) to $\sigma_\epsilon = 0$, we obtain $E_a = 0.10V_c$, which is more than 3 times larger than the value $E_a^{(0)} = 0.032V_c$ in the ordered system (see the dashed and dotted lines in Fig. 5). This shows again the interplay between the effects of interaction and the static energy fluctuations: A small energy fluctuation $\sigma_\epsilon \ll E_a$ (but still larger than the thermal energy, $\sigma_\epsilon > k_B T$) increases the activation energy E_a by a factor of 3.

The agreement between model and experiment shown in Fig. 4, enables us to give a coherent explanation of the non-Arrhenius behavior and the associated problem of optimized ionic conduction. The non-Arrhenius behavior occurs as a result of a crossover between two different temperature regimes: For $T \ll T_x$, the activation energy E_a is governed by the combined effect of the energetic disorder and the Coulomb repulsion between the mobile ions (as shown in Eq. (7)), while for $T \gg T_x$ the apparent activation energy becomes smaller, because the disorder becomes irrelevant and the ionic mobility is governed by the effects of the Coulomb interaction alone. Accordingly, for $T \gg T_x$ all curves in Fig. 4a approach the dashed line with slope $E_a^{(0)}/V_c$, which corresponds to the values obtained in the ordered system. The dashed line thus sets an upper bound for the ionic mobility, which by reducing σ_ϵ is approached at lower temperatures. This way the conductivity is optimized by extending the high-temperature regime to lower temperatures, i.e. by decreasing T_x.

The analogous behavior occurs in the experiment when z is increased (see Fig. 4b), which suggests that the disorder of the ionic pathways is decreased by increasing the iodine content. (This does, however, not nescessarily imply that the iodine ions form some regularly structured α-AgI-like microdomains in the glassy network, as it was conjectured e.g. in [35] but later questioned in [11].) One can roughly estimate the location of the experimental limiting curve by extrapolating the conductivity data in Fig. 4b to higher temperatures. The resulting curve is drawn as dashed line in Fig. 4b and has an activation energy $E_0 = 0.11\,\mathrm{eV}$. In order to check for consistency, we compare this value with the activation energy $E_a^{(0)} = 0.032V_c$ of the limiting curve in the model. As a result we get $V_c = 3.4\,\mathrm{eV}$, which is an acceptable value. For example, if we assume that the number density of Ag^+ ions is about $2 \times 10^{22}\mathrm{cm}^{-3}$ as in

xAgI+$(1-x)$Ag$_2$O-B$_2$O$_3$ [36], and that the high-frequency dielectric constant is $\epsilon_\infty \approx 2$, we obtain $V_c \simeq 3.3$ eV.

4 Conclusions

We have studied the transport of ions in an energetically disordered structure and found that in the presence of both disorder and Coulomb interaction the dispersive behavior of the ac-conductivity and the non-Arrhenius behavior of the dc-conductivity observed in glassy fast ion conductors is correctly reproduced. We have concentrated on the dynamic conductivity and found that both Coulomb interaction and structural disorder are essential for describing the experimental situation properly. Deviations from the standard behavior occur also in other quantities, e.g. in spin lattice relaxation and quaielastic neutron scattering, and can be explained also within the framework of the present model [20, 29]

While the model describes well the dispersive transport behavior, it does not aim to explain those transport anomalies that occur, when the composition of the glass is changed. Examples are the mixed-alkali effect or the drastic increase of the dc conductivity with the ionic content in single ionic glasses. For a description of these effects one has to take into account explicitely the couplings of the mobile ions to the underlying structural network [26, 37, 38].

References

1. H. Jain, N.L. Peterson, and H.L. Downing, *J. Non-Cryst. Solids* **55**, 283 (1983).
2. K. Funke, *Prog. Solid St. Chem.* **22**, 111 (1993).
3. C.T. Moynihan, L.P. Boesch, and N.L. Laberge *Phys. Chem. Glasses* **14**, 122 (1973).
4. M.D. Ingram, *Phys. Chem. Glasses* **28**, 215 (1987).
5. O. Kanert, J. Steinert, H. Jain, and K.L. Ngai, *J. Non-Cryst. Solids* **131-133**, 1001 (1991); O. Kanert, R. Küchler, K.L. Ngai, and H. Jain, *Phys. Rev.* B **49**, 76 (1994).
6. O. Kanert, R. Küchler, J. Diekhöfer, X. Lu, and H. Jain, *Phys. Rev.* B **49**, 629 (1994).
7. P. Heitjans, W. Faber, and A. Schirmer, *J. Non-Cryst. Solids* **131-133**, 1053 (1991); P. Heitjans, *Solid State Ionics* **18/19**, 50 (1986); W. Franke and P. Heitjans, *Ber. Bunsenges. Phys. Chem.* **96**, 1674 (1992).
8. K. Funke and I. Riess, *Z. Phys. Chem. Neue Folge* **140**, 217 (1984); K. Funke, *Z. Phys. Chem. Neue Folge* **154**, 251 (1987).

9. T. Suemoto and M. Ishigame, *Phys. Rev.* B **33**, 2757 (1986).
10. L. Börjesson *Phys. Rev.* B **36**, 4600 (1987).
11. L. Börjesson, L.M. Torrel, and W.S. Howells, *Phil. Mag.* B **59**, 105 (1989).
12. G. Carini, M. Federico, and G. Tripodo, *Phil. Mag.* B **65**, 153 (1992).
13. D.P. Almond, G.K. Duncan, and A.R. West, *Solid State Ionics* **8**, 159 (1983).
14. A.K. Jonscher, *Nature* **267**, 673 (1977).
15. J. Kincs and S. W. Martin, *Phys. Rev. Lett.* **76**, 70 (1996).
16. K. Ngai and A. K. Rizos, *Phys. Rev. Lett.* **76**, 1296 (1996).
17. M. D. Ingram, C. Vincent, and A. Wandless, *J. Non-Cryst. Solids* **53**, 73 (1982).
18. A. Bunde, D.K. Chaturvedi, and W. Dieterich, *Z. Physik* B **47**, 209 (1982); A. Bunde and W. Dieterich, *Phys. Rev.* B **31**, 6012 (1985).
19. P. Maass, J. Petersen, A. Bunde, W. Dieterich, and H.E. Roman, *Phys. Rev. Lett.* **66**, 52 (1991).
20. M. Meyer, P. Maass, and A. Bunde, *Phys. Rev. Lett.* **71** 573 (1993).
21. K.L. Ngai, *Comments Solid State Phys.* **9**, 127, 141 (1979); K.L. Ngai, in: *Fast Ion Transport in Solids*, ed. by P. Vashishta, J.N. Mundy, and G.K. Shenoy (North Holland, New York, 1979); K.L. Ngai and O. Kanert, *Solid State Ionics* **53-56**, 936 (1992).
22. S. R. Elliott and A. P. Owens, *Phys. Rev.* B **44**, 47 (1991).
23. W. Schirmacher, M. Prem, J.-B. Suck, and A. Heidemann, *Europhys. Lett.* **13**, 523 (1990).
24. J.C. Dyre, *Phys. Rev.* B **47**, 9128 (1993); **48**, 12511 (1993).
25. A. Hunt, *J. Phys.* C **3**, 7831 (1991); **4**, 5371 (1992); *J. Non-Cryst. Solids* **160**, 183 (1993).
26. D. Knödler and W. Dieterich, *Physica* A **191**, 426 (1992); W. Dieterich, D. Knödler, and P. Pendzig, *J. Non. Cryst. Solids* **172-174**, 1237 (1994).
27. M. P. Allen and D. J. Tildesley, *Computer Simulation of Liquids*, (Clarendon Press, Oxford 1987).
28. R. Kubo, M. Toda, and N. Hashitsume, *Statistical Physics II* (Springer-Verlag, Berlin, Heidelberg 1985).
29. P. Maass, M. Meyer, and A. Bunde, *Phys. Rev.* B **51**, 8164 (1995).
30. J. C. Dyre, *Phys. Rev.* B **49**, 11709 (1994).
31. V. Ambegoakar, B. I. Halperin, and J. S. Langer, *Phys. Rev.* B **4**, 2612 (1971); B. I. Shklovskii and A. L. Efros, *Zh. Eksp. Teor. Fiz.* **60**, 867 (1971) [*Sov. Phys. JETP* **33**, 468 (1971)].
32. H. Bässler, *Phys. Rev. Lett.* **58**, 767 (1987).
33. A. Bunde and S. Havlin, in *Fractals and Disordered Systems*, ed. by

A. Bunde and S. Havlin, 2nd ed., (Springer, Heidelberg 1996).
34. C. Cramer, K. Funke und T. Saatkamp: *Phil. Mag.* B **71**, 701 (1995).
35. M. Tachez, R. Mercier, J. P. Malugani, and A. J. Dianoux, *Solid State Ionics* **20**, 93 (1986).
36. T. Minami, T. Shimizu, and M. Tanaka, *Solid State Ionics* **9** & **10**, 577 (1983).
37. P. Maass, A. Bunde, and M. D. Ingram, *Phys. Rev. Lett.* **68**, 3064 (1992).
38. A. Bunde, M.D. Ingram, and P. Maass, *J. Non-Cryst. Solids* **172–174**, 1222 (1994).

MOLECULAR WIRES: OPERATIONAL PRINCIPLES AND APPLICATIONS

J.R. REIMERS
School of Chemistry, University of Sydney
Sydney NSW, 2006, Australia

N.S. HUSH
School of Chemistry and Department of Biochemistry, University of Sydney
Sydney NSW, 2006, Australia

Our work over the past decade in determining the basic operational principles for potential elements of molecular electronic devices such as molecular wires and molecular switches is briefly reviewed. This includes theoretical treatments for kinetic and spectroscopic processes and their relationship to fundamental molecular properties such as the effective donor to acceptor coupling. Analytical formalue considerably more accurate than the standard one due to McConnell are constructed for several paradigmatic model hamiltonians including those pertinent to chemical bonding; reasons for the widespread occurrence of exponential bridge-length dependences are advanced, as are means (eg, manipulation of bridge resonances) for overcoming this phenomenon. Applications to six different chemical systems are discussed including organic and inorganic molecular wires. Our recently developed methodology for the treatment of interactions between molecular electronic devices and their condensed-phase environments (including interactions with solvent molecules and local/external electric fields) is also discussed, as are a number of applications.

1 Introduction

At the Adriatico Conference on Electron and Ion Transfer in Condensed Media (Trieste, 15-19 July, 1996), we spoke about recent progress in analytical models for long-range electron-transfer processes. This prompted discussion as to our broader interests in molecular wires (and associated materials such as molecular switches and even non-linear optical devices), their operational principles and applications. It was concluded that this book provides an appropriate medium for a brief summary of our work in this area over the previous decade. Unfortunately, it does not provide scope for a major review of this vast field, and little opportunity to properly acknowledge the theoretical contributions of others including Taube, McConnell, Newton, Ratner, Larsson, Meyer, Ulstrup, Beratan, Joachim, Kutznetsov, and many others that have been especially influential to our thinking, as well as the experimental work to which most of this paper is focused. Readers are referred to the cited references in order to ascertain a wider view of the current status of work in this area. The material originally covered in the talk is discussed in Sections 2.5-2.6.

Conceptually, molecular wires are envisaged as comprising terminal donor (D) and acceptor (A) functionalities joined via a molecular bridge (B). The donor is seen as donating an electron (or hole, or vibrational or electronic excitation, etc.) to the bridge which conducts this disturbance, sinking it in the acceptor. A central quantity of interest is the rate at which this process occurs, varied as a function of bridge length and/or composition and as a function of the nature of the donor and acceptor. We have developed a general theory for this rate process, obtained analytical solutions as paradigms appropriate in particular limits, developed numerical applications modelling specific experimental situations, and investigated the interaction of model chemical systems with its environment and with external applied electric fields. While a wide range of bridge materials have been studied, our work to date has concentrated on donor and acceptor moieties which are organic or inorganic molecular fragments. Today, there is much interest in using metallic electrodes as the donor and/or acceptor; all of the methods which we have developed are equally as applicable to these systems as well, the actual treatments requiring at most simple modification to reflect the different chemistry. In overview, this paper is partitioned into the following sections:

2 General operational principles
- 2.1 Quantum dynamics formalism
- 2.2 Analytical solution for the rate
- 2.3 The Slow Drain, Rabi, and Golden Rule limits
- 2.4 Simplifications involving effective hamiltonians
- 2.5 McConnell's model hamiltonian for intermolecular interactions
- 2.6 Model hamiltonians for intramolecular interactions
- 2.7 Relating spectroscopic and kinetic processes

3 Selected Applications
- 3.1 Spiro-type bridges and switches
- 3.2 Norbornologue-bridges organic molecules
- 3.3 Even-polyene-bridged inorganic complexes
- 3.4 Polymethinecyanine (Brooker-dye) bridged systems
- 3.5 Fused rigid oligoporphyrinic bridges
- 3.6 Bacterial photosynthetic reaction centres

4 The interactions of molecular electronic devices with their environment

5 Conclusions

2 General Operational Principles

The function of any system is described in terms of a quantum-mechanical hamiltonian $\mathbf{H_T}$ for the D-B-A system in its environment. This operator can conveniently be split in to three parts,

$$\mathbf{H_T} = \mathbf{H_g} + \mathbf{H_S} - i\lambda|\psi_A\rangle\langle\psi_A| \quad (1)$$

where $\mathbf{H_g}$ is the gas-phase Hamiltonian for the system, $\mathbf{H_S}$ is the environment-induced perturbation, and the last term, an imaginary contribution to the energy of the acceptor state $|\psi_A\rangle$ which dictates the rate $\lambda/\hbar$ at which the transferred charge or excitation leaves the acceptor (possibly to a connected functional unit, or simply by leakage to the environment). It is usual to combine $\mathbf{H_g}$ and $\mathbf{H_S}$ into one term $\mathbf{H}$ which simply describes the effective interactions between the donor, bridge and acceptor moieties found *in situ*. In practical calculations it may be determined using whatever means is both reliable and computationally feasible, eg., empirically (using, say, a parameterized Hückel or soliton (etc.) Hamiltonian), semiempirically, or by using one of the many density functional or *ab initio* techniques; means of including environmental effects are becoming generally available[1]. We have found that while empirical approaches are qualitatively indicative, they are difficult to apply quantitatively; they lead to analytically solvable equations, however, and this we exploit in Section 2.5-2.6.

2.1 *Quantum Dynamics Formalism*[2]

The basic physical model envisages that either a particle (electron or hole) or excitation is injected into the donor. This disturbance is then coherently transferred to and from the acceptor via the bridge but, as the acceptor is connected to its surrounds, this disturbance slowly leaks out at the other end. Using the techniques of time-dependent quantum mechanics, the flow of the initial disturbance can be followed and the probability P that the disturbance remains on the D-B-A system determined. The time dependence of P will, in general, be quite complex consisting of both exponentially decaying and oscillatory components. We have found a useful single parameter describing the kinetics to be the "rate constant" k defined as

$$k = \left[\int_0^\infty P(t)\,\mathrm{d}t\right]^{-1} \quad (2)$$

This function reduces to the expected result if $P(t)$ is dominated by a single exponentially decaying component. If more than one component is important then this formula averages over the significant components; however, it also includes contributions from any small-amplitude long-lived components. Such long-lived components are likely in a real chemical system to lead to alternative reaction products and hence not contribute to the primary process. Empirically, to exclude such contributions we introduce[2] a Gaussian damping function of standard deviation σ into the dynamics which effectively reduces the quantum yield of the process from unity to ϕ. In practical applications, σ (or, alternatively, ϕ) should be set to match the physical interactions between the system and its environment; if such information is not available then we find[2-4] that a reasonable value of σ is 0.07, a value which typically produces $\phi \approx 0.99$; setting $\sigma = 0$ produces the full yield $\phi = 1$.

2.2 *Analytical Solution for the Rate*[2]

The quantum dynamics equations are straightforward to solve numerically for general hamiltonians and hence the rate constant may be determined. We have shown[2-4] that an excellent approximation to k (exact for $\sigma = 0$) is

$$\hbar k = \frac{2\lambda \bar{V}^2}{\bar{V}^2 n_r' n_a' / \phi' + \lambda^2} \tag{3}$$

where, $\bar{V}$ n_r', n_a', and ϕ' are all determined *without* solving the quantum dynamics numerically simply[2] from **H** itself. These parameters are interpreted as: $\bar{V}$ - the net donor-acceptor coupling, n_r'- the approximate number of resonant states, n_a'- an asymmetry index which is 1 if the donor and acceptor are symmetrically related, and ϕ'- the approximate quantum yield. If no damping is used ($\sigma = 0$) then $\phi' = 1$ and all states must be treated as being resonantly connected so that $n_r' = n+2$ where n is the number of bridge states. Formally, n_r', n_a', and ϕ' are obtained by diagonalizing **H** producing eigenvectors $\mathbf{U}_j$ with eigenvalues E_j:

$$\phi' = \sum_{j=1}^{n+2} U_{Dj}^2 \exp \frac{-\sigma^2}{2U_{Dj}^4} + \sum_{k=1}^{j-1} p_{jk} \exp \frac{-\sigma^2}{8p_{jk}^2} ,$$
$$n_r' = \frac{1}{\phi'} \sum_{j=1}^{n+2} \exp \frac{-\sigma^2}{2U_{Dj}^4} , \quad n_a' = \frac{1}{n_r'\phi'} \sum_{j=1}^{n+2} \left(\frac{U_{Dj}}{U_{Aj}} \right)^2 \exp \frac{-\sigma^2}{2U_{Dj}^4} \tag{4}$$

where

$$p_{jk} = \frac{2\, U_{Dj}\, U_{Dk}\, U_{Aj}\, U_{Ak}}{E_j - E_k} \tag{5}$$

$\bar{V}$ is obtained by writing $\mathbf{H} = \mathbf{H^0} + \mathbf{\Delta H}$ where $H_{ij}^0 = H_{ij}(1-\delta_{iA})\ (1-\delta_{jA})$ ie, essentially the donor and bridge part of **H**. $\mathbf{H^0}$ is then diagonalized producing eigenvectors $\mathbf{U}_j'$, allowing the coupling to be expressed as

$$\bar{V}^2 = \left[\sum_{j=1}^{n+2} (U'_{Dj})^2 \exp \frac{-\sigma^2}{2\,(U'_{Dj})^4} \right]^{-1} \sum_{j=1}^{n+2} (U_{Dj})^2 \left(\sum_{i=1}^{n+2} H_{Ai}\, U'_{ij} \right)^{-2} \exp \frac{-\sigma^2}{2\,(U'_{Dj})^4} \quad (6)$$

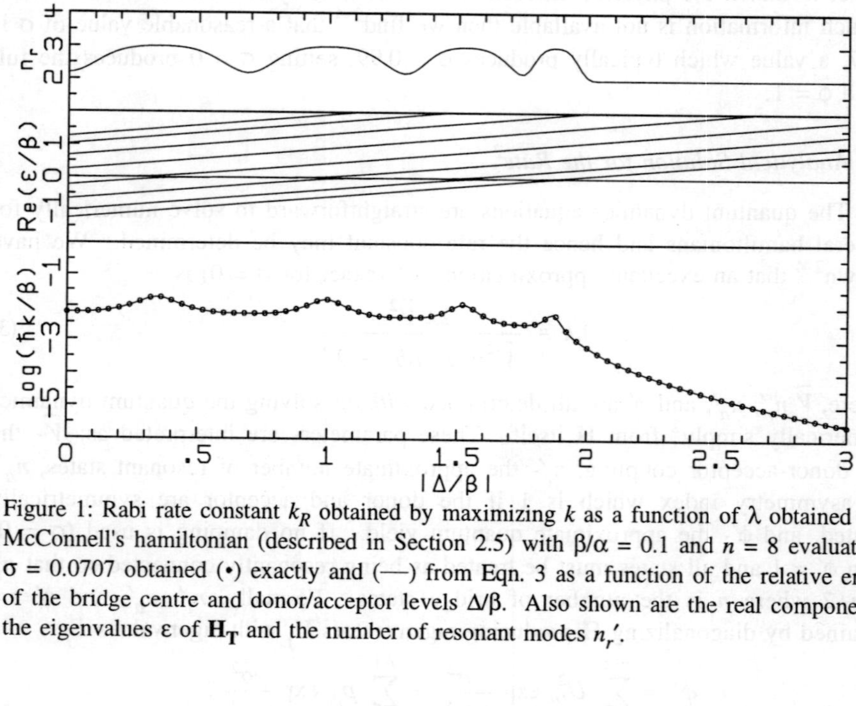

Figure 1: Rabi rate constant k_R obtained by maximizing k as a function of λ obtained from McConnell's hamiltonian (described in Section 2.5) with $\beta/\alpha = 0.1$ and $n = 8$ evaluated at $\sigma = 0.0707$ obtained (•) exactly and (—) from Eqn. 3 as a function of the relative energy of the bridge centre and donor/acceptor levels Δ/β. Also shown are the real component of the eigenvalues ε of $\mathbf{H_T}$ and the number of resonant modes n_r'.

An example of the net coupling obtained through a model bridge of length $n = 8$ is shown in Fig. 1. This figure shows the effect of raising the bridge energy levels with respect to the (equal) donor and acceptor levels. Four bridge resonances are crossed (another four resonances exist arising from symmetry-related lowering the bridge energy levels), as evidenced from the plots of the real parts of the eigenvalues ε of $\mathbf{H_T}$ shown in this figure. Each resonance gives rise to a kink in the net coupling and an increase in n_r' from 2 to 3. Results obtained numerically from Eqn. 2 as well as those obtained from Eqn. 3 are shown and are essentially identical, indicating the quality of the analytical formula.

2.3 *The Slow Drain, Rabi, and Golden Rule Limits*[2]

Equation 3 indicates the central role played by both λ and the key system parameter $\bar{V}$ in controlling the kinetics: as shown in Fig. 2, three kinetic regimes are evident, depending on whether the leak rate $\lambda/\hbar$ is slower than, equal to, or faster than the rate of transfer to/from the donor $\bar{V}/\hbar$. The first case is called the *Slow Drain* regime, the second case forms the *Rabi* regime in which the maximum overall rate

$$\hbar k_{\max} = (n_r' \, n_a')^{-1/2} \, |\bar{V}| \tag{7}$$

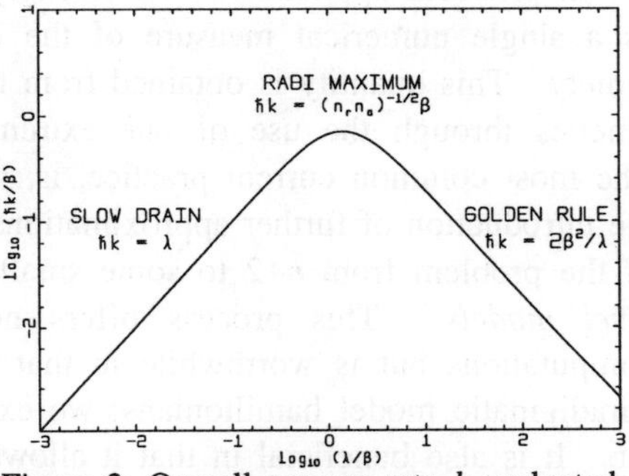

Figure 2: Overall process rate k evaluated as a function of bath coupling λ and net donor-acceptor coupling $\beta = |V|$.

is produced and the third case forms the overdamped or *Golden Rule* regime. Practical devices could be devised which operated in any of these regimes, with the slow drain regime being attractive in that the net coupling through the wire is so large that its magnitude is irrelevant and the wire appears as a perfect conductor, the properties of the output stage being the only determinant of the kinetics.

Most research up to the present time has concentrated not on producing functional devices but rather on the simpler question of what materials constitute the most efficient bridges (ie, which bridges give rise to the largest values for $\bar{V}$)? We[2] have fully characterized the situations in which there are no bridge states present ($n = 0$, directly connected donor and acceptor) or just one. For $n = 0$, the hamiltonian can be expressed as

$$\mathbf{H} = \begin{bmatrix} H_{DD} & H_{DA} \\ H_{DA} & H_{AA} \end{bmatrix} \tag{8}$$

and Eqns. 3 and 7 for $\sigma = 0$ evaluates to give

$$\hbar k = \frac{2\lambda H_{DA}^2}{2H_{DA}^2 + (H_{AA} - H_{DD})^2 + \lambda^2}, \quad \hbar k_{\max} = \left[2 + \frac{(H_{AA} - H_{DD})^2}{H_{DA}^2} \right]^{-1/2} |H_{DA}| \tag{9}$$

The fastest rate possible is $2^{-1/2}|H_{DA}|$ and occurs when $H_{DD} = H_{AA}$, ie., the when donor and acceptor are resonant; in this scenario $n_r' = 2$ and $n_a' = 1$, but universally $\bar{V} = H_{DA}$.

2.4 *Simplifications Involving Effective Hamiltonians*[2,4]

In practical applications, we strongly recommend the direct evaluation of $\bar{V}$ as a single numerical measure of the coupling between a donor and acceptor moiety. This quantity is obtained from the exact general solution of the quantum kinetics through the use of one extremely generally applicable approximation. The most common current practice, is, however, not this but rather proceeds via the introduction of further approximations which seek to reduce the dimensionality of the problem from n+2 to some smaller value m; these are called *effective* m-*level models.* This process offers no advantages in actual numerical rate computations but is worthwhile in that it leads to analytical solutions for some paradigmatic model hamiltonians; we explore this feature further in Section 2.5-2.6. It is also beneficial in that it allows the ready connection between observed donor → acceptor electronic transition moments and rate constants in kinetics applications; this is explored in Section 2.7.

The first approximation introduced is the (integral) value to be used for the dimension m. Our general formalism is useful here in that m must clearly exceed n_r', the number of resonant modes[2-4]. Almost universally, effective 2-level models are used in which the bridge is not explicitly represented at all. Hence, the well-known kinetics solutions for hamiltonians of the form of Eqn. 8 can immediately be applied, but note that the matrix elements are now to be interpreted as effective ones, which implicitly include all of the effects of the bridge:

$$\mathbf{H}' = \begin{bmatrix} H'_{AA} & H'_{DA} \\ H'_{DA} & H'_{AA} \end{bmatrix} \tag{10}$$

We have found the quality of this approach to exceed that which would naively be expected[2-4]. Effective 3-level models, in which one (resonant) bridge level is retained in order to describe the kinetics more accurately, have also had their parameter space characterized[2] and are underutilised, however. In either case, to proceed further, generally a second assumption[4,5] is necessary concerning the nature of the eigenvalues of the corresponding bridge Green's function **G**; this works well in all regions away from resonance[4] and, in the case of an effective2-level model, leads to Larsson's equation[5,6]

$$H'_{DA} = -\sum_{i,j=2}^{n+1} H_{Di} G_{ij} H_{Aj} \tag{11}$$

2.5 *McConnell's Model Hamiltonian*[4]

A simple but widely applicable model Hamiltonian is the modified symmetric Hückel hamiltonian introduced originally by McConnell[7] in the context of long-range magnetic interactions. This hamiltonian represents a linear chain with nearest-neighbour connections only: the donor is coupled to the first bridge element with strength β, as is the acceptor coupled to the last bridge element, while adjacent bridge elements are coupled with strength α and the isolated bridge elements are displaced by Δ with respect to the donor and acceptor levels. Joachim[8] and us[3] have considered numerical the available parameter space while Evenson and Karplus[6] have derived an analytical solution to Eqn. 11. This we[4] expressed as

$$H'_{DA} = \frac{(-1)^n \alpha^2}{\beta} \frac{y - \dfrac{1}{y}}{y^{n+1} - \dfrac{1}{y^{n+1}}} \quad \mathbf{(12)}$$

where

$$y = \frac{\Delta}{2\beta} + \left(\frac{\Delta^2}{4\beta^2} - 1 \right)^{1/2} \quad \mathbf{(13)}$$

is real if $\Delta^2 > 4\beta^2$ and complex otherwise. Complex y leads to the effective coupling oscillating in magnitude as the bridge length n increases; these oscillations arise from the tuning of resonances between the bridge levels and the donor and acceptor levels. Near these resonances an effective 2-level model is not appropriate, and neither is the use of the Green's function approximation[4]. Figure 3 shows calculated Rabi rate constants evaluated (at σ = 0) exactly using Eqn. 7 and approximately via an effective 2-level model using Eqns. 9 and 12, and the deficiencies in the approximate treatment are clearly exposed. As soon as y becomes real, an event which corresponds physically to the donor and acceptor levels moving outside of the band spanned by the bridge levels, Eqn. 12 may be

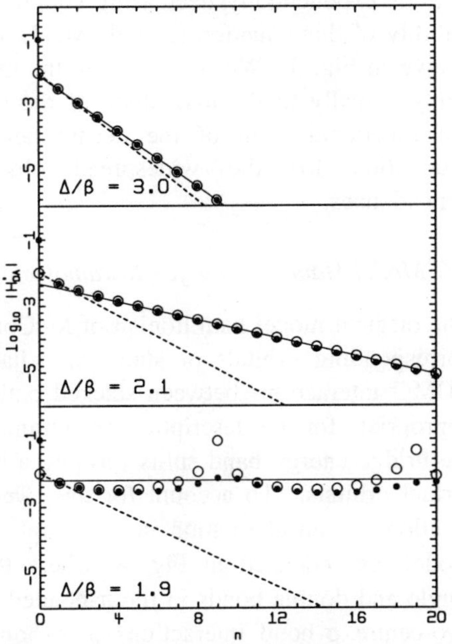

Figure 3: Plot of the Rabi rate constant obtained evaluated: exactly (•), 2-level weak-coupling approximation (o), our exponential-decay law Eqn. 14 (—), and from McConnell's law Eqn. 15 (– –).

accurately approximated by a coupling which decreases exponentially with increasing bridge length

$$H_{DA}^{\prime} = \frac{(-1)^n \alpha^2}{\beta} \left(1 - \frac{1}{y^2}\right) y^{-n} \qquad \textbf{(14)}$$

This function is also shown in Fig. 3 and is clearly a very good approximation away from resonance. When $\Delta^2 \gg 4\beta^2 n^2$, this equation reduces to McConnell's equation[7]

$$H_{DA}^{\prime} = \frac{\alpha^2}{\beta} \left(\frac{-\beta}{\Delta}\right)^n \qquad \textbf{(15)}$$

This equation is of fundamental importance in this field in that it predicts the ubituos exponentially decreasing coupling with bridge length. The condition for validity of this equation is rarely valid, however, and the consequences of this are shown in Fig. 3. We see here that the major assumptions in McConnell's formula apply actually to the magnitude of the exponent and the prefactor rather than to the functional form of the decay; hence, we in fact construct a more solid foundation for the widespread observation of exponential bridge-length dependences.

2.6 *Model Hamiltonians for Intramolecular Interactions*[2]

The original model hamiltonian of McConnell is appropriate to situations in which corresponding orbitals or states on adjacent molecules interact- eg., HOMO or LUMO interactions between stacked phthalocyanines[3] porphyrin rings[9]. It is not appropriate for the description on intramolecular σ or π bonding as in these cases the bridge energy band splits forming a large band gap between the occupied and virtual orbitals. To account for this, we have modified McConnell's hamiltonian to allow for an alternation in the coupling β from β_1 to β_2. Applications of this model are indicated in Fig. 4, where the two β's reflect either the alternating single and double bonds in the π-bonded systems or the alternating one-centre and two-centre σ-bond interactions in σ-bonded system. Three classes of alternate systems occur:

(i) the bridge is even in length (*n* = 2*m*+1) with $|\beta_2| > |\beta_1|$. This is the usual case and is appropriate for most chemical systems. The bridge forms occupied and virtual bands with a large band gap.

(ii) the bridge is odd in length (*n* = 2*m*). This occurs for odd polyenes ions and related molecules such as the Brooker dye ions[10]. In model calculations it is characterized by the presence of a single non-bonding level in the centre of the donor-acceptor band gap; in practical applications, electron-phonon coupling

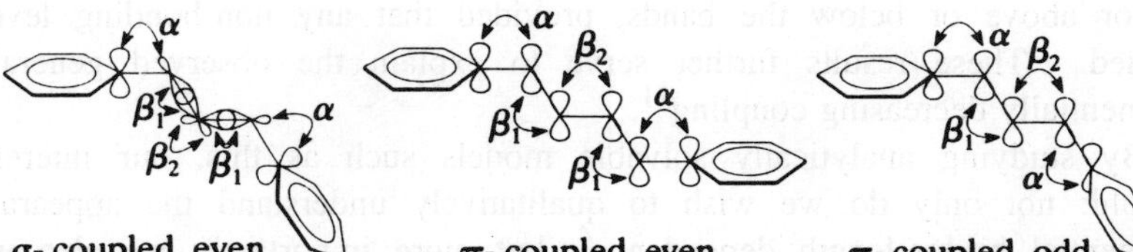

Figure 4: Typical intramolecular coupling schemes with odd and even bridge lengths; all have $|\beta_1| > |\beta_2|$.

initiates soliton formation which significantly delocalizes the molecular charge and lowers the energy of the non-bonding level.
(iii) the bridge is even in length ($n = 2m+1$) with $|\beta_2| < |\beta_1|$. We know of no examples of such a system. Our model hamiltonian indicates that two non-bonding levels should form in the band gap. This offers[4] the exciting consequence that, if the donor and acceptor are located between these two levels, the effective coupling would increase exponentially with increasing bridge length!

Specifically, the effective donor-acceptor coupling is given by

$$H'_{DA} = \frac{-\alpha^2}{\Delta}\,\frac{y-y^{-1}}{y^{m+1}-y^{-m-1}} \quad \text{or} \quad \frac{\alpha^2}{\beta_2}\,\frac{y-y^{-1}}{y^{m+1}-y^{-m-1}+\frac{\beta_2}{\beta_1}(y^m-y^{-m})} \tag{16}$$

when the bridge length n is odd or even, respectively, where

$$y = \frac{1}{2\beta_1\beta_2}\left\{\Delta^2-\beta_1^2-\beta_2^2+[\Delta^2-(\beta_1+\beta_2)^2]^{1/2}[\Delta^2-(\beta_1-\beta_2)^2]^{1/2}\right\} \tag{17}$$

Choosing $y' = y$ outside the bands and $y' = 1/y$ inside the band gap so that $y' > 1$ always, the coupling (away from resonance with any possible non-bonding level) is given to high accuracy by[4]

$$H'_{DA} = \frac{-\alpha^2}{\Delta}(y'-1/y')(y')^{-m} \quad \text{or} \quad \frac{\alpha^2}{\beta}\,\frac{y'-1/y'}{y'+\beta_2/\beta_1}(y')^{-m} \tag{18}$$

for odd or even n, respectively. These results are directly analogous to those obtained for the simpler McConnell Hamiltonian, and indeed can be reduced to a form directly analogous to McConnell's equation, Eqn. 15. Oscillatory behaviour is predicted when the donor and acceptor bands lie within the bridge occupied or virtual orbital spaces, but this may be difficult to realize as crossing a resonance would result in bridge oxidation or reduction. Exponentially decreasing coupling is predicted as soon as the donor and acceptor orbitals move either into the band gap or above or below the bands, provided that any non-bonding levels are avoided. These results further serve to explain the observed generality of exponentially decreasing coupling[11].

By studying analytically solvable models such as this, our interests are twofold: not only do we wish to qualitatively understand the appearance of exponential bridge-length dependence[4], but more importantly we also wish to exploit the more general solutions to predict scenarios in which the bridge-length dependence is either weak or non-exponential[2,3,11].

2.7 *Relating Spectroscopic and Kinetic Processes*

From the early days it was recognized by Hush[12,13] that the molecular processes which control kinetic processes such as the flow of an electron or hole through a molecule are the same processes which facilitate molecular charge-transfer absorption spectroscopy. By measuring the charge-transfer transition moment M it is thus possible to determine H'_{DA} and hence estimate electron-transfer rate constant. Using the adiabatic approximation, if an electronic transition moves an electron a macroscopic distance r (assumed to be much greater than the spatial extent of the initial and final orbitals) and requires energy $h\nu$, the effective donor (initial orbital) to acceptor (final orbital) coupling is given by

$$H'_{DA} = \frac{M h\nu}{m^{1/2} r} \tag{19}$$

where $m = 1$ or 2 is the number of electrons which can make the transition. The change in dipole moment $\Delta\mu$ between the initial and final states provides another measure of the electron transfer, with[14]

$$r^2 = (\Delta\mu)^2 + 4M^2/m \tag{20}$$

Determination of the coupling from experimental data has traditionally relied on the use of estimated values for r in Eqn. 19, a process which may be unreliable[14-16]. Cave and Newton[17] have obtained improved results, however, using Eqn. 20 to first determine r; this they term the "Generalized Mulliken-Hush (GMH) approach, and it can be applied to both theoretical and experimental data provided that both $\Delta\mu$ and M are known.

3 Selected Applications

3.1 *Spiro-Type Wires and Switches Switches*[18,19]

Aviram[20] suggested that spiro-linkages may be useful as molecular switches, and an postulated application of this is shown in Fig 5 in which the spiro switch links two polythiol molecular wires. Their is a net +*ve* charge on the spiro switch and, depending on its location, different conduction paths are formed. The location of the switch, which is bistable, can be adjusted using an applied electric field. We[18] considered using *ab initio* calculation methods the magnitude of the electric field strengths required to flip the switch, and the length of time for which the new setting would be remembered. The final result was the elucidation of the general nature of such molecular switch devices. We have also considered the possible use of spiro linkages as molecular wires[19] and have elucidated their through-bond interaction mechanisms; a sample system is shown in Fig. 6 in which a four-ring bridge links ruthenium donors and acceptors. A feature of the spiro-type wires is that there is finite communication between donors and acceptors only when they lie out of the plane of the ring to which they are respectively attached.

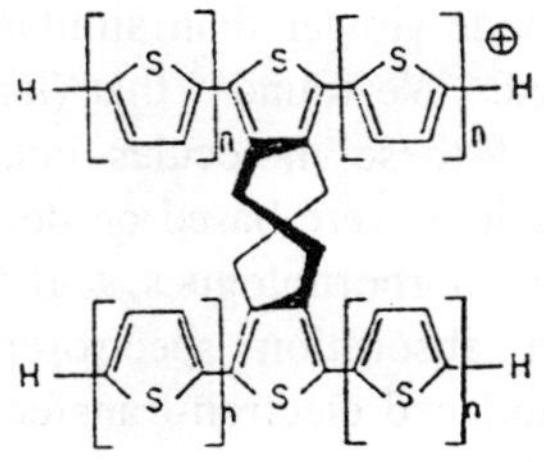

Figure 5: Example of two molecular wires controlled by an Aviram-type spiro switch.

Figure 6: A spiro-bridged complex

3.2 *Norbornologue-Bridged Organic Molecules*

The series of rigid norbornolog bridges synthesised by Paddon-Row[21] form a useful systems for studying electron transfer. These are organic even σ-bonded systems with π donors and acceptors much as in Fig. 4; a specific example is shown in Fig. 7, but bridges of up to 20 σ-bond lengths have been synthesised. Our interest lay in the characterization of the charge-transfer pro-

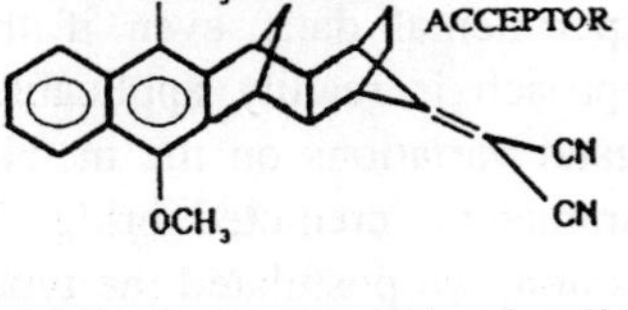

Figure 7: A 4-σ bond functionalized norbornologue molecular wire.

cess[22]. We were particularly concerned with the role of electronic symmetry selection rules in determining the electron-transfer rate constant: molecules in which the process is forbidden were found to occur at rates only ca. an order of magnitude slower than similar processes in molecules for which the process was allowed. We found[23] that (large) vibronic coupling within the naphthalene donor units of these molecules could account for the observed rate constants. Our conclusions were based on detailed analysis of the UV electronic absorption bands of these norbornologues, and this analysis was later explicitly verified using two-photon absorption spectroscopy[24]. This was the first example of a fast photoinduced electron-transfer for which the D-A *electronic* coupling was zero by symmetry.

3.3 *Even-Polyene-Bridged Inorganic Complexes*[11,25]

Polyacetylene is an extremely important material which, typical of conducting polymers in general, contains long lengths of conjugated material with alternating single and double bonds. Small segments with molecular formula $(CH)_{2m}$ have also been popular as short molecular wires. Woitellier, Launay, and Spangler[26] synthesised a series of mixed-valence bis-ruthenium complexes in which the ruthenium donor and acceptor centres were bridged by polyenes of various length, attached via para-pyridyl "alligator clips", see Fig. 8. The intervalence charge transfer band intensity was used to monitor the strength of the intermetallic interaction. In 1989 we modelled their results[11,25], finding that the coupling decreased in what was crudely an exponential manner, but that the changing chemical nature of the ligand caused discernible differences. At the time, it was not feasible to perform any *a priori* calculation of the coupling through such molecules *in situ*, and we constructed a hybrid model based on the *ab initio* electronic structure of each individual ligand combined with a two-parameter empirical model of the metal d_π orbital. Fully empirical models were found to provide little insight as they could easily be adjusted to fit any experimental data, even if their physical basis was without foundation. Our approach is readily applicable, however, as standard tables list the effects of ligand variations on the metal d_π redox potential: as this is one of our empirical parameters, predicted bridge-length dependences are easily generated. In this fashion, we postulated the types of ligand most likely to induce minimal decay of the coupling with increasing bridge length.

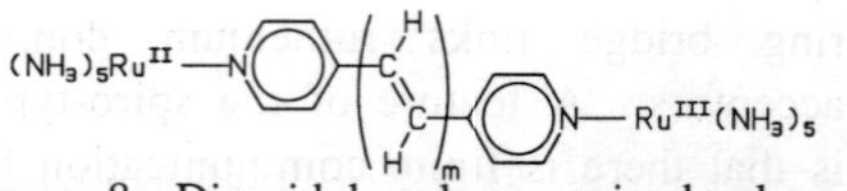

Figure 8: Dipyridyl polyene mixed-valence complexes

3.4 *Polymethinecyanine (Brooker-Dye) Bridged Systems*

In our numerical studies[3,11] on the polyene wires, we observed that the net donor-acceptor coupling decreased less rapidly and sometimes non-exponentially with bridge length when the band gap was narrow. Based on the historical perspective that odd polyene ions $[R\text{-}C_{2m+1}H_{2m+1}\text{-}R]^{\pm}$ and the isoelectronic Brooker dye cations $[NR_2\text{-}C_{2m\text{-}1}H_{2m\text{-}1}\text{-}NR_2]^+$ had all CC bonds equivalent, a highly dispersed charge, and little or no asymptotic band gap, we postulated that they would facilitate dramatically increased coupling when compared to standard even-length polyenes of similar length. This initiated a major project in which the properties of molecules of length up to m = 80 units were evaluated by a combination of *ab initio*, semi-empirical and empirical techniques[10,27]. We found that soliton formation resulted in the concentration of the charge in one small region (the "soliton") of the chain, and in this region the bond-length alternation becomes quite small; typical results are shown in Fig. 9. However, outside of this region, Peierls distortion set in producing large strands of "standard" even polyacetylene. Hence, the asymptotic band gap is unmodified but the gap is populated with a finite number of discrete soliton levels[28,29], one of which is the "non-bonding" level which is central to the results obtained from model hamiltonians, see Section 2.6. While the soliton levels could promote resonance interactions with specific donors and acceptors which may have profound consequences, we found only a modest yet significant increase in the end to end coupling with the use of odd-length molecules.

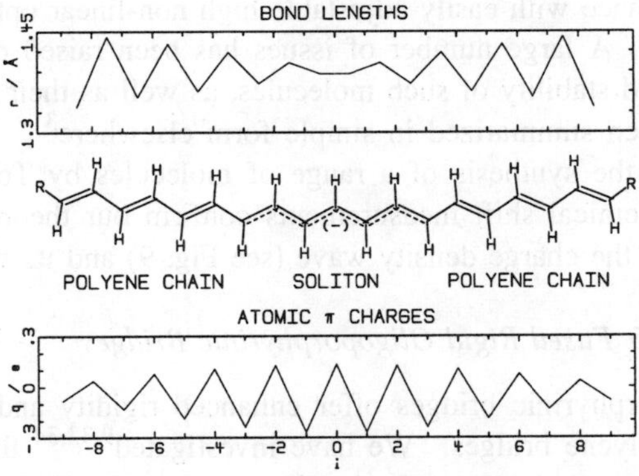

Figure 9: Sample soliton development in the odd polyene chain $R\text{-}C_{19}H_{19}\text{-}R^-$ with R = H; r is the CC bond length, q the carbon charge, and i the number of carbon atoms from the chain centre.

Solitons are localized geometrical disturbances which entrap a charge (electron or hole). As is often exploited in conducting polymers, they interact strongly with their environment and can translate along a polyene chain at the cost of very modest amounts of energy. It became clear that we must be able to model these interactions in order to describe qualitatively the motion of solitons. This in part motivated our studies of solvent effects on the simple, well-characterized systems described in Section 4. In the meantime we considered the potential surface for the motion of solitons in Brooker dyes in the gas phase. Both *ab initio* and AM1 SCF calculations predicted that, after a critical length of about 19 CH units, this potential contains two minima corresponding to solitons localized at each end of the chain. It is not easy to determine the reliability of this prediction for real molecules in condensed phase: eg., electron correlation effects will reduce the tendency to localize, while environmental interactions could dramatically enhance it. If localization occurs then the Brooker dye would act as a molecular switch[30]. Application of external electric fields could force the soliton to one end, and then the intrinsic soliton double-well potential could cause the soliton to remain in that position after the applied field is removed. We have considered the operational characteristics for such a device, and found that enhanced properties compared to modern dynamic memory is feasible[30]. Such a device is also interesting in that it constitutes a highly polarizable cation and as such would show large non-linear optical properties. These properties would change dramatically as a function of the location of the soliton, however, promising a device with easily adjustable high non-linear optical responses[30].

A large number of issues has been raised concerning the synthetic feasibility and stability of such molecules, as well as their basic properties. The results have been summarized in simple form elsewhere[31]. The most important aspect of this is the synthesis of a range of molecules by Tolbert and Ogle[32]; their ^{13}C NMR chemical shift measurements confirm our theoretical calculations of the structure of the charge density wave (see Fig. 9) and its variation with bridge length.

3.5 *Fused Rigid Oligoporphyrinic Bridges*

Porphyrinic bridges offer enhanced rigidity and synthetic flexibility compared to polyene bridges. We have investigated[9,33,34] the electronic properties of a series of a series of porphyrin oligomers synthesised by Crossley *et al.*[35-37] such as the

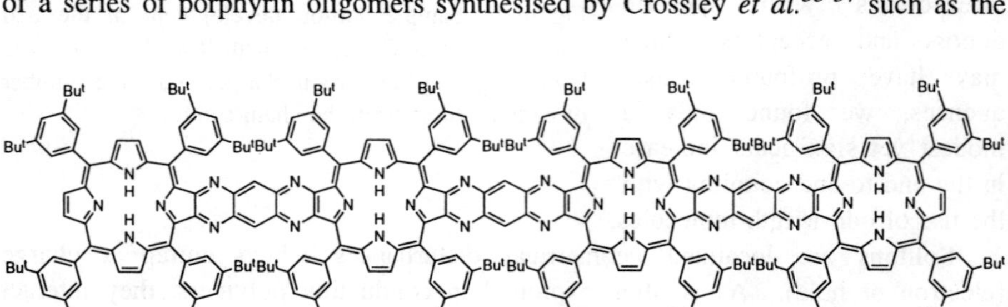

Figure 10: Linear fused oligoporphyrin tetramer

linear tetramer shown in Fig. 10. While this molecule has the topology of a linear molecular wire, each porphyrin has four attachment sites and a large variety of topologies are possible[33]; different topologies could have different molecular electronic functions. Their synthetic flexibility can be manipulated by adding a central metal atom, the effects of which can be quite profound eg., switching the porphyrin HOMO and SHOMO orbitals, and by substitution at the beta and meso sites or by modifying the inter-porphyrin link. In this way, materials with desired electronic and chemical (eg., solubility, stability) properties can be constructed. Compared to polyene molecular wires, we found that these materials have considerably lower inter-unit coupling but this is more than adequately compensated by the very large distance (ca., 14 Å) spanned by a single unit[9]. Hence, much larger effective coupling over long distances can be sustained using the oligoporphyrins. The linear tetramer shown in Fig. 10 is 56 Å long, sufficient for spanning membranes, while the linear octamer is ca. 110 Å long, sufficient for spanning gaps between modern nanoelectrodes etc..

3.6 *Bacterial Photosynthetic Reaction Centres*[38,39]

These are naturally occurring light-harvesting systems that involve long-distance electron transfer involving bacteriochlorophyll and bacteriopheophytin units. Our interest has concentrated on the role and properties of the "special pair" radical cation P^+ formed during the initial electron transfer step; the special pair is a system of two overlapping bacteriochlorophyll molecules which forms the donor unit for the electron-transfer process. The electronic spectra of P^+ are interesting in that they provide key information about the coupling between the two bacteriochlorophylls. To interpret the spectra, important aspects of which had remained unassigned for 25 years[39], we needed to understand the general description for the electronic states of interacting radicals[40] and the role of vibronic coupling in electron-transfer devices[15]. The principles learned are generally applicable, and even include methodologies necessary to implement semiempirical electronic structure methods such as INDO to electron transfer problems: in general, attempts have been made to use semiempirical parameterizations to implicitly include core reorganization effects, but we see that no such general parameterization is possible and that, just as in *ab initio* calculations, multi-reference CI calculations are necessary in order to provide an unbiased description of both the initial and final electronic states. With this and other improvements, we were able to identify the physical significance of the INDO results and hence assign transitions that had previously been calculated using INDO but for which no interpretation was forthcoming[39].

4 The Interaction of Molecular Electronic Devices with their Environment

In all of the studies described in Section 3 it became apparent that it is essential to calculate properties of molecular electronic devices *in situ* in a condensed phase. This is because the basic electronic properties, which often involve long-range Coulomb interactions, are strongly effected by the environment. Also, functional devices will usually involve as a central part the interaction of the device with electric fields emanating from the surroundings.

Experimentally, fundamental progress was made by Boxer *et al.*[41] who applied large external electric fields to inorganic molecular wires are their prototypes and observed the resulting change in their electronic absorption spectrum. From their results it is possible to obtain information concerning how the electric fields emanating from solvent molecules affect the electronic transition energy (the "solvent shift"), and how the molecular dipole moment changes as a result of the transition (the effective amount of charge transferred).

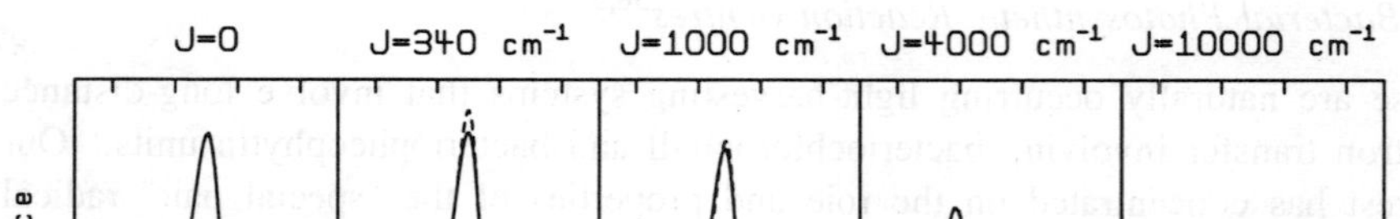

Figure 11: Simulated MLCT absorption spectra and electric-field difference spectra for $[Ru^{2+}(NH_3)_5]_2$pyrazine (reorganization energy $\lambda = 2500\ cm^{-1}$) as a function of frequency ν referenced to the monomer absorption band maximum for various values of the excited-state net intermetallic coupling *J*. (—) simulated band ; (– –) best-fit band contour to the standard Liptay functional form.

Qualitatively, the results of Boxer *et al.* are very illuminating in that from them it is immediately obvious[42] whether the charge transfer is between localized donor and acceptor centres, as is envisaged in our basic physical description, or whether it is between symmetric and antisymmetric global delocalized wavefunctions (the extreme strong coupling limit). This is indicated in Fig. 11 in which the change in the electronic absorption spectrum of the metal to ligand charge transfer (MLCT) transition in $Ru^{2+}(NH_3)_5$-pyrazine-$Ru^{2+}(NH_3)_5$ is shown synthesised using various ratios of the net donor-acceptor coupling J to the reorganization energy λ associated with hole localization in the excited state. In the strong coupling ($J/2\lambda \gg 1$) limit, delocalization occurs and the spectral change is first-derivative in nature[14] (ie., weak with one node) whereas in the weak coupling ($J/2\lambda \ll 1$) limit, localization occurs and the spectral change is second-derivative in nature (ie., strong with two nodes); for intermediate coupling, the spectral change is intermediate and non-Liptay[14,42,43] in nature.

Quantitatively, their results were puzzling, however, as they indicated, eg., that for the MLCT transition in $Ru^{2+}(NH_3)_5$-pyrazine actually transferred only 20% of an electron, contrary to standard chemical intuition. We set out to interpret this result, and it was clear from preliminary studies[14] that a high-quality description of all ligands was necessary, as was a reliable calculation for the interaction of the complex with the solvent.

We developed a method[44] for estimating solvent shifts and dipole moment changes and tested it on the spectra of azines in water[44-47]. These molecules, which involve only a small amount of charge transfer, were chosen as a significant amount of detailed experimental data is available, their relatively small internal charge transfer makes them difficult test cases, they are themselves often ligands in important charge-transfer systems, for them also issues of localization and delocalization are important and often difficult to resolve[46], and they have been historically important in the development of the theory of hydrogen bonding. Our method was able to interpret all available experimental data. It contains no arbitrarily adjustable parameters, and is based on the determination of an explicit solvent structure using numerical simulation techniques.

Our method for treating solvation was also tested[48] by considering transitions in $Fe^{2+}(H_2O)_6$ in aqueous solution. This included charge transfer to solvent transitions which produce solvated electrons and involve huge solvent shifts. While the primary purpose of this study was to determine the most appropriate way to treat long-range and other effects which make simulation of inorganic systems technically difficult, the results indicated that our method is accurate over an extremely large range of chemical situation.

Finally, we applied this method to investigate MLCT transitions in $Ru^{2+}(NH_3)_5$-pyrazine[16] and $Ru^{2+}(NH_3)_5$-pyridine[49,50]. Using both *ab initio* and

INDO (applied as per Section 3.6) methods to describe the electronic structure of the chromophores in the gas phase, we were able to reproduce the observed solution absorption frequencies as well as the observed small values of the dipole moment change accompanying the electronic MLCT transition. The text-book view of this transition moving a whole electron (less the fraction associated with back bonding) from the donor to the acceptor was substantiated, and the observed small dipole change results from the polarizability of the solvent and solute.

5 Conclusions

We have developed a conceptual framework in which the operation of molecular electronic devices can be viewed. This explains why most bridge materials display exponential or near-exponential decreases in the end to end coupling with increasing bridge length, and point to ways by which either non-exponential or weak-exponential dependences could be obtained. It also points to the importance of bridge resonances (cf. Figs. 1 and 3). There is considerable current interest (see eg. Ratner *et al.*[51,52]) in the interaction of molecules or bridges with metal electrodes in which these will be particularly important, and our analytical methods are capable of describing this interaction. We have developed numerical methods for the investigation of specific practical problems, and intend to develop these to handle connections to metal electrodes.

Proper treatment of the environment will play an important part in the future development of research in this area. We have laid a foundation on which this could be based. Currently, we are exploring environmental effects on vibrational spectra[53,54], using these are a probe for local electric fields near molecular wires[55], and are also investigating weak non-polar dispersive solvent-solute interactions.

6 Acknowledgments

J.R.R. is indebted to the Australian Research Council for the provision of a Senior Research Fellowship.

7 References

1 J.R. Reimers and G.B. Bacskay in Encyclopaedia of computational chemistry, ed. P. von R. Schleyer (Wiley, London, 1997) submitted.
2 J.R. Reimers and N.S. Hush, *Chem. Phys.* 134, 323 (1989).
3 J.R. Reimers and N.S. Hush, *Chem. Phys.* 146, 89 (1990).

4 J.R. Reimers and N.S. Hush, *J. Photochem. Photobiol. A* 82, 31 (1994).
5 S. Larsson, *J. Am. Chem. Soc.* 103, 4034 (1981).
6 J.W. Evenson and M. Karplus, *J. Chem. Phys.* 96, 5272 (1992).
7 H.M. McConnell, *J. Chem. Phys.* 35, 508 (1961).
8 C. Joachim, *Chem. Phys.* 116, 339 (1987).
9 J.R. Reimers *et al*, *Chem. Phys. Lett.* 256, 353 (1996).
10 J.S. Craw *et al*, *Chem. Phys.* 167, 77 (1992).
11 J.R. Reimers and N.S. Hush, *Inorg. Chem.* 29, 3686 (1990).
12 N.S. Hush, *Prog. Inorg. Chem.* 8, 391 (1967).
13 N.S. Hush, *Electrochim. Acta* 13, 1005 (1968).
14 J.R. Reimers and N.S. Hush, *J. Phys. Chem.* 95, 9773 (1991).
15 J.R. Reimers and N.S. Hush, *Chem. Phys.* 208, 177 (1996).
16 J. Zeng *et al*, *J. Am. Chem. Soc.* 118, 2059 (1996).
17 R.J. Cave and M.D. Newton, *Chem. Phys. Lett.* 249, 15 (1996).
18 N.S. Hush *et al*, *J. Am. Chem. Soc.* 112, 4192 (1990).
19 A.P.L. Rendell *et al*, *J. Am. Chem. Soc.* 110, 8343 (1988).
20 A. Aviram, *J. Am. Chem. Soc.* 110, 5687 (1988).
21 J.M. Warman *et al*, *Nature* 320, 615 (1986).
22 H. Oevering *et al*, *J. Am. Chem. Soc.* 109, 3258 (1987).
23 J.R. Reimers and N.S. Hush, *Chem. Phys.* 146, 105 (1990).
24 J.R. Reimers *et al*, *Chem. Phys. Lett.* 169, 622 (1990).
25 J.R. Reimers and N.S. Hush, *Inorg. Chem.* 29, 4510 (1990).
26 S. Woitellier *et al*, *Inorg. Chem.* 28, 758 (1989).
27 J.S. Craw *et al*, *Chem. Phys.* 167, 101 (1992).
28 J.R. Reimers *et al* in Molecular Electronics- Science and Technology Vol II, ed. A. Aviram (AIP, New York, 1992), p. 11.
29 J.R. Reimers *et al*, *Mol. Cryst. Liq. Cryst. A* 234, 51 (1993).
30 J.R. Reimers and N.S. Hush, *Chem. Phys.* 176, 407 (1993).
31 J.R. Reimers *et al*, *Biosytems* 35, 107 (1995).
32 L.M. Tolbert and M.E. Ogle, *J. Am. Chem. Soc.* 111, 5958 (1989).
33 T.X. Lü *et al*, *J. Phys. Chem.* 98, 11878 (1994).
34 J.R. Reimers *et al*, *J. Am. Chem. Soc.* 117, 2855 (1995).
35 M.J. Crossley and P.L. Burn, *J. Chem. Soc., Chem. Commun.* 1569 (1991).
36 M.J. Crossley *et al*, *J. Chem. Soc., Chem. Commun.* 2379 (1995).
37 M.J. Crossley *et al*, *J. Chem. Soc., Chem. Commun.* 1921 (1995).
38 J.R. Reimers and N.S. Hush, *Chem. Phys.* 197, 323 (1995).
39 J.R. Reimers and N.S. Hush, *J. Am. Chem. Soc.* 117, 1302 (1995).
40 J.R. Reimers and N.S. Hush, *Inorg. Chim. Acta* 226, 33 (1994).
41 D.H. Oh *et al*, *J. Am. Chem. Soc.* 113, 6880 (1991).

42 J.R. Reimers and N.S. Hush in Mixed Valence Systems: Applications in Chemistry, Physics, and Biology, ed. K. Prassides (Kluwer Acad. Publishers, Dordrecht, 1991), p. 29.
43 W. Liptay, *Z. Naturforsch. Teil. A* 20, 272 (1965).
44 J. Zeng *et al*, *J. Chem. Phys.* 99, 1508 (1993).
45 J. Zeng *et al*, *Chem. Phys. Lett.* 206, 323 (1993).
46 J. Zeng *et al*, *J. Chem. Phys.* 99, 1482 (1993).
47 J. Zeng *et al*, *J. Phys. Chem.* 100, 9561 (1996).
48 J. Zeng *et al*, *J. Phys. Chem.* 98, 11075 (1994).
49 J. Zeng *et al*, *J. Phys. Chem.* 99, 10459 (1995).
50 J. Zeng *et al*, *J. Phys. Chem.* (1996) in press..
51 V. Mujica *et al*, *J. Chem. Phys.* 104, 7296 (1996).
52 M. Kemp *et al*, *J. Phys. Chem.* 100, 8349 (1996).
53 N.S. Hush and J.R. Reimers, *J. Phys. Chem.* 99, 15798 (1995).
54 J.R. Reimers *et al*, *J. Phys. Chem.* 100, 1498 (1996).
55 L. Hall *et al*, *J. Phys. Chem.* (1997) submitted.

Tunneling conduction in molecular wires

by

Sven Larsson, Daniel Karlsson, and Violetta Cepevska

Department of Physical Chemistry
Chalmers University of Technology
S 412 96 Gothenburg, Sweden

Abstract: The Green function method is used to compare the electron transfer (ET) capability Θ of a molecular wire to that of aliphatic support molecules. Due to the π system, Θ^2 for the conjugated molecule is a factor 40,000 times larger than Θ^2 for the shorter aliphatic molecule. The consistency of the results has been checked by calculating the distance dependence of Θ^2 for the corresponding shorter molecules. We find an exponential decrease in both cases but a much slower one for the wire than for the support molecule.

1. Introduction

It was recently demonstrated by scanning tunneling microscopy (STM) that linear conjugated molecules may appear conducting and act as wires [1] (fig.1). In the present paper we study theoretically the connection between structure of the molecule and (nonresonant) tunneling capability.

The tunneling or electron transfer (ET) capability of a molecular wire may be defined as the square Θ^2 of a Green function [2–4]:

$$\Theta(E) = \sum_{i=1}^{N} \frac{C_{1i} C_{Ni}}{E - E_i} \tag{1}$$

where the summation extends over the molecular orbitals $\{\phi_i\}$ of the bridge, occupied or unoccupied, with orbital energies $\{E_i\}$ and coefficient $C_{\nu i}$ for atom orbital ν. E is the energy of the electron to be transferred. The atomic orbitals are counted so that orbitals 1 and N are at the end of the bridge, overlapping with the donor and acceptor orbitals in STM tip and metal. Eq.(1) is useful if the donor and acceptor do not oxidize or reduce the bridge and if the interaction with the bridge is not exceptionally large.

In actual calculations on large systems one is usually forced to apply 'Extended Hückel' (EH) [2] or semi–empirical methods based on the Hartree–Fock method [5]. In simpler systems ab initio methods including symmetry breaking and other correlation effects have been applied, to a large extent confirming the results by the simpler methods [6,7]. The valence orbitals (minimal LCAO set) are sufficient to describe tunnelling through the molecule and existence of virtual orbitals, for example C3d orbitals [8], need not be invoked, at least not for a rough description of the essential physics of the phenomenon.

If the solvent itself acts as a bridge one may rewrite eq.(1) as a product of ratios of interactions between the solvent molecules to the energy gap between solvent and transferred electron, demonstrating an exponential decrease of the coupling with distance between donor and acceptor [8,9]. In the case of π systems, closed but approximate expressions may be obtained which show that the coupling decreases slower than exponential [10–14].

When donor and acceptor are metals, a derivation based on the golden rule is possible as has been shown by Mujica et al. [4]. Eq. (1) still applies if the applied voltage is small and is useful in the comparison between different types of bridges when everything else is the same.

2. Calculational method

Eq.(1) may be derived from any eigenvalue problem for the molecular orbitals (MO) for the donor–bridge–acceptor system [2]. In the original derivation [2] the EH model was used, but this method has several drawbacks. One is that there is no theorem which relates the orbital energies to total energies. In Hartree–Fock (HF) methods we may use Koopmans' theorem in the following way [15]:

$$I_i=E^{j}_{N-1}-E_N=-\epsilon_i \Rightarrow I_i-I_j=E^{i}_{N-1}-E^{j}_{N-1}=\epsilon_j-\epsilon_i$$

$$A_i=E_N-E^{a}_{N+1}=-\epsilon_a \Rightarrow A_a-A_b=E^{b}_{N+1}-E^{a}_{N+1}=\epsilon_b-\epsilon_a \qquad (2)$$

The semi–empirical HF method used here, CNDO/S, was invented to describe spectra of organic π systems [16]. The method was modified and extended in an earlier paper by Larsson et al. [17]. The method has been found to give reasonable estimations of ionization energies and electron affinities via the orbital energies and of excitation energies in the configuration interaction step. However, since the CNDO/S method is considered as less useful for geometry optimization, the geometries for n–dodecanethiol (DT) and the molecular wire (MW) 4,4'– di(phenylene)benzothiolate were optimized piecewise using AM1 [18]. The all–*trans* conformation of DT was found to have the lowest energy. A single DT or MW molecule was then calculated using the CNDO/S method and the results used in eq.(1).

We will now shortly repeat how eq.(1) is derived. The eigenvalue problem (d=donor, a=acceptor, b=bridge) is partitioned [19]:

$$\begin{bmatrix} H_{dd} & H_{da} & H_{db} \\ H_{ad} & H_{aa} & H_{ab} \\ H_{bd} & H_{ba} & H_{bb} \end{bmatrix} \begin{bmatrix} C_d \\ C_a \\ C_b \end{bmatrix} = E \begin{bmatrix} C_d \\ C_a \\ C_b \end{bmatrix}. \qquad (3)$$

C_b is obtained from the third equation and inserted in the first and second equations implicit in eq.(3). The latter two equations then read:

$$\overline{H}\begin{bmatrix} C_d \\ C_a \end{bmatrix} = \begin{bmatrix} H_{dd}-H_{db}(H_{bb}-E)^{-1}H_{bd} & H_{da}-H_{db}(H_{bb}-E)^{-1}H_{ba} \\ H_{ad}-H_{ab}(H_{bb}-E)^{-1}H_{bd} & H_{aa}-H_{ab}(H_{bb}-E)^{-1}H_{ba} \end{bmatrix}\begin{bmatrix} C_d \\ C_a \end{bmatrix}$$

$$= E\begin{bmatrix} C_d \\ C_a \end{bmatrix} \qquad (4)$$

We assume for simplicity that a single atomic orbital (AO) on the metal connects to an AO χ_1 of the bridge and a single AO of the tip connects to a single AO χ_N on the bridge. The interaction matrix element between the two atomic orbitals on metal and tip connecting to the bridge is then obtained from eq.(4):

$$\overline{H}_{da}(E) = H_{da} - [H_{d1}\ 0\ 0\ \ldots\ 0](H_{bb}-E)^{-1}\begin{bmatrix} 0 \\ 0 \\ \vdots \\ H_{Na} \end{bmatrix} \qquad (5)$$

The 'direct' matrix element H_{da} can be set equal to zero in the present context in view of the large distance between tip and metal. Diagonalizing $\mathbf{H_{bb}}$, the coupling matrix elements of eq.(5) may be written as [2]:

$$\overline{H}_{da}(E) = H_{d1}\, H_{Na} \sum_{i=1}^{N} \frac{C_{1i}C_{Ni}}{E-E_i} = H_{d1}\, H_{Na}\, \Theta(E) \qquad (6)$$

In the case of DT and its shorter counterparts we have calculated Θ between the $S2p_z$ orbital and the H1s orbital closest to the tip. In the case of MW we have calculated Θ between the S3p orbital parallel to the conjugated π orbitals and the C2p–π orbital closest to the STM tip, since H atom has no 2p valence orbitals. To be able to compare to results for the aliphatic molecules we have considered the transfer across the final H atom as transfer through free space. For this transfer β has been set to 3.2 $Å^{-1}$ [20], which means that Θ is multiplied by a factor 0.03 to account for the decrease over the CH distance of 1.1 Å.

It is possible to get further insight into the problem of tunneling through a molecule between a surface and an STM tip if instead of diagonalizing the martix H_{bb} we use a series expansion of the inverse matrix [5]. We write:

$$\mathbf{H}_{bb} - \mathbf{E} = \mathbf{D} + \mathbf{B} \tag{7}$$

where **D** is a diagonal matrix and **B** a matrix with zeroes in the diagonal. The inverse may then be written [2,5]:

$$(\mathbf{D}+\mathbf{B})^{-1} = \mathbf{D}^{-1} - \mathbf{D}^{-1}\mathbf{B}\mathbf{D}^{-1} + \mathbf{D}^{-1}\mathbf{B}\mathbf{D}^{-1}\mathbf{B}\mathbf{D}^{-1} - .... \tag{8}$$

provided **B** is 'small' compared to **D**. The effective coupling matrix element is then, using eq.(5):

$$\overline{H}_{da}(E) =$$

$$H_{da} - \mathbf{H}_{db}\mathbf{D}^{-1}\mathbf{H}_{ba} + \mathbf{H}_{db}\mathbf{D}^{-1}\mathbf{B}\mathbf{D}^{-1}\mathbf{H}_{ba} - \mathbf{H}_{db}\mathbf{D}^{-1}\mathbf{B}\mathbf{D}^{-1}\mathbf{B}\mathbf{D}^{-1}\mathbf{H}_{ba} + \ldots \quad (9)$$

The only important matrix element of $(\mathbf{H}_{bb} - \mathbf{E})^{-1} = (\mathbf{D}+\mathbf{B})^{-1}$ is then the one which couples to the closest AO's of metal and tip. The first term in eq.(9) contributes only the negligible direct interaction between metal and tip. Since the couplings between the bridge atomic orbitals are roughly proportional to the overlap, the following terms in eq.(9) contribute significantly only when the couplings B_{ij} are between neighbouring atoms. The second term in eq.(8) for example can also be ignored since it connects a single atom on the bridge to the metal and tip atoms. The large contributions in eq.(9) are those which connect atomic orbitals in a pathway all the way from tip to metal. If the energy denominators are large compared to the numerators we further conclude that the contributions for long pathways are smaller the longer the pathway. There are comparatively few large terms in eq.(9) and due to the product form of the term, with the number of factors proportional to the donor–acceptor distance, the dependence on distance should be exponential. However, since the number of terms giving a large contribution in eq.(9) is still formidable, the smaller terms cannot be safely ignored. The use of eq.(9) as a basis for an interpretation of the very different behaviours for different types of bridge, therefore, is not completely straightforward.

3. Results

We present results for the systems DT and MW used in the experiments of ref.[1]. E in $\Theta(E)$ of eq.(1) was chosen to be in the middle of the HOMO–LUMO gap. The transfer capabilities $\Theta(E)^2$ for this value are given in fig.2. We find that Θ^2 is 44,000 times larger for the conjugated molecule than for DT and the notation 'molecular wire' for the former molecule hence appropriate.

The decrease of Θ^2 with R for a saturated bridge (varying the number of CH_2 groups) is exponential: $\Theta^2 = \exp(-\beta R)$ with β=1.14 $Å^{-1}$. Acetylene and phenyl groups were successively replaced by an H atom to examine the distance dependence in a conjugated system of this type (table 1). We found a much slower decrease with distance (β=0.303 $Å^{-1}$) than for the saturated series of molecules.

The function Θ varies slowly in the HOMO–LUMO gap except of course close to the energies of HOMO and LUMO where it tends to infinity. We made some other choices of E but found the same qualitative result. A more detailed study has been made by Samanta et al. for related, symmetric bridges [21].

Since fullerenes have been found to mediate charge transport in lipid bilayers [22] we calculated Θ for C_{60}. We obtained a value which is not much different from the value for a saturated chain of the same length. To find out whether the contribution comes from direct interaction between apex C2p orbitals in the interior of the C_{60} molecule or from interaction through the CC bonds, we replaced the twenty carbon atoms along the equator by He atoms. The coupling was then much reduced. Since this result may be explained as due to the absence of π orbitals on the He atoms we may conclude that the best pathway is indeed through the fullerene π system. This result is interesting since in most other cases the shortest pathway has turned out to be the best.

The transfer capability of the fullerene is not larger than for the aliphatic DT molecules (fig.1) of corresponding length. We believe that the good conduction properties of the fullerenes, consistent with the conclusions in ref.[22], is that resonance tunnelling is involved in the sense that the fullerene bridge acts as donor or acceptor during the transfer process. The low value of the Coulomb repulsion integral in extended molecules like the fullerenes is of direct importance to decrease the effect of the 'Coulomb blockade'.

For C_{60} as for all other molecules the large contributions in eq.(1) do not come from the orbitals around the HOMO–LUMO gap. There are many positive and negative contributions from many MO's and the result is therefore hard to give a physical interpretation in terms of the contributions in eq.(1).

4. Discussion

The results show that π bridges are better bridges for electrons than aliphatic ones. Evenson and Karplus [11] and Reimers and Hush [10,12] defined a dimensionless ratio between energy difference to the transferring orbital and coupling between orbitals (assumed to be the same between all atoms). The decrease is exponential for a large ratio, which is consistent with the present results. However, it still remains to be explained why a saturated bridge gives a slower transfer than a conjugated bridge, since in the aliphatic chains there is also good overlap between the orbitals with only slightly larger energy denominators compared to conjugated systems. The contributions from other tunnelling 'pathways' than the most obvious, connected ones may be large in many cases, which makes a detailed interpretation difficult.

The conclusion that π bridges are better than σ bridges seems to contradict the experimental [23] and theoretical [24] finding that π systems do not improve tunnelling in proteins. Proteins have even been described as organic glass with uniform ET properties [23]. On the other hand Gray et al. [24] conclude that the natural pathways are in fact better than arbitrary pathways through a protein. Our results here show that if conjugated systems are directed in the right way, there will be an enhanced ET capability of the bridge, but this advantage is apparently hard to achieve in natural selection.

It is also of interest that in both aliphatic and conjugated systems, the coupling is provided by the valence orbitals (corresponding to a minimal basis set in ab initio calculations) of the bridge. Virtual atomic orbitals, as for example the C3d orbitals in ref.[8] are not involved to any large extent. This might be tested in ab initio calculations where 3d functions are included as polarization functions. The latter are important for the total energy, but, since their expansion coefficients are much smaller than for the valence orbitals, the contribution to Θ negligible.

In an aliphatic chain the all–trans conformation is adopted by the present support molecules. Saturated CH_2 chains in this conformation form a relatively good tunnelling medium as has been predicted theoretically [6] and shown experimentally by Waldeck et al.[27]. The calculations show a decrease of Θ if the aliphatic chain is turned away from the all–trans conformation [6,17]. In fact conformations can be found which give zero Θ [6,17]. Interestingly enough, in the results given in fig.1, the all–trans chains give a much better tunnelling than the σ orbitals of the conjugated molecules.

In many experiments on molecular wires is is obvious that the wire acts as an electron acceptor or donor itself [27]. π systems have an important role to play also in this case since they are easily reduced or oxidized. If the wire is involved as a redox system each ET step in the three level system has to be treated as an ET system. For example it will be important to calculate the reorganization energy for each possible localization of the electron.

Acknowledgement: We are greatful for support from the Swedish Natural Science Research Council (NFR).

References

1. L.A. Bumm, J.J. Arnold, M.T. Cygan, T.P. Dunbar, T.P. Burgin, L. Jones, II, D.L. Allara, J.M. Tour, P.S. Weiss, Science **271**, 1705 (1996); S.J. Stranick, A.N. Parikh, Y.–T. Tao, D.L. Allara, and P.S. Weiss, J. Phys. Chem. **98**, 7636 (1994).

2. S. Larsson, J. Am. Chem. Soc. **103**, 4034 (1981); S. Larsson, J. Chem. Soc. Faraday Trans. 2, **79**, 1375 (1983).

3. M.A. Ratner, J. Phys. Chem. **94**, 4877 (1990); M.D. Todd, A. Nitzan, and M.A. Ratner, J. Phys. Chem. **97**, 29 (1993).

4. V. Mujica, M. Kemp, M.A. Ratner, J. Chem. Phys. **101**, 6849, 6856 (1994); M. Kemp, A. Roitberg, V. Mujica, T. Wanta, and M.A. Ratner, J. Phys. Chem. **100**, 8349 (1996)

5. S. Larsson and M. Braga, Int. J. Quant. Chem. QBS **20**, 65 (1993).

6. A. Broo and S. Larsson, Chem. Phys. **148**, 103 (1990).

7. M. Braga and S. Larsson, Chem. Phys. Letters **213**, 217 (1993); M. Braga and S. Larsson, Int. J. Quant. Chem. **44**, 839 (1992); J. Phys. Chem. **97**, 8929 (1993).

8. H. McConnell, J. Chem. Phys. **35**, 508 (1961).

9. S. Larsson, J. Phys. Chem. **88**, 1321 (1984).

10. J.R. Reimers and N.S. Hush, Chem. Phys. **146**, 89 (1990).

11. J.W. Evenson and M. Karplus, J. Chem. Phys. **96**, 5272 (1992).

12. J.R. Reimers and N.S. Hush, J. Photochem. Photobiol. A Chem **82**, 31 (1994).

13. S. Larsson, Chem. Phys. Letters **90**, 136 (1982).

14. C. Joachim, Chem. Phys. **116**, 339 (1987).

15. T. Koopmans, Physica **1**, 104 (1934).

16. J. Del Bene and H.H. Jaffé, J. Chem. Phys. **48**, 1807, 4050 (1968); **49**, 1221 (1968).

17. S. Larsson, A. Broo, B. Källebring, and A. Volosov, Int. J. Quant. Chem. QBS **15**, 1 (1988).

18. M.J.S. Dewar, E.G. Zoebisch, E.F. Healy, and J.J.P. Stewart, J. Am. Chem. Soc. **107**, 3902 (1985).

19. E. Gora, Z. Physik 120, 121 (1942–1943); P.–O. Löwdin, J. Mol. Spectr. **10**, 12 (1963).

20. S. Larsson, M. Braga, A. Broo, and B. Källebring, Int. J. Quant. Chem. QBS **18**, 99 (1991).

21. M.P. Samantha, W. Tian, S. Datta, J.I. Henderson, and C.P. Kubiak, Phys. Rev. B **53**, R7626 (1996).

22. K.C. Hwang and D. Mauzerall, J. Am. Chem. Soc. **114**, 9705 (1992); S. Niu and D. Mauzerall, J. Am. Chem. Soc. **118**, 5791 (1996).

23. C.C. Moser, J.M. Keske, K. Warncke, R.S. Farid, and P.L. Dutton, Nature **355**, 796 (1992). R.S. Farid, C.C. Moser, P.L. Dutton, Curr. Opinion Struct. Biol. **3**, 225 (1993).

24. A. Broo and S. Larsson, Int. J. Quant. Chem. QBS **16**, 185 (1989); A. Broo and S. Larsson, J. Phys. Chem. **95**, 4925 (1991).

25. H.B. Gray, D.N. Beratan, J.N. Onuchic, and J.R. Winkler, Science **258**, 1740 (1992); D.R. Casimiro, J.H. Richards, J.R. Winkler, and H.B. Gray, J. Phys. Chem. **97**, 13073 (1993).

26. A. Haran, D.H. Waldeck, R. Naaman, E. Moons, and D. Cahen, Science **263**, 948 (1994).

27. J.R. Reimers, J.S. Craw, G.B. Bacskay, and N.S. Hush, Chem. Phys. Letters **256**, 353 (1996).

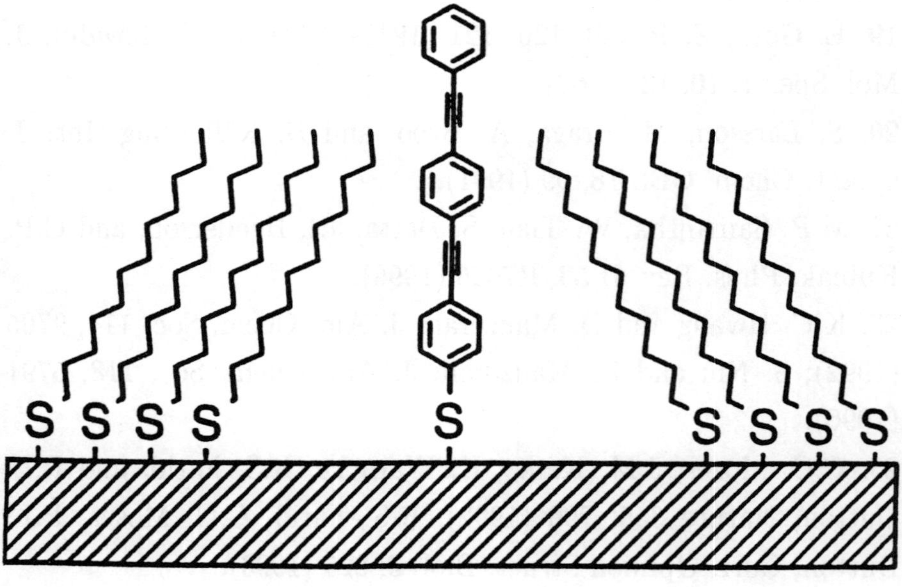

Fig. 1. System studied in the present paper. Further information is given in ref.[1].

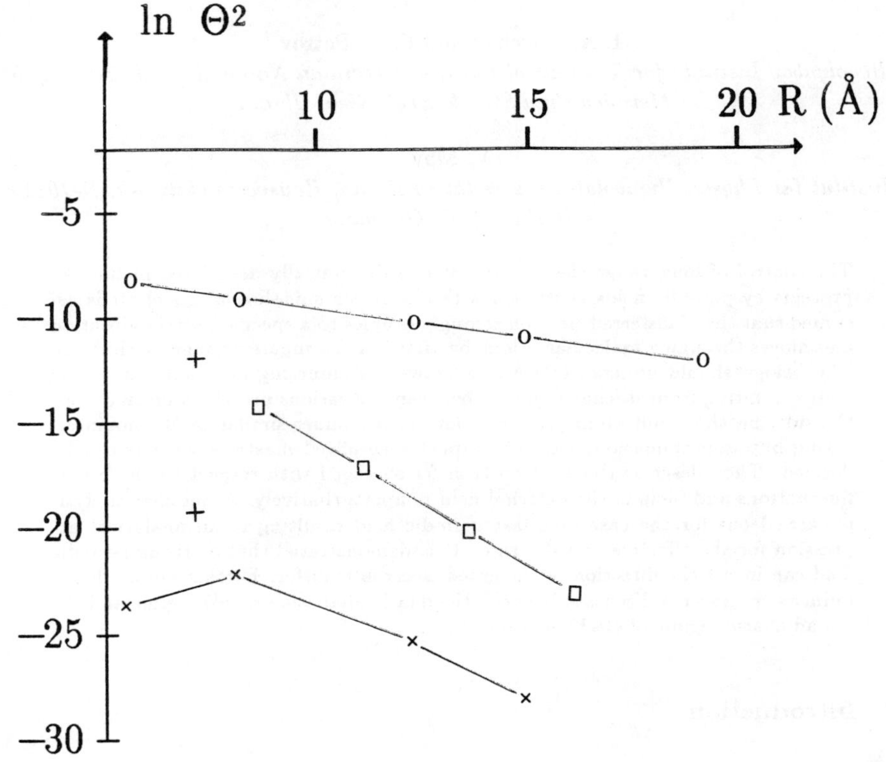

Fig. 2. $\ln|\Theta|^2$ as a function of the length of the molecule. — **o** — molecules wire; — □ — aliphatic DT molecules; — **x** — transfer in the σ bonds of molecular wire; upper + for C_{60}; lower + for $C_{50}He_{10}$ (see text).

PERIODIC FIELD CONTROL OF LONG RANGE ELECTRON TRANSFER IN DYNAMICALLY DISORDERED MEDIA

I. A. Goychuk and E. G. Petrov
Bogolyubov Institute for Theoretical Physics, Ukrainian National Acad. Sci., 14-b Metrologichna Str., 252143 Kiev, Ukraine

V. May
Institut für Physik, Humboldt-Universität zu Berlin, Hausvogteiplatz 5-7, D-10117 Berlin, F. R. Germany

The control of long-range electron transfer in dynamically disordered media like proteins by periodic fields is studied within a driven spin-boson model. It is assumed that the transferred electron strongly couples to a specific protein vibration and moves through a molecular bridge by virtue of the superexchange mechanism. The bridge should undergo transitions between conducting and non-conducting states resulting from stochastic jumps between the various protein conformations. Considering the equilibrium protein modes in a manner similar to the noninteracting blip approximation, a set of coupled generalized master equations can be derived. They describe the electron transfer averaged with respect to the bridge fluctuations and include the external field nonperturbatively. A detailed analysis is carried out for the case of a fast periodic field resulting in an analytical expression for the effective transfer rate. It is demonstrated that a strong periodic field can invert the direction of the gated electron transfer. Furthermore, a field-induced crossover is discussed between the quasi-adiabatic (gated) regime and the nonadiabatic regime of electron transfer.

1 Introduction

Long-range electron transfer reactions (ET) in complex molecular systems like proteins attracted much attention over the years [1–8]. Especially, the control of such reactions by protein dynamics has been in the focus of interest. The dynamics of a protein involves different kinds of internal motions extending from a femtosecond regime up to milliseconds.

Fast and small-amplitude vibrations assisting the ET can be modeled by harmonic oscillators staying in thermal equilibrium with the rest of the protein modes. With respect to these fast vibrations long-range ET takes place in the nonadiabatic regime and can be characterized by nonadiabatic transfer rates derived, e.g., within Fermi's Golden Rule [2]. In contrast, slow protein dynamics (diffusion or flipping motions of molecular groups, concerted conformational transitions, etc.) can be modeled by a stochastic process. Such types of dynamics leads to a random time-dependence of the nonadiabatic transfer rate and a description is possible within stochastic theories of dynamic disorder[9].

An approach based on a Fokker-Plank like equation was originally proposed by *Agmon* and *Hopfield*[10], *Sumi* and *Marcus*[11], and by *Shaitan* and *Rubin*[12]. In all these studies, the slow reaction coordinate was modeled as an overdamped Brownian oscillator (or by the Markovian Gaussian process in other terms). The dichotomic model may serve as alternative description if protein dynamics mainly determined by jumps between two different conformations (see, e.g. [13,14,15]).

It could be demonstrated that slow molecular motions strongly modify ET kinetics. They could change the transfer dynamics from the single–exponential case to the non–exponential one, and could completely control (gate) the ET by a turnover from the nonadiabatic to the adiabatic transfer regime. Additionally, external fields may alter long–range ET in a remarkable manner. Related studies have recently acquired importance in the general context of quantum control of chemical reactions in condensed media. In view of possible interference effects in the case of a simultaneous influence of an external field and the medium fluctuations, an *a priori* treatment based on the Golden Rule formula combined with a stochastic balance equations would fail and a more sophisticated treatment is strongly desirable.

Special interest was dedicated to the driven spin–boson model[16–21] and its application to the control of ET. Especially, it was recognized that by means of a strong periodic field, which is fast compared to the transfer rate, one can change the rates over orders of magnitude. The possibilities to accelerate or to suppress ET reactions [16,17,21], and even to invert the direction of the transfer [21] have been theoretically predicted.

The work in this direction was mainly concentrated on nonadiabatic ET. In the case of dynamically disordered media, new effects stemming from the combined influence of a strong periodic field and dynamic disorder can come into play. Below we give an example for such a combined influence in using a sufficient simple and illustrative model. To meet this goal we proceed with the study of the model of stochastically interrupted dissipative tunneling [19].

2 Model and theory

Let us describe the driven long-range ET in a model including two electronic states, i.e. the donor (D) and acceptor (A) state with energies E_1 and E_2, respectively, which are separated in space. Bridging states of the macromolecule provide the effective electronic coupling $V_{eff}(t)$ between the donor and acceptor states as a result of the so–called superexchange mechanism (for some details see, e.g., Ref. [22]). Such an approximation is possible if the bridging states are energetically well separated from the donor and acceptor states, and any sequential transfer can be completely neglected.

The coupling to the macromolecular vibrations introduces fluctuations of the energy gap $E_g = E_1 - E_2$, which will be treated quantum mechanically. Fluctuations of the conformation of the bridging part may introduce a time–dependence the effective coupling $V_{eff}(t)$. Presumably, this may appear in a protein where the bridging molecular groups of the electronic pathway contain flipping groups, e.g., a tyrosine ring [23]. In the simplest cases if the switching occurs between two energetically equivalent conformations the resulting fluctuations of V_{eff} can be modeled by a symmetric dichotomic Markovian process (DMP, or telegraph noise [24,25]). In such a case $V_{eff}(t)$ jumps between the values V_1 and V_2 ($V_1 > V_2 > 0$) with equal escape frequencies $\nu_1 = \nu_2 = \nu$. The dichotomic type of fluctuations is a very important example for the stochastic influence stemming from an inhomogeneous medium. Such fluctuations were revealed, for instance, in the molecular dynamics simulations of the time–dependent energy gap of the photosynthetic reaction center [15] (see also [13,14]).

Furthermore, we suppose that a strong and time–dependent electric field $\mathcal{E}(t)$ is applied to the DA system, which couples to the ET via the dipole moment $d = er$ of the ionic pair D^+A^- created by the ET (r is the transfer distance).

The Hamiltonian of the used model has the following form

$$H(t) = \frac{1}{2}\varepsilon(t)\hat{\sigma}_z + V_{eff}(t)\hat{\sigma}_x + \frac{1}{2}\hat{\sigma}_z \sum_\lambda \kappa_\lambda (b_\lambda^\dagger + b_\lambda) + \sum_\lambda \hbar\omega_\lambda (b_\lambda^\dagger b_\lambda + \frac{1}{2}). \quad (1)$$

The pseudo–spin operators $\hat{\sigma}_z = |1\rangle\langle 1| - |2\rangle\langle 2|$ and $\hat{\sigma}_x = |1\rangle\langle 2| + |2\rangle\langle 1|$ are defined by the two electronic states, $|1\rangle$ and $|2\rangle$. They are coupled one to another via the fluctuating effective matrix element $V_{eff}(t)$. The time–dependent energy gap $\varepsilon(t) = E_g - d\,\mathcal{E}(t)$ accounts for the time–dependent external field. b_λ and $b_\lambda^\dagger$ are the boson annihilation and creation operators of protein vibrations considered as a heat bath. The interaction of the transferring electron with protein vibrations can be represented by the stochastic force operator $\hat{F}(t) = \sum_\lambda \kappa_\lambda (b_\lambda^\dagger e^{i\omega_\lambda t} + b_\lambda e^{-i\omega_\lambda t})$ (written here in the Heisenberg representation). This stochastic force controls the quantum fluctuations of the energy bias. It obeys Gaussian statistics, and is completely characterized by the zero mean, $\langle \hat{F}(t)\rangle_T = 0$, and the complex second moment [26,27],

$$K(t) = \langle \hat{F}(t)\hat{F}(0)\rangle_T = \frac{1}{2\pi}\int_0^\infty J(\omega)\frac{\cosh(\hbar\omega/2k_BT - i\omega t)}{\sinh(\hbar\omega/2k_BT)} d\omega \,. \quad (2)$$

The bath spectral function is given by $J(\omega) = 2\pi\hbar^{-2}\sum_\lambda \kappa_\lambda^2 \delta(\omega-\omega_\lambda)$, and the brackets $\langle..\rangle_T$ denote the average over the heat bath with temperature T.

In the case of long–range ET the effective electronic D–A coupling, $V_{eff}(t)$, is weak and can be treated as a perturbation. This circumstance justifies the derivation of a master equation for the difference of the donor and acceptor level population $\sigma_z(t)$ in the Born approximation with respect to the dressed electronic coupling (for more details see Refs. [19,20,28]). In line with Refs. [19,20] we chose the *initial* decoupling between the transferring electron and the heat bath at $t=0$, and apply the polaron transformation [29] to the dressed states. Than, the following master equation can be obtained

$$\dot{\sigma}_z(t) = -\int_0^t V_{eff}(t)V_{eff}(t')f(t,t')\sigma_z(t')dt' - \int_0^t V_{eff}(t)V_{eff}(t')g(t,t')dt' \ . \quad (3)$$

The equation contains the functions

$$f(t,t') = \frac{4}{\hbar^2}\exp[-G_s(t-t')]\cos[G_a(t-t')]\cos(\frac{1}{\hbar}\int_{t'}^t \varepsilon(\tau)d\tau) \ ,$$

$$g(t,t') = \frac{4}{\hbar^2}\exp[-G_s(t-t')]\sin[G_a(t-t')]\sin(\frac{1}{\hbar}\int_{t'}^t \varepsilon(\tau)d\tau) \ , \quad (4)$$

which are expressed via the bath influence function $G(t) = G_s(t) + i\,G_a(t) = \int_0^t dt_1 \int_0^{t_1} K(t_2)dt_2 + i\frac{E_r}{\hbar}t$ [27]. The last quantity is determined via the double–integrated second moment (2) and via the bath reorganization energy, $E_r = \frac{\hbar}{2\pi}\int_0^\infty \frac{J(\omega)}{\omega}d\omega$.

The master equation (3) is the generalization of the equation resulting from the non–interacting blip approximation (NIBA) to the case of a fluctuating electronic coupling and a time–dependent free–energy gap. It contains the external field $\mathcal{E}(t)$ in a nonperturbative manner and is a direct generalization of the master equation used in Refs.[16,17,20] to the case of a fluctuating electronic coupling. Eq. (3) can be also considered as an extension of the master equation (9) of Ref. [19] to the case of a strong external field. It can be also obtained in carrying out the NIBA within a path integral description [30].

Until this point the time–dependence of $V_{eff}(t)$ and $\mathcal{E}(t)$ has been not fixed. In a first step, let us specify $V_{eff}(t)$ in accordance with the above mentioned dichotomic model as $V_{eff}(t) = V_0 + \Delta\alpha(t)$, where $V_0 = \langle V_{eff}\rangle =$

$(V_1 + V_2)/2$ and $\Delta = (V_1 - V_2)/2$. The quantity $\alpha(t) = \pm 1$ characterizes the DMP as a process with zero mean $\langle \alpha(t) \rangle = 0$, and with the exponentially decaying autocorrelation function $\langle \alpha(t)\alpha(t') \rangle = \exp[-\nu|t-t'|]$ [25]. The noise–averaging has been denoted by $\langle ... \rangle$.

In this case of a DMP, the noise–averaging of the master equation (3) can be carried out *exactly* using one of the theorems of *Bourret, Frisch*, and *Pouquet*[25] and the theorem of *Shapiro and Loginov*[31] (for details see[19]). As a result, a set of master equations for the noise–averaged difference of level populations, $\langle \sigma_z(t) \rangle$, and the correlator, $\langle \alpha(t)\sigma_z(t) \rangle$ can be obtained.

Next, let us consider the case of a fast oscillating field,

$$\mathcal{E}(t) = \mathcal{E}_0 \cos(\Omega\, t)\ . \tag{5}$$

such that $\Omega >> \nu,\ 1/\langle \tau \rangle$. (The characteristic time $\langle \tau \rangle$ of the ET reaction will be defined below). According to this inequality, we can use the approach derived in Ref. [18,21] to obtain master equations coarse grained with respect to the fast oscillations of the field. Hence, in line with Ref. [19,21] the following set of reduced master equations is obtained

$$\begin{aligned}
\frac{d}{dt}\langle \overline{\sigma}_z(t) \rangle = & - \int_0^t (V_0^2 + \Delta^2 e^{-\nu(t-t')}) \overline{f}(t-t') \langle \overline{\sigma}_z(t') \rangle dt' \\
& - V_0 \Delta \int_0^t (1 + e^{-\nu(t-t')}) \overline{f}(t-t') \langle \alpha(t') \overline{\sigma}_z(t') \rangle dt' \\
& - \int_0^t (V_0^2 + \Delta^2 e^{-\nu(t-t')}) \overline{g}(t-t') dt', \\
\frac{d}{dt}\langle \alpha(t) \overline{\sigma}_z(t) \rangle = & - \nu \langle \alpha(t) \overline{\sigma}_z(t) \rangle - V_0 \Delta \int_0^t (1 + e^{-\nu(t-t')}) \overline{g}(t-t') dt' \\
& - \int_0^t (\Delta^2 + V_0^2 e^{-\nu(t-t')}) \overline{f}(t-t') \langle \alpha(t') \overline{\sigma}_z(t') \rangle dt' \\
& - V_0 \Delta \int_0^t (1 + e^{-\nu(t-t')}) \overline{f}(t-t') \langle \overline{\sigma}_z(t') \rangle dt'\ .
\end{aligned} \tag{6}$$

Note the introduction of the functions

$\overline{f}(t) = \frac{4}{\hbar^2} \sum_{n=-\infty}^{\infty} J_n^2(\zeta) \exp[-G_s(t)] \cos[G_a(t)] \cos[(n\Omega + E_g/\hbar)t]\ ,$

$$\overline{g}(t) = \frac{4}{\hbar^2}\sum_{n=-\infty}^{\infty} J_n^2(\zeta)\exp[-G_s(t)]\sin[G_a(t)]\sin[(n\Omega + E_g/\hbar)t]\ , \quad (7)$$

which are the quantities of Eq. (4) averaged over the fast oscillations. They contain the Bessel functions $J_n(\zeta)$ of the first kind at the argument $\zeta = d\mathcal{E}_0/\hbar\Omega$, which gives a measure for the coupling strength of the external field to the ET.

The obtained set of equations (6) generalizes the master equation used in recent studies [16,17,21] to the case of a dichotomically fluctuating electronic coupling. Furthermore, the Eqs. (6) also extend the master equation used in our previous studies dealing with the influence of dynamic disorder on ET reactions [19] to the case of the action of a monochromatic field.

Note that in the general case the set (6) describes non-exponential transfer kinetics what may be characterized by a mean relaxation time[32],

$$\langle\tau\rangle = \frac{\int_0^\infty [\langle\overline{\sigma}_z(t)\rangle - \langle\overline{\sigma}_z(\infty)\rangle]dt}{\langle\overline{\sigma}_z(0)\rangle - \langle\overline{\sigma}_z(\infty)\rangle}. \quad (8)$$

Accordingly, an effective rate constant can be defined as $\Gamma = 1/\langle\tau\rangle$.

Now, let us consider a model of stochastically interrupted tunneling where the bridge fluctuates between the conducting state ($V_1 \neq 0$) and the non-conducting state ($V_2 = 0$). In the used approximation scheme any non–Markovian contributions to the averaged master equations (6) can be safely neglected. Thus, ET kinetics turn out to be double–exponential in the general case. Providing an *initial* decoupling between the transferring electron and the bridge fluctuations we obtain the effective transfer rate as

$$\Gamma = \frac{\nu k(\nu)}{\nu + k(\nu)}\,. \quad (9)$$

This expression contains the nonadiabatic transfer rate $k(\nu) = V_0^2(\tilde{f}(0)+\tilde{f}(\nu))$, with $\tilde{f}(\nu) = \tilde{f}_+(\nu) + \tilde{f}_-(\nu)$ and with

$$\tilde{f}_\pm(\nu) = \frac{2}{\hbar^2}\sum_{n=-\infty}^{\infty} J_n^2(\zeta)\mathrm{Re}\int_0^\infty dt\exp[-G(t) + i\,n\Omega t \pm iE_g t/\hbar - \nu t]\,. \quad (10)$$

Eq. (9) gives the typical form of a rate expression derived in theories which connect the limits of the adiabatic and nonadiabatic ET (see, e.g., Refs.[8,33,34,35]). If the condition $\nu \ll k(\nu)$ holds, the transfer is non–exponential and takes place in the gated (adiabatic) regime with the effective transfer rate $\Gamma \approx \nu$. In the opposite limit, $\nu \gg k(\nu)$, single–exponential nonadiabatic kinetics appear with the nonadiabatic transfer rate $k(\nu)$. Note, that the rate

(9) strongly depends on the frequency of the external field, Ω, the coupling strength ζ, and on the switching rate ν. It can be changed by external field over orders of magnitude [16,17,21]. In the context of the present model this means than one can switch the ET between gated and nonadiabatic regimes applying a periodic field. This is one possible effect of external fields on ET kinetics in media with dynamic disorder.

Furthermore, it should be stressed that the steady–state populations of the donor and acceptor levels, $P_D^s = (1 + \langle\overline{\sigma}_z(\infty)\rangle)/2$ and $P_A^s = (1 - \langle\overline{\sigma}_z(\infty)\rangle)/2$, respectively, can drastically deviate from their equilibrium values in the absence of the external field. This pronounced nonequilibrium effect has been already demonstrated for the nonadiabatic electron transfer [17,21]. It can be characterized by the effective energy

$$\tilde{E}_g = k_B\, T\, \ln \frac{P_A^s}{P_D^s}. \tag{11}$$

This quantity coincides with the actual energy gap E_g in the absence of the external field ($\mathcal{E}_0 = 0$) and the bridge fluctuations ($\Delta = 0$). But, in the presence of an external field it can be reduced drastically, and even can become negative. The last effect may be interpreted as the inversion of the ET direction. For the considered model the calculation of $\tilde{E}_g$ yields

$$\tilde{E}_g = k_B\, T\, \ln \frac{\tilde{f}_+(0) + \tilde{f}_+(\nu)}{\tilde{f}_-(0) + \tilde{f}_-(\nu)}. \tag{12}$$

Note, that expression (12) has the same structure for the gated as well as for the nonadiabatic regime. This observation allows to predict the possibility to invert the direction of the gated transfer by an external field despite the fact that the transfer rate $\Gamma \approx \nu$ could be the same as in the absence of the field.

3 Some illustrative calculations

To illustrate the above mentioned effects we will consider the case of a strong coupling to a single specific mode of the medium. This specific mode can be represented by a damped oscillator which feels the remaining vibrational degrees of freedom as friction. Such a model is of special interest for the ET in proteins. It is given by the spectral function $J(\omega) = 8\kappa_0^2/\hbar^2\gamma\omega_0\omega/[(\omega_0^2 - \omega^2)^2 + 4\gamma^2\omega^2]$, where ω_0 is the frequency of the specific mode, κ_0 denotes the coupling constant to the transferred electron, and γ is the broadening of the oscillator levels due to friction [8]. In the strong coupling limit ($\kappa_0 >> \hbar\omega_0$)

one can perform the short-time approximation for $G_{s(a)}(t)$ [8]. This yields $G(t) \approx E_r k_B T_{eff} t^2/\hbar^2 + iE_r t/\hbar$, with the reorganization energy $E_r = \kappa_0^2/\hbar\omega_0$. The effective temperature $T_{eff} = \hbar^2(4\pi k_B E_r)^{-1} \int_0^\infty \coth(\hbar\omega/2k_BT)J(\omega)d\omega$ depends on the absolute temperature T and the friction [8] and coincides with T in the high–temperature limit. In the given approximation we get from equation (10)

$$\tilde{f}_\pm(\nu) = \frac{2\pi}{\hbar}\frac{1}{\sqrt{4\pi E_r k_B T_{eff}}} \sum_{n=-\infty}^{\infty} J_n^2(\zeta) Re[w(\frac{E_g \mp E_r + n\hbar\Omega + i\hbar\nu)}{2\sqrt{E_r k_B T_{eff}}})] \,, \quad (13)$$

where $w(z) = \exp(-z^2)\mathrm{erfc}(-iz)$ is the error function of a complex variable, and $\mathrm{erfc}(z)$ denotes the complementary error function. In the case of $\nu = 0$, Eq. (13) reduces to the form

$$\tilde{f}_\pm(0) = \frac{2\pi}{\hbar}\frac{1}{\sqrt{4\pi E_r k_B T_{eff}}} \sum_{n=-\infty}^{\infty} J_n^2(\zeta) e^{-(E_g + n\hbar\Omega \mp E_r)^2/4E_r k_B T_{eff}} \,. \quad (14)$$

This expression has been already derived in the theory of periodically driven ET [17,21]. The only difference here and in Eq. (12) is the use of the effective temperature T_{eff} instead the temperature T. This difference reflects the factual utilization of *Hopfield's* model of the ET [2]. (To neglect the friction effects one has to assume that $\gamma \ll \hbar\omega_0$.)

Now, the effective transfer rate Γ, Eq. (9), and the effective energy gap $\tilde{E}_g$, Eq.(12), are completely defined. To perform numerical calculations we chosen the following set of parameters $E_g = 3600\mathrm{cm}^{-1}$, $E_r = 1200\mathrm{cm}^{-1}$, $\hbar\omega_0 = 250\mathrm{cm}^{-1}$, $V_0 = 1\mathrm{cm}^{-1}$ and put $\gamma = 0$. These parameters are typical for the long–range electron tunneling in metalloproteins like cytochrom c (see, e.g., [36]). Results of the numerical calculations are presented in Fig. 1(a) and Fig.1 (b). They demonstrate the effect of the induced crossover from the nonadiabatic regime to the gated regime (Fig. 1(a)) and the effect of the space inversion of the gated electron (Fig. 1(b)) which were discussed above. Assuming a tunneling distance of $r \approx 20$Å one can show that for the given set of parameters that the effect of ET inversion requires a field strength of the order of $\mathcal{E}_0 \sim 5\cdot 10^5$V/cm or larger. Such fields are quite accessible in the experiment.

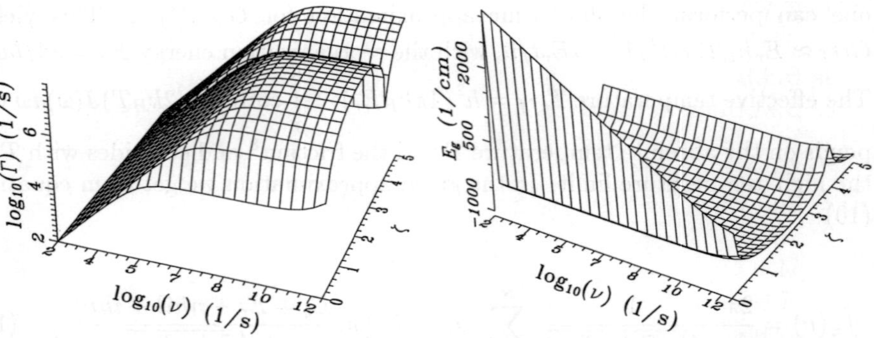

Figure 1: Dependence of the effective transfer rate Γ (left part) and of the effective energy gap (right part) on the noise frequency (reverse autocorrelation time) ν and on the strength of coupling $\zeta = e r \mathcal{E}_0 / \hbar\Omega$ to the external periodic field at the external frequency $\hbar\Omega_0 = 4800\ \mathrm{cm}^{-1}$. T=100 K. Other parameters are given in the text.

Acknowledgments

Support of this work by the *Volkswagen–Stiftung* (Federal Republic of Germany) is gratefully acknowledged.

References

1. J.Ulstrup, *Charge Transfer Processes in Condensed Media* (Springer, Berlin, 1979).
2. J.J.Hopfield, Proc. Natl. Acad.Sci. USA. **71**, 3640 (1974).
3. J.Jortner, J. Chem. Phys. **64**, 4860 (1976).
4. R.A.Marcus and N.Sutin, Biochim. Biophys. Acta **811**, 265 (1985).
5. D.De Vault, *Quantum Mechanical Tunneling in Biological Systems, 2nd ed.* (Cambridge University, New York, 1984).
6. *Long-range Electron Transfer in Biology*, ed.G.Palmer (Springer, Berlin, 1991).
7. E.G.Petrov, *Physics of Charge Transfer in Biosystems* (Naukova Dumka, Kiev, 1984) (in Russian).
8. A.Garg, J.N.Onuchic, and V.Ambegaokar, J.Chem.Phys. **83**, 4491 (1985).
9. R.Zwanzig, Acc.Chem.Res. **23**, 148 (1990).
10. N.Agmon and J.J.Hopfield, J. Chem. Phys. **78**, 6947 (1983).
11. H.Sumi and R.A.Marcus, J.Chem. Phys. **84**, 4849 (1986).

12. K.V.Shaitan and A.B.Rubin, Molekul. Biol. **16**, 1004 (1982) (in Russian).
13. B.S.Brunschwig and N.Sutin, J. Am. Chem. Soc. **111**, 7454 (1987).
14. B.M.Hoffman and M.R.Ratner, J. Am. Chem. Soc. **109**, 6237 (1987).
15. J.H.Gehlen, M.Marchi, and D.Chandler, Science **263**, 499 (1994).
16. Yu.Dakhnovskii, J. Chem. Phys. **100**, 6492 (1994).
17. Yu. Dakhnovskii and R.D.Coalson, J. Chem. Phys. **103**, 2908 (1995); Yu. Dakhnovskii, D.G.Evans, H.J.Kim, and R.D.Coalson, *ibid.* **103**, 5459 (1995).
18. M.Grifoni, M.Sassetti, P.Hänggi, and U.Weiss, Phys. Rev. E **52**, 3596 (1995).
19. I.A. Goychuk, E.G. Petrov and V. May, J. Chem. Phys. **103**, 4937 (1995).
20. I.A.Goychuk, E.G.Petrov, and V.May, Phys. Rev. E **52** 2392 (1995).
21. I.A.Goychuk, E.G.Petrov, and V.May, Chem. Phys. Lett. **253**, 428 (1996).
22. H.McConnell, J.Chem.Phys.**35**, 508 (1961).
23. R.F.Goldstein and W.Bialek, Comments Mol. Cell. Biophys.**3**, 407 (1986). J.A.McCammon, Rep. Prog. Phys **47**, 1 (1984).
24. N.G.Van Kampen,*Stochastic Processes in Physics and Chemistry* (North-Holland, Amsterdam, 1984).
25. R.C.Bourret, U.Frisch, and A.Pouquet, Physica **65**, 303(1973).
26. A.J.Leggett *et al.*, Rev. Mod. Phys. **59**, 1 (1987).
27. U.Weiss, *Quantum Dissipative Systems*, Series in Modern Condensed Matter Physics Vol.2 (World Scientific, Singapore, 1993).
28. E.G. Petrov, Teor. Mat. Fiz. **68**, 117 (1986); E.G.Petrov, V.I.Teslenko, and I.A.Goychuk, Phys.Rev.E **49**, 3894(1994).
29. T.Holstein, Ann.Phys. (N.Y.) **8**, 343 (1959).
30. M.Grifoni, Phys. Rev. E **54** (in press).
31. V.E.Shapiro and V.M.Loginov, Physica A **91**, 563(1978).
32. Such a description is often used for the complicated kinetic processes[11,24], see also H.L.Friedman and M.D.Newton, Faraday Disc. Chem. Soc. **74**,73 (1982); D.Y.Yang, J. Stat.Phys. **74**, 631 (1994).
33. L.D.Zusman, Chem.Phys.**49**, 295 (1980).
34. I.Rips and J.Jortner, J. Chem. Phys. **87**, 2090 (1987).
35. S.Roy and B.Bagchi, J.Chem. Phys. **100**, 8802 (1994).
36. J.N.Onuchic, J. Chem. Phys. **86**, 3925 (1987); R.F.Goldstein and A.Bearden, Proc. Natl. Acad.Sci. USA **81**, 135 (1984).

PROTEINS AS SPECIFIC MEDIA FOR CHARGE TRANSFER REACTIONS

L.I. KRISHTALIK, E.L. MERTZ, V.V. TOPOLEV
A.N.Frumkin Institute of Electrochemistry, Russian Academy of Sciences, Leninskii prosp 31, 117071, Moscow, Russia

The highly-organized spatial structure for proteins results in three important properties of them as polar media of charge-transfer reactions. They are: the existence of a permanent intraprotein electric field, a low dielectric constant, and the time evolution of their dielectric response. The first property influences strongly the reaction equilibrium energetics. The second one manifests itself in a low reorganization energy. The latter effect has been confirmed by an independent method, *viz.* by the study of Stokes shifts of a chromophore bound to a protein. Only combination of the effects of intraprotein field and of low reorganization energy makes proteins effective catalysts of charge-transfer reactions. The third property affects substantially the kinetics of very fast processes as revealed by an analysis of electron transfers in photosynthetic reaction centers.

1 Introduction. The specific properties of proteins: an outline

In the present paper, we will consider enzymatic charge transfer reactions, namely the transfer of electrons and/or protons. For these particles with a strongly expressed quantum behavior, a substantial contribution to their activation energy is due to the polar medium reorganization. [1-5] Our task is to follow how the general principles of the charge transfer theory manifest themselves in such specific media as proteins.

For the systems of so high complexity as proteins, it is hopeless to obtain the strict quantitative results basing on the first principles only. Hence, in the theoretical analysis of enzymatic reactions we have to resort to some approximations, and the problem is to choose approximations adequate to the purpose of the analysis. A self-evident property of proteins is their structural specificity making a given protein appropriate to perform a definite function, and not the others. Therefore, a full analysis of an enzymatic process shall employ a structural information on the protein. On the other hand, there is much common in behavior of many enzymes of rather different structures, and, therefore, we can expect that some substantial features of enzymes are due to some general principles of their architecture rather than to specific details of the structure of a concrete protein.

All proteins are linear polymers formed by peptide bonds and folded in a definite secondary and tertiary structures. The peptide groups are highly polar ones. Their dipole moment in free liquid amides is close to 4 D, in proteins this value increases due to the mutual polarization of peptide groups, especially the hydrogen-bonded ones. Dipole moments of about 5-6 D can be inferred from the experimental data on dissolved α-helical peptides [6, 7] and on increase in C-O and decrease in C-N bond lengths in proteins as compared to monomolecular amides. [8] Quantum-chemical calculations [9] have also shown the rise of the dipole moment to about 6 D after closing the hydrogen bond link in the helical tetrapeptide.

The concentration of peptide groups in proteins is very high. The low-molecular-weight liquids with a similar concentration of these groups, *viz.* amides, are high-dielectric media (in particular, N-methylformamide is a solvent with one of the highest dielectric constants $\varepsilon = 180$). However, in proteins, in contrast to liquid solvents, the peptide dipoles are incorporated in a large and rather rigid structure, and hence their mobility, their ability to reorient under the action of an external electric field is drastically restricted. They have, of course, some mobility, as it is revealed by various experimental techniques and molecular dynamics simulations, but the mutual orientation of dipoles can change only in relatively narrow limits, in the opposite case the spatial structure of protein would be destroyed.

The low ability of the protein dipoles to reorient in an external electric field means a weak dielectric response, *i.e.* a low dielectric constant. Experimental measurements of static dielectric constants of dry proteins give the value of ε about 3.5-4. [10-14] Similar quantities were obtained, at least for the bulk of the protein, by molecular simulations taking an explicit account of the mobility of peptide groups. [15-20]

In a liquid solvent, at any arbitrary point of its bulk, and at any arbitrary moment of time, there does exist some electric field set up by surrounding dipoles. Due to random fluctuations of the orientation of these dipoles, the field fluctuates also randomly, and its value averaged over a long enough time interval equals to zero. In the case of an ion introduced into a solvent, the electric field of the ion will cause a preferential orientation of dipoles creating some organization of the medium and hence causing a permanent non-zero field of dipoles. In a protein, dipoles fluctuate too but their vectors remain in some definite bounds of spatial angle. Hence, their electric field fluctuates around some non-zero value, this value depending upon the protein structure and upon the position of the point under consideration. We see that, inside the protein, there does exist some permanent (averaged in time) structure-dependent field of dipoles pre-existing before the introduction of any extraneous charge. Therefore, the protein can be called a "preorganized medium". [21]

We see that the organization of dipoles into a definite structure has two interconnected cosequences: the existence of some permanent intraprotein electric field and the low dielectric constants of these media.

These considerations show that proteins can be defined as highly-polar low-dielectric media. This definition stresses the principal difference between proteins (and, possibly, some other macromolecular bodies) and low-molecular solvents. In the latter, due to practically free rotation of their molecules, "polarity" and dielectric constant change in parallel. In proteins, "polarity" (dipole moment and concentration of polar groups) and dielectric constant are not interconnected in such a direct way. From this point of view it is no sense to oppose the notions of "highly-polar medium" against "low-dielectric medium" as applied to proteins. They are both at the same time, and these different (and seemingly contradicting) properties manifest themselves, as will be shown below, largely in different reaction parameters.

Due to a very high complexity of proteins structures, there exists a hierarchy of different motions, with their characteristic times lying in the interval of several orders of magnitude wide. [22-25] Correspondingly, a wide distribution of characteristic times of their dielectric responses should be typical of proteins. Evidently, for very short reaction times some kinds of relatively slow polarization modes have not enough time to manifest themselves in the course of the charge transfer, and hence the effective medium dielectric response is lower than the response corresponding to the static dielectric constant. So, the third property of proteins as dielectica is their different effective dielectric constants corresponding to different time scales. [26, 27]

2 The activation energy of enzymatic reactions

The activation energy of the elementary act of electron or proton transfer $\Delta G^{\neq}$ is determined, according to a well-known equation, by two parameters, namely the reorganization energy λ and the free energy of the reaction elementary act ΔG (see, *e.g.*, [1-5])

$$\Delta G^{\neq} = \frac{(\lambda + \Delta G)^2}{4\lambda} \qquad (1)$$

The reorganization energy λ can be considered as a sum of the inner-sphere λ_i and medium λ_S contributions. In its turn, ΔG also includes two constituents. First is the "structural" component ΔG_{str} which consists of the "chemical" term depending on the inner structure of the reactants, and of an electrostatic term, namely the energy of the charge transfer in the pre-existing intraprotein electric field depending on the protein structure. The second contribution is the energy of the charge - medium polarization interaction ΔG_{pol} (in the case of two charged particles emerging in the course of reaction, ΔG_{pol} includes also the energy of their mutual interaction).

2.1 The equilibrium reaction energy

The importance of an intraprotein electric field of peptide dipoles (more strictly, multipoles, because we deal here with a neutral system of several atomic partial charges) was recognized first in early seventies. [21, 28] Later in the same decade, several other groups explored the problem. [29-33] Nowadays, the activity in the field (including other electrostatic aspects of the equilibrium energetics) is very high, and we have to restrict ourselves to referring to some recent reviews. [34-47] However, one problem deserves to be discussed especially, *viz.* the problem of dielectric constant in these electrostatic calculations.

The notion of the dielectric constant of proteins in the framework of continuum formalism has been discussed many times. [15, 34, 48, 49] The main conclusion of these considerations is that the effective dielectric constant that should be used in calculations depends on the problem, namely on the approximations used to describe the charge density distribution in the protein. Let us now consider briefly how the dielectric constant(s) should be used consistently in the semi-continuum calculations of the enzymatic reaction energetics. [50]

Knowing the exact electronic density distribution and the nuclei positions, we could calculate the electric field at any point of the system considering the charge density distribution in vacuum, *i.e.*, at dielectric constant $\varepsilon = 1$. However, such a complete knowledge of the protein's electronic structure is practically inaccessible, and therefore we have to restrict ourselves to an approximate description of the charge density distribution, *e.g.*, ascribing to each atom (or each bond) a definite partial charge evaluated on the basis of experimental data and quantum-chemical calculations for rather small model molecules. While transferring these partial charges to protein macromolecule, we should take into account the effect of atoms' charges on the electronic clouds of atoms of other groups, that is the effect of their electronic polarization. Averaging this effect, we come to description of the system as a structure formed by fixed charges embedded into a medium with an optical (electronic) dielectric constant ε_O. It should be noted that the mutual interaction of partial charges influences, of course, the atoms' equilibrium positions, but this effect should not be accounted for as some quasi-orientational polarization because we start with known equilibrium coordinates of an already folded protein.

Let us consider now formation of an ion inside a protein molecule, the ion which was absent in the equilibrium X-ray structure. From one hand, this ion appears at a definite place in the structure, and the energy of the newly-formed charge is influenced by the field pre-existing at this point. This field, as it was discussed above, can be calculated with the help of the optical dielectric constant ε_O. From the other hand, the field of the new ion interacts not only with electrons of the medium, but produces also some shifts of its fully or partially charged atoms resulting in a new equilibrium structure, but with the new atomic coordinates being unknown. The situation can be described as a superposition of a set of induced dipoles upon the original network of partial charges. This means, the new charge interacts with all kinds of medium polarization, and hence the energy of this interaction depends on the total value of the static dielectric constant ε_S.

Proteins are often surrounded by another medium, in particular water. In this case, the electric fields should be calculated with account of the heterogeneity. This means that, in calculating the pre-existing field, we should use optical ε_O for protein but static ε_{SW} for water because water is not preorganized.

We can conclude that, for the correct semi-continuum calculation of the energy of ions in protein, one should employ two protein's dielectric constants simultaneously, the optical and static ones. This differs from the widely applied protocol, when a single value of ε is chosen. This will give rise to substantially different quantitative results. It is well known that, for instance, the field of protein dipoles brings a marked contribution into the reaction energy compensating usually a large part of the loss of the ion solvation energy (see, *e.g.*, the reviews quoted above). With ε_O used for the calculation of this field, the effect will increase about 1.5 times.

2.2 *The reorganization energy*

Let us go over to discussion of the medium reorganization energy λ_S. According to the Marcus formula, in the simple case of the transfer of charge e between two spherical reactants of radii a_1, a_2 and an inter-center distance R_{12} situated in a homogeneous medium

$$\lambda_s = e^2\left(\frac{1}{\varepsilon_o} - \frac{1}{\varepsilon_s}\right)\left(\frac{1}{2a_1} + \frac{1}{2a_2} - \frac{1}{R_{12}}\right) \qquad (2)$$

Low value of the dielectric constant of proteins affects strongly the reorganization energy of any intraprotein charge transfer. For common solvents having high or medium dielectric constants $\varepsilon_S >> \varepsilon_O$, and hence the polaron coupling constant $C = 1/\varepsilon_O - 1/\varepsilon_S$ only slightly varyes with a change of ε_S. On the contrary, the transition to a medium with a very low dielectric permittivity changes the value of λ_S drastically. For water, with $\varepsilon_O = 2.1$ and $\varepsilon_S = 78$, $C = 0.46$ (the optical, purely electronic ε_O of water equals to 1.8; however, in the calculations of the reorganization energy a so called "quantum boundary" of the dielectric constant should be employed which accounts for the high-frequency vibrational modes displaying a quantum behavior [51]). For proteins, "quantum" $\varepsilon_O \approx 2.5$ (the usual value for liquid amides corrected for higher density of proteins) and $\varepsilon_S = 4$, hence

$C \approx 0.15$, *i.e.* about three times smaller. Therefore, as it was shown in [52, 53], a low reorganization energy is typical of proteins as the media for charge transfer reactions, and hence it is one of the physical reasons of their catalytic effect.

The same conclusion was inferred also from a qualitative microscopic considerations: the dipoles inside the protein globule are rather rigidly fixed and cannot undergo a noticeable reorientation, hence the corresponding reorganization energy is low. [21] Quantitative molecular dynamics simulations for some biological charge transfer reactions confirmed low reorganization energies for particular proteins studied in these simulations. [54-63] [a]

Real enzymatic systems are heterogeneous ones, *e.g.*, protein globule is surrounded by aqueous solution. In literature, two different approaches to calculations of the reorganization energy have been used, so called "fixed-displacement-field" and "fixed-charge-density" methods. [64] They gave coinciding results for homogeneous, but quantitatively different ones for inhomogeneous media. Quite recently, the convincing arguments in favor of the "fixed-charge-density" formalism as the only rigorous one have been given. [64-66] This means that for obtaining the reorganization energy the general Eq. 3 should be used:

$$\lambda_s = \frac{1}{8\pi} \int_V \left[\frac{1}{\varepsilon_0} (\mathbf{D}_i^o - \mathbf{D}_f^o)^2 - \frac{1}{\varepsilon_s} (\mathbf{D}_i^s - \mathbf{D}_f^s)^2 \right] dV \qquad (3)$$

Here $\mathbf{D}_i$ and $\mathbf{D}_f$ are the dielectric displacement vectors corresponding to the initial and final charges' distributions, $\mathbf{D}^o$ and $\mathbf{D}^s$ refer to the cases when all dielectric phases have their optical or, correspondingly, static ε's.

As it will be shown below, the effects of ΔG_{pol} and ΔG_{str} can be analyzed, to some extent, separately considering ΔG_{str} as an independent parameter. This is due to the fact that ΔG_{str} is strongly dependent on the protein structure while ΔG_{pol} and λ_s are connected with the protein's general dielectric response, much less sensitive to structural details. It can be shown quite rigorously that the charging energies, *i.e.*

---

[a] The role of a low reorganization energy of enzymes was especially emphasized in one of the papers by Warshel and co-workers. [60] Unfortunately, they presented this qutie correct conclusion, fully coinciding with our previous results, as something different in principle to our treatment. Their point of view is that the low dielectric constant of the medium is irrelevant to the catalytic activity of enzymes because it would reduce λ but at the same time "nonpolar protein sites would slow rather than accelerate charge-transfer reactions due to the loss in self energy (solvation energy) associated with bringing the charged fragments from water to the nonpolar active site". It is easy to see, that this reasoning is based on the identity of notions "low-dielectric" and "nonpolar". Our position is that only an account of the combined action of low dielectric constant and of intraprotein electric fields can explain all sides of the energetics of enzymatic catalysis (see subsection 2.4). Therefore, as a matter of fact, there are no fundamental contradictions between the results obtained by our semi-continuum approach and Warshel's microscopic analysis.

ΔG_{pol} and λ_s, are independent of the pre-existing electric field. [50] Therefore, these energies can be considered in a rather general form using the model of a structureless dielectric.

Let us consider first simple enzymes represented by globular proteins. As in our previous papers, [52, 53, 67, 68] we model them by a sphere with two spherical reactants immersed into the globule.

The solution of an electrostatic problem for a spherical globule can be represented in the form of a series of the Legendre polynomials [69] (the other, fully equivalent, way is to describe the electric field inside the sphere as a superposition of fields of real charges inside the sphere and of a distribution of image charges outside the globule [70]). The result of our analysis [71] is

$$\lambda = e^2\left(\frac{1}{\varepsilon_p^o} - \frac{1}{\varepsilon_p^s}\right)\left(\frac{1}{2a_1} + \frac{1}{2a_2} - \frac{1}{R_{12}}\right) +$$

$$+(A_1^o + A_2^o - A_3^o) - (A_1^s + A_2^s - A_3^s) \tag{4}$$

Here the first term identical with the Marcus equation (2) gives the reorganization energy in an infinite protein, the quantities $A_{1,2}$ are the contributions of the images (Legendre polynomials) to the charging energies of each reactant separately, A_3 - to their interaction energies, the quantities calculated for both media having optical (A_i^o) or static (A_i^s) dielectric constants. The correspoding expressions are

$$A_1 = \frac{e^2}{2}\sum_{n=0}^{\infty}\left(\frac{r_{01}}{R}\right)^{2n}\left(\frac{1}{R}\right)\frac{n+1\ \varepsilon_p - \varepsilon_w}{\varepsilon_p[n\varepsilon_p + \ n+1\ \varepsilon_w]} \tag{5}$$

$$A_2 = \frac{e^2}{2}\sum_{n=0}^{\infty}\left(\frac{r_{02}}{R}\right)^{2n}\left(\frac{1}{R}\right)\frac{n+1\ \varepsilon_p - \varepsilon_w}{\varepsilon_p[n\varepsilon_p + \ n+1\ \varepsilon_w]} \tag{6}$$

$$A_3 = e^2\sum_{n=0}^{\infty}\frac{r_{01}{}^n r_{02}{}^n}{R^{2n+1}}\frac{n+1\ \varepsilon_p - \varepsilon_w}{\varepsilon_p[n\varepsilon_p + \ n+1\ \varepsilon_w]}P_n\ cos\theta \tag{7}$$

$P_n(cos\theta)$ is the Legendre polynimial, ε_p and ε_w are the dielectric constants (optical or static) of protein and water, respectively, r_{01} and r_{02} are the moduli of the radius-vectors of two reactants, θ is the angle between these vectors; in (5), (6) the equality $P_n(1) = 1$ has been accounted for.

To calculate ΔG_{pol}, *i.e.* the equilibrium charging energy, the proper combination of static terms should be used (a sum or difference of charging energies with or without interaction energy).

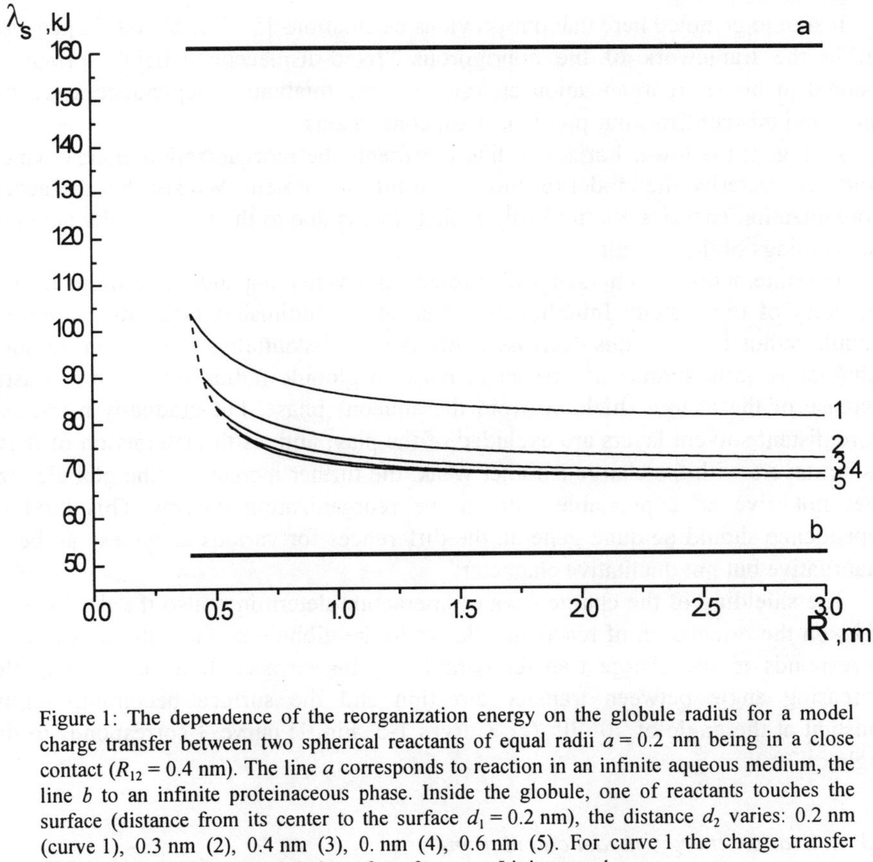

Figure 1: The dependence of the reorganization energy on the globule radius for a model charge transfer between two spherical reactants of equal radii $a = 0.2$ nm being in a close contact ($R_{12} = 0.4$ nm). The line a corresponds to reaction in an infinite aqueous medium, the line b to an infinite proteinaceous phase. Inside the globule, one of reactants touches the surface (distance from its center to the surface $d_1 = 0.2$ nm), the distance d_2 varies: 0.2 nm (curve 1), 0.3 nm (2), 0.4 nm (3), 0. nm (4), 0.6 nm (5). For curve 1 the charge transfer direction is parallel to the globule surface, for curve 5 it is normal.

Some results of these calculations for one set of geometric parameters (reactants' radii, interreactant distances, a constant depth of immersion of one of the reactants) are presented at Fig. 1. They give much lower (by, *e.g.*, 70-90 kJ/mol) reorganization energies for an enzymatic charge transfer than for the same reaction in an aqueous solution (cf. curves 1-5 and the upper horizontal line a). The physical reason for that is quite clear: some part of the solvent in the vicinity of the reactants is replaced by protein, *i.e.* the low-reorganization medium, and that decreases the total reorganization energy. Shielding by protein reduces the interaction of the charge being transferred with aqueous surroundings, decreasing in such a way the reorganization of the latter. The better is the shielding, the lower is the reorganization energy.

It should be noted here that our previous calculations [52, 53, 67, 68, 72] carried out in the framework of the nonrigorous "fixed-displacement-field" formalism resulted in higher reorganization energies but the qualitative dependences are the same, and this confirms our previous main conclusions.

At Fig. 1, the lower horizontal line *b* presents the reorganization energy which would characterize the model reaction in an infinite protein. We see that the actual reorganization energies are markedly higher: this is due to the highly-polar aqueous surroundings of the globule.

The interaction of charges with protein and water depends, naturally, on the geometry of the system. Initially, the value of λ_s diminishes fast with increasing globule radius but later this decrease slows down substantially. The reason for such behavior is quite simple: at first an increase in globule radius results in a drastic increase of the charge shielding from the aqueous phase, but gradually more and more distant solvent layers are excluded of the play, and, as the interaction of these distant layers with the charge is rather weak, the further increase of the globule size does not give an appreciable gain in the reorganization energy. This kind of dependence should be quite general, the differences for various enzymes can be of quantitative but not qualitative character.

The shielding of the charge - water interaction determines also the dependence of λ_s on the orientation of reactants relative to the globule surface: the maximal λ_s corresponds to the charge transfer parallel to the surface, it decreases with the increasing angle between transfer direction and the surface becoming almost constant at the angle of 20°-30° (cf. Curves 1-5, Fig. 1; curve 3 corresponds to the angle of 30°).

2.3 *Activation energy: model calculations*

In discussing the regularities for the activation energy we shall account for both reorganization energy λ (a sum of the medium and inner-sphere contributions) and the equilibrium energy of the elementary act ΔG. The latter consists, as we have mentioned before, of some component dependent on the structures of both reacting groups and protein ΔG_{str} and of an "polarization" component ΔG_{pol}. At the present stage of our discussion, we will consider the "structural" component as a parameter (ΔG_{str}).

The charge - polarization interactions manifest themselves in the influence of the system geometry on the Born solvation energies (and on the charge - charge interaction energies). This leads to a qualitatively new effect, absent in homogeneous systems, namely to the dependence of the reaction energy on the orientation of reactants with respect to the interface. [73]

Depending on the reaction type, these effects will affect the ΔG differently. Let us consider first the process of exchange of the charge position (charge shift), *e.g.*,

$$A^- + B = A + B^-$$

In the case when B is closer to the globule surface than A, the charge is transferred from interior of the globule towards the boundary. The increase in the charge interaction with the aqueous phase makes the charge transfer in this direction favorable. The influence of geometric parameters on ΔG_{pol} are qualitatively similar to their influence on λ_s. Therefore,the dependence of the activation energy on the globule radius, reactants' orientation, depth of their immersion, interreactant distance are similar to the corresponding dependences of λ_s (cf. Figs. 1 and 2).

In any catalytic process, the system must return to its initial state after completion of the catalytic cycle, *i.e.* the charge transfer from A to B should be followed by a step with the reverse transfer direction. Therefore, at any disposition of the reactants, we have considered both possible directions of the charge transfer. The charge transfer outwards, *i.e.*, closer to the high-dielectric medium, is favorable and, hence, decreases ΔG_{pol}. The comparison of Figs. 2 and 3 shows that the activation energy for outward charge transfer is always lower than for the inward direction.

At Figs. 2, 3 we give sets of curves corresponding to different ΔG_{str}, namely 0 and ±41.8 kJ/mol (0 and ±10 kcal/mol). These values are close by the order of magnitude to the usual range of energies for different steps of enzymatic catalysis. As can be seen from a comparison of these two figures, the activation energy for the endoergonic transfer even in the outward direction remains higher than for the inward exoergonic process. Therefore, the outward endoergonic process is here the rate-determining one.

The preference of the outward transfer holds true only for the reactants of similar size. When one of them is substantially larger than the other, the charge shift to the former may become more favorable.

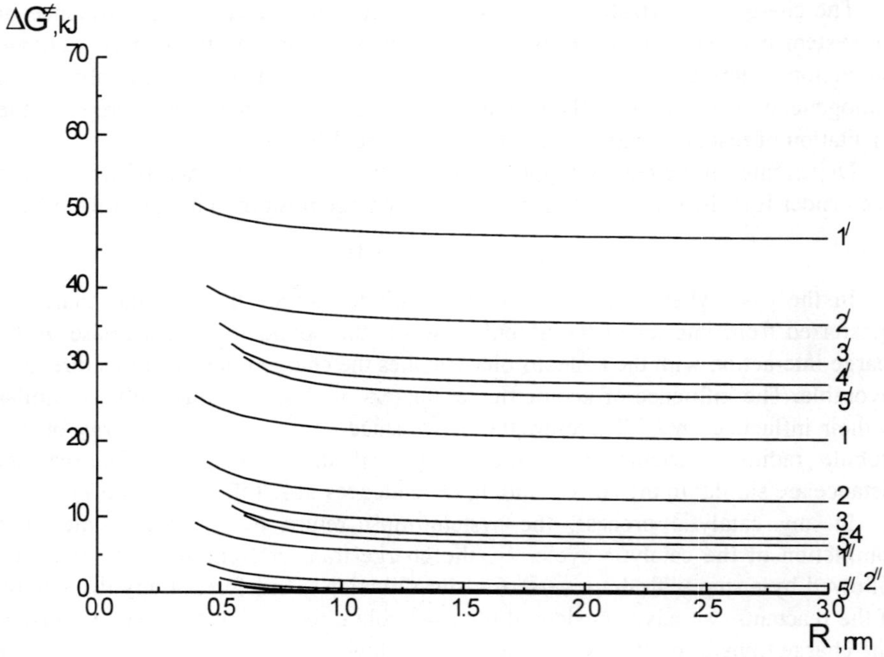

Figure 2: Activation energy for the same model reaction as at Fig. 1. The charge acceptor (1) is touching the surface, the charge donor (2) is situated at different distances (curves 1-5, as at Fig. 1): an outward charge transfer. Nonprimed curves correspond to $\Delta G_{str} = 0$, one prime refers to an endergonic reaction ($\Delta G_{str} = +41.8$ kJ/mole), two primes to an exergonic one ($\Delta G_{str} = -41.8$ kJ/mole).

We can formulate now a general conclusion that, in the rate-determining step, the charge displacement should preferably proceed from the reactant with a higher charging energy to the reactant with a lower one; this can be achieved at larger radius of the charge acceptor or/and its position closer to the high-dielectric aqueous surrroundings (the transfer from inside of the globule to its surface).

We can conclude also that the favorable disposition of the reactant should provide the charge transfer at an acute angle to the globule surface: under this condition the activation energy is rather low, and a relatively small additional decrease which could be achieved in some systems at a charge transfer almost perpendicular to the surface would be connected with much deeper immersion of the initial ion into globule, the immersion demanding a large energy expenditure.

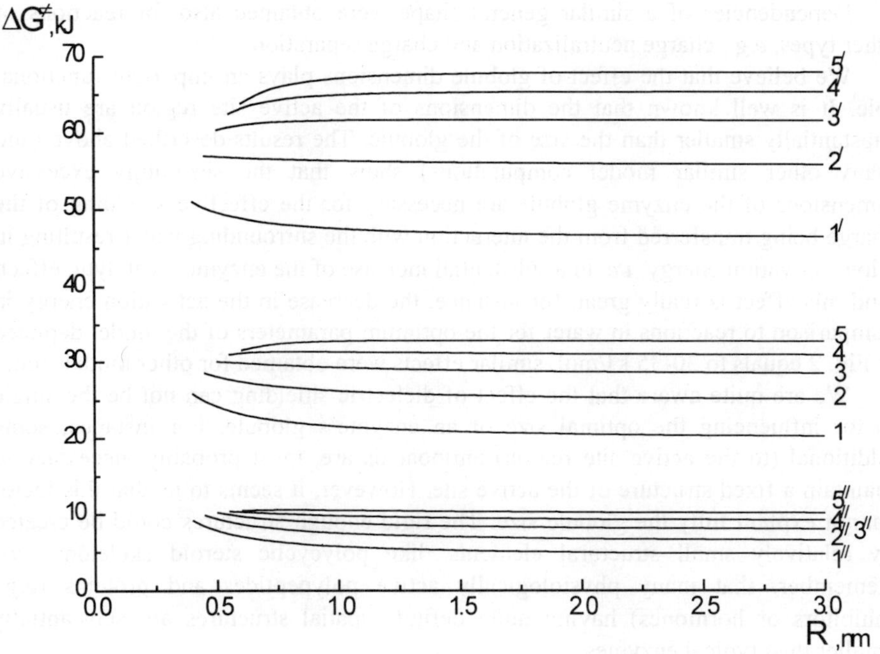

Figure 3: Activation energy for the same reaction but in opposite direction (an inward charge transfer). Note that for an endergonic reaction of Fig. 2 the reverse charge transfer means an exergonic process, and vice versa.

The third important conclusion follows from the form of the dependence of the activation energy for the rate-determining step on the globule radius. We observe a stronger decrease followed by a weaker dependence. As it is shown at Fig. 2, for the curves corresponding to the optimum reactants configuration (curves 3-5, 3'-5'), at the globule radius of about 0.95-1.0 nm ΔG is only 2 kJ higher than at R = 3 nm (at R = 1.35-1.45 nm increase in ΔG equals to 1 kJ/mole). The radius close to 1 nm (or to 1.4 nm) can be considered in a sense as optimum. Indeed, its further increase would require a large additional consumption of protein (proportional to R^3) but will not bring about an appreciable gain in the reaction rate (only by about 2.2 or 1.5 times). The optimal size of the globule increases markedly with an increasing interreactant distances, at a deeper immersion of reactants into globule, *etc.* (examples of real globulae with much larger optimum radii will be given below).

Dependencies of a similar general shape were obtained also for reactions of other types, *e.g.*, charge neutralization and charge separation.

We believe that the effect of globule dimensions plays an important functional role. It is well known that the dimensions of the active site region are usually substantially smaller than the size of the globule. The results described above (and many other similar model computations) show that the seemingly excessive dimensions of the enzyme globule are necessary for the effective shielding of the charge being transferred from the interaction with the surrounding water resulting in a low activation energy, *i.e.* in a substantial increase of the enzyme's catalytic effect. And this effect is really great: for instance, the decrease in the activation energy in comparison to reactions in water for the optimum parameters of the model depicted at Fig. 2 equals to 30-35 kJ/mol, similar effects were obtained for other models too.

We are quite aware that the effect of dielectric shielding can not be the single factor influencing the optimal size of an enzyme's globule. For instance, some additional (to the active site region) aminoacids are, most probably, necessary to maintain a fixed structure of the active site. However, it seems to us that this factor cannot explain fully the globule size. The rigid enough structures could be created by relatively small structural elements, like polycyclic steroid skeletons, *etc.* Remember, that many physiologically active polypeptides and proteins (*e.g.*, inhibitors or hormones) having quite definite spatial structures are substantially smaller than typical enzymes.

In our calculations, we have accepted the structural component ΔG_{str} as some parameter, constant for the series of the curves. If we compare the optimum radii for $\Delta G_{str} = 0$ and $\Delta G_{str} = +41.8$ kJ, *i.e.*, practically in a full range of the rate-determining endoergic steps, we will see that the difference between them is relatively small. For example, for curves 3-5 and 3'-5', Fig. 2, the difference in optimum radii does not exceed 0.05 nm. Therefore, we can conclude that the relative radii evaluated by this method are not very sensitive to the value of the structural component ΔG_{str}. This can serve as an intuitive rationale to analyse separately the influence of polarization and structure factors, as a verification of the possibility to analyse, in the first approximation, in a separate way the effects of polarization interactions and the structural component of the energy.

Below, we will apply the same formalism to an analysis of several real enzymatic reactions. Our approach will be as follows: we will start with the experimentally established reaction mechanism in order to find out what the groupings are involved in charge transfer. Then, on the basis of X-ray crystallographic data, we will define the main geometric parameters: the size of reacting groups, their mutual disposition and the distances between them. The reactants in their exact geometry will be positioned inside a spherical globule. The distances of the reactants from the globule surface will be also estimated from the structural data. The latter parameters will be the approximate ones because the surface of a real globule is, strictly speaking, neither spherical nor a smooth one, and we have to adjust the real geometries to the model of spherical globule. Therefore, we consider these distances as some parameters varying within reasonable limits consistent with the crystallographic data available.

First of all, we should notice that the general disposition of charges in the active sites corresponds to the optimum configurations following from the analysis of the idealized models given above. So, for ribonuclease with the reactions of a single charge shift, we observe, as predicted above, the charge transfer in the outward direction at an acute angle to the globule surface. For α-chymotrypsin and lysozyme with the reactions of charge separation in the presence of the third charge, three ions form an obtuse triangle with all sides being at acute angles to the surface (the geometry which also follows from model calculations). Charge neutralization preceeded by an ion formation (papain) takes place with two reactants on the line forming again an acute angle to the surface.

Let us now consider the globulae's optimum radii. The values, calculated under condition that the activation energy is by 2 kJ/mole higher than the minimal possible value, are as follows (in parentheses the data for the activation energy increase by 1 kJ/mole): for ribonuclease 1.1 (1.6) nm, lysozyme 0.95 (1.6) nm, papain 1.5 (2.3-2.4) nm, α–chymotrypsin 1.5-1.6 (2.0-2.4) nm. The average radii of these molecules calculated for a sphere of a given mass and the density 1.33 g/ml are: for ribonuclease 1.57, lysozyme 1.63, papain 1.84, α–chymotrypsin 1.9 nm. We see that these two sets of data are, in principle, similar. The calculated radii are conditional ones, and therefore it is better to speak only about the correct order of magnitude. However, more important is that the calculated relative values of the optimum globule sizes reflect reasonably the tendency in variation of the experimental radii. This shows that the effect of the general dielectric shielding plays indeed an important role in determination of the optimum enzyme's size.

2.4 Enzymes' high catalytic activity as a result of combined polarization and pre-existing field effects

As it was shown above, the structureless-dielectric model results in conclusion that the reorganization energy for protein is lower than for water. As a consequence, the activation energy for many charge transfer processes in a protein is also lower. This conclusion refers to the true activation energy of an elementary act of reaction supposing that all the charged reactants exist already inside the protein globule or that, for initially neutral reacting species, the charge separation is not strongly disfavored from the energetic point of view.

However, the transfer of an ion from water into a low-dielectric medium demands substantial energy expenditure. Similarly, the charge separation in aqueous solution is much more favorable than in a low-dielectric surroundings. Therefore, to compare the efficiency of these two media we have to confront the activation energy in an aqueous phase and the activation energy in a low-dielectric medium plus the energy necessary to transfer the reactants from water to this medium (or an excessive energy needed for the charge separation in the second case). We will not dwell on the relevant calculations, [74, 75] giving only their final conclusion: at usual size of enzyme globules, for all types of reactions considered the energy expenditure due to the loss of solvation energy is higher than the decrease in the activation energy caused by lower reorganization energy in a model “structureless protein”. This means that the structureless low-dielectric medium is less effective than water, and hence such a model cannot explain high catalytic activity of enzymes.

We have noticed before that the intraprotein electric field set up by the protein dipoles organized in a definite structure largely compensates for the energy loss due to a low Born solvation energy. This compensation cannot take place at any point inside the globule, but at definite reactants’ positions inside the active site of enzyme. In living nature, an intricate catalytic complex with a rather fixed structure is necessary to overcome the entropy restraints, in particular, to put together two or more particles in a proper orientation, to ensure the high specificity of binding, to shield the intermediates from undesirable attacks from aqueous solution, *etc.* However, such a complex is inevitably fairly large in size, and, due to its rigidity, has a rather low dielectric constant. All this leads to the shielding of the charge - water interaction and, hence, to an additional energy consumption in transfer of charged particles into the macromolecule. The medium preorganization, *i.e.*, the internal electric field compensates for this unfavorable effect. From the other side, the shielding effects on the equilibrium energy and on the reorganization energy should be optimized, and this results, as it was discussed above, in an optimization of the globule size.

We have defined proteins as low-dielectric highly-polar media, both these properties being due to a common cause: an arrangement of a great many of polar groups in a fixed structure. Low dielectric constant results in low reorganization energy, which is an essentially nonequilibrium parameter of reaction. High polarity of the protein constituents manifests itself in a significant, structure-dependent intraprotein electric field acting favorably on the process equilibrium energy (making it at least not exceeding substantially the same quantity in aqueous solution). These two effects together provide a marked decrease in the activation free energy, and this presents one of important general physical reasons for the high catalytic activity of enzymes.

The intraprotein electric potential may compensate the loss of the Born solvation energy in a various extent. An exact compensation is not necessary: in the case when the equilibrium ΔG of the reaction elementary act becomes comparable with the corresponding value in aqueous solution, a substantial acceleration will take place due to decrease in the reorganization energy. An additional acceleration can be achieved by decreasing ΔG to the level lower than in water. However, there should be some limit in lowering the energy of the charge transfer: when electric field produces a very strong stabilization of a charged state then its subsequent transformation into a neutral product (or release of a free ion into solution) will be strongly retarded; therefore, the next step of the total process can become the rate-determining one, and the overall kinetics may turn out to be markedly slowed.

The combination of high polarity of monomeric units and their fixation inside relatively rigid structures distinguishes proteins from other types of polymers like polyhydrocarbons or polycarbohydrates. Thus we can hypothesize that precisely this feature of proteins gave them an advantage over some other types of biopolymers in the natural selection of the "construction material" for enzymes.

Contradicting to this conclusion seems to be the existence of a quite different class of biological catalysts, namely the ribozymes discovered by Cech and his coworkers. [76] These ribonucleic acids catalyze effectively transesterification and hydrolysis, *i.e.*, reactions accompanied by a substantial charge transfer. However, some analogy can be found between RNA and proteins: the backbone of RNA contains a regular arrangement of highly polar phosphodiester groupings, and the polar side-chains of purine and pyrimidine bases are linked to this backbone. The hypothetical structure of the core of *Tetrahymena* ribozyme implies a rather dense packing of these structural elements. [77] So, in ribozymes we may meet with the same principle as in proteins: a high concentration of strongly polar groups substantially restricted in their orientational mobility. With the further accumulation of the relevant structural information, it would be extremely interesting to analyze the ribozymes function on the same physical basis as outlined above for the proteinaceous enzymes.

3 Spectroscopic estimates of the reorganization parameters

As it was shown before, the idea of a low reorganization energy of proteins permits to understand many features of the enzymatic charge transfer kinetics. At the same time, it seems to be important to prove the idea by another way, independent of kinetic data. We are attempting now at this direction using the measurements of the Stokes shifts of fluorescence spectra. [78]

For a purely classical description of polar modes, the difference of the energies of the maxima of the long-wavelength absorption and emission (fluorescence) spectra, a Stokes shift, equals to the reorganization energy doubled. [79, 80] With a due account of a discrete vibrational structure of the spectra caused by the quantum modes, this equality applies strictly not to the difference of the maxima but to the first momenta (centers of the spectra). [81] Our analysis has shown that the corresponding correction, though substantial from the quantitative point of view, does not change qualitatively the picture and, what is most important, does not prevent one from drawing correlation between the difference of the maxima and the medium dielectric properties.

To study this correlation, we have chosen the complex of α-chymotrypsin with a fluorescent dye proflavine. Proflavine is a well-known competitive inhibitor of chymotrypsin. It sorbs tightly in its binding pocket, in close vicinity of the active site. [82, 83] We have studied formation of this complex in more details to define the conditions ensuring the full binding of the dye, and excluding thereby any appreciable concentration of the free proflavine. [84]

As a reference system, we have started with the proflavine solutions in various solvents. Our experimental data are presented at Fig. 4. They show a tendency to a roughly linear correlation of λ with the solvents coupling parameter C as could be expected from the Marcus equation (2). Substantial deviations from the apparent linearity were observed for the solvents having substantially different static dielectric constants ε_S. These deviations have two reasons. First, the Marcus equation is an approximate one since it neglects the dielectric inhomogeneity of the solvent - solute system due to the difference of the effective ε's of the solvent and the bulky solute molecule. Second, the electronic density distribution in the ground and excited states depends on the surroundings' dielectric constant, and therefore the charge transferred effectively at the optical excitation is different for different media.

We have evaluated the first effect using the model of a dielectric oblate spheroid cavity. [85]. Variations of the electronic density distributions were calculated quantum-chemically using the modification [86] of the self-consistent reaction field theory. [87] A combination of these two effects results in evaluation of some conditional "reorganization energy" that can be considered as an integral parameter describing the dielectric properties ("reorganization ability") of the medium and replacing the coupling constant of the more approximate theory.

At Fig. 5, the experimental values of a half of the Stokes shifts are plotted against this integral parameter. For all aprotic solvents (circles), we observe a good correlation conforming the theory. [88] This correlation was used as a basis for further analysis.

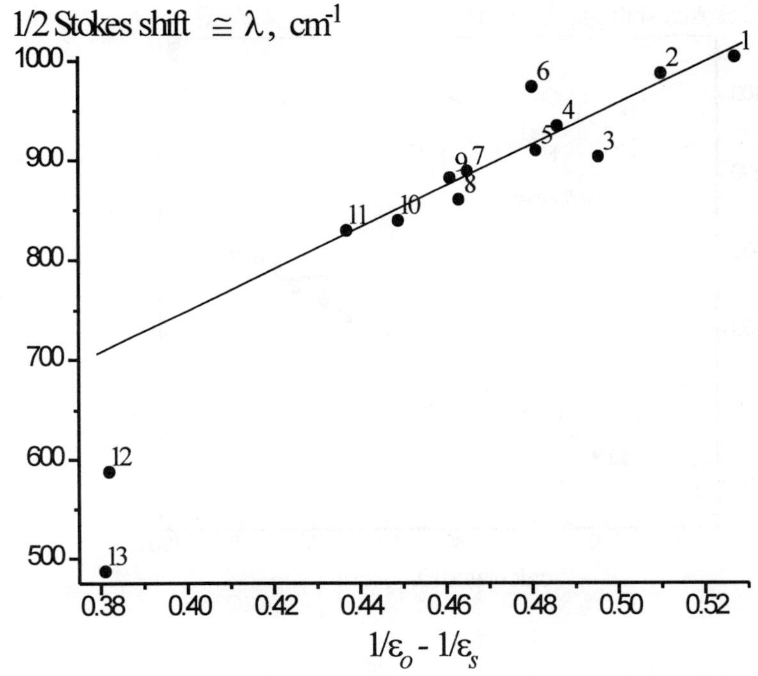

Figure 4: Stokes shift of a dicationic dye proflavine in various polar aprotic solvents and their mixtures plotted versus the coupling parameter of the solvent. ***1***- acetonitrile; ***2***- γ-butyrolactone (25% vol.) - acetonitrile mixture; ***3***- acetone; ***4***- propylene carbonate (62%) - acetone mixture; ***5***- acetonitrile (32%) - γ-butyrolactone mixture; ***6***- propylene carbonate; ***7***- γ-butyrolactone (86%) - acetone mixture; ***8***- *N,N*-dimethylformamide; ***9***- γ-butyrolactone; ***10***- dimethyl sulfoxide (77%) - acetone mixture; ***11***- dimethyl sulfoxide; ***12***- 1,2-dichloroethane; ***13***- dichloromethane.

The experimental point for water (triangle) lies far higher than it could be expected for the medium with similar dielectric properties. We believe that this increase can be ascribed to the additional energy necessary for reorganization of the network of water's hydrogen bonds in the close vicinity of the dye molecule. As an argument in favor of this interpretation we can consider a marked difference in reorganization energy for heavy water (square) whose macroscopic dielectric properties are very close to those of the light water.

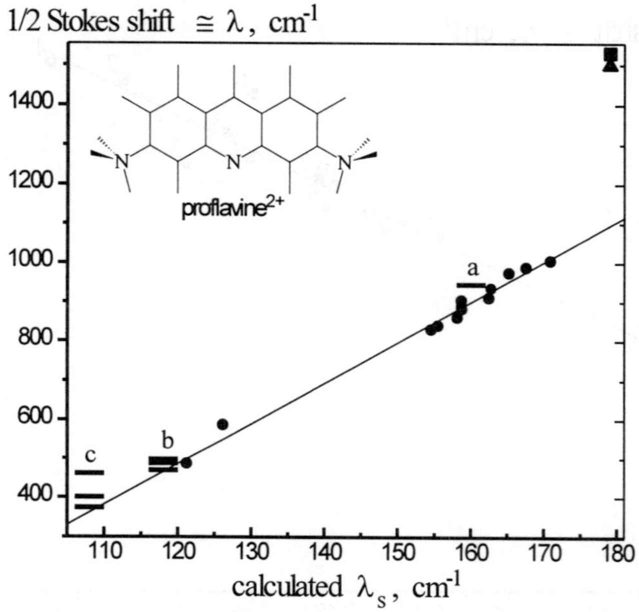

Figure 5: Stokes shift of a dicationic dye proflavine in various media. The data for the free dye solutions in various polar aprotic solvents (circles) and in light (triangle) and heavy (square) waters are plotted versus the conditional "reorganization energy" of the solvent. The data for the dye-chymotrypsin complex are marked by lines. The *a* line stands for the complex in aqueous solution at pH 8.5. The *b* group of lines stands for the complex in protein films (made of the excess of the same protein chymotrypsin) dried from the complex-protein pH 8.5 water solution at humidity of 0, 85 and 45% (from the upper to the lower lines). The *c* group of lines stands for the complex in the same protein films but impregnated by low-polar solvent chloroform (the humidity is, from upper to lower lines, 0, 45 and 85%).

The horizontal line *a* at the Fig. 5 corresponds to the Stokes shift of proflavine in the complex with chymotrypsin dissolved in aqueous solution (pH 8.5). First of all, this value is substantially lower than for the free dye in water. It should be stressed that for the aqueous solution of the complex the solvent isotope effect is negligibly small (of order of few cm^{-1}). This shows that in the complex with chymotrypsin the proflavine molecule is shielded enough from a direct contact with water (or, may be, it contacts water only by a part of the molecule) to prevent a serious reorganization of the hydrogen bonds network. However, not only the effect of hydrogen bonds is responsible for lowering the reorganization energy. The Stokes shift is substantially lower here than for an aprotic solvent with water's macroscopic dielectric parameters. This is in full agreement with the picture outlined in this paper: protein has a lower reorganization parameter than water, and therefore the corresponding quantity for the dissolved complex should have the value intermediate between characteristics of the protein and water.

The reorganization energy for the dye-protein complex in aqueous solution reveals to be rather high, close to values for moderately polar solvents. This unexpectedly high value can be explained by proximity of the rearranging charges to the external medium: the large molecule of proflavine cannot fit completely the binding pocket of α-chymotrypsin designed to sorb much smaller aromatics, and therefore the dye molecule should protrude towards aqueous phase.

We attempted also at obtaining the characteristics of the protein itself measuring the Stokes shifts in dry protein films of the proflavine-chymotrypsin complex. These data can give us only the upper limit of the desirable quantity. The reasons are that, due to inhomogeneity of the samples obtained, the emission spectra are broadened, and, even at the lowest concentrations of the dye we could experimentate with, we could not fully exclude some excitation energy migration from species with higher energy to the species with the lower one. Nevertheless, the results (the group of lines *b* at Fig. 5) show that the protein indeed has a very low reorganization energy, lower than for most polar solvents.

For a protein in solution, there are ionized groups on its surface, and their counter-ions are dissolved in the bulk. The field of both these ions is well screened by the aqueous surroundings. Upon drying a protein solution, even free of extraneous electrolytes, ionized groups and counter-ions transform into surface-attached ionic pairs. Therefore, there appears, on the surface of globule, a marked amount of rather mobile and poorly screened large dipoles, and their reorientation should contribute to the reorganization energy.

Generally, there are three kinds of effects of the films' humidity on the reorganization energy. First, water molecules solvate polar surface groups and reduce their electric field which results in decreasing the dielectric response and, thus, the reorganization energy. Second, the effective size of solvated groups is higher. This should decrease the mobility of the groups. Slowed down reorientation of the enlarged groups does allow these degrees of freedom to participate in the reorganization during the lifetime of the excited state of the dye (we found the lifetime for the complex in aqueous solution to be about 0.3 ns). Third, increasing amount of water molecules in the film gives rise to increasing the reorganization due to water' own reorientation ability. The three effects influence the reorganization energy in different directions which explains the non-monotonous dependence of the reorganization energy on the humidity (Fig. 5, the *b* group of lines).

The effect of the side polar groups can be reduced by their dielectric screening and viscous friction. We have impregnated the films with a low-dielectric solvent, chloroform ($\varepsilon = 4.8$). The reorganization energies are indeed lower (group *c*) and more sensitive to humidity. The strongest reduction of the reorganization energy is observed at the highest humidity (of 85%). In this case, the effective size of the side groups is the largest due to better solvation by water. This means that the reduction of the reorganization energy due to impregnation is caused predominantly by decreasing the mobility of the ion pairs, and their contribution to reorganization energy for the dye bound close to the protein surface is large. Comparing these data with the experimental points for homogeneous liquid solvents we can conclude that the protein by itself behaves as a medium substantially less polar than dichloromethane ($\varepsilon_O = 2.03$, $\varepsilon_S = 8.9$, $C = 0.38$): compare the lowest circle point for dichloromethane with the lowest line of the *c* group at Fig. 5.

4 Effect of the time-evolution of the proteins' dielectric response

At the beginning of this paper, we have stated that one of specific features which can be expected for proteins as polar media is a very wide spectrum of their relaxation times. This property affects especially strongly the fastest reactions like the electron transfers in photosynthetic reaction centers, *etc.* In these cases, the polarization processes, which are many orders of magnitude slower than the reaction rate, have no time to develop in a full extent in the course of the charge transfer (see, *e.g.*, [89-92]). The situation can be described phenomenologically in terms of a set of effective dielectric constants ε operative in the corresponding time intervals. [26, 27]

For the photosynthetic reaction centers, the experiments with the different characteristic time of measurement have given different values of ΔG, the free energy of the charge separation reaction: $-0.18 \div -0.19$ eV for picoseconds, [93] $-0.23 \div -0.24$ eV at 10 ns, [91-93] -0.26 eV at 100 μs, [94] -0.30 eV in millisecond time span, [95] the equilibrium value -0.34 eV or less (by the absolute magnitude) [93] (see [26, 27] for more detailed discussion; we do not consider here very small initial gaps [91] due, most probably, to the biographical inhomogeneity of reaction centers [96]). This time-evolution of the free energy can be ascribed to the temporal variation of the dielectric response and, correspondingly, to the change of the reactants Born solvation energies and their Coulombic interaction.

We have analyzed the effect of the variable ε_S on the free energy of reaction and on its reorganization energy (in [26] and, in a more rigorous way, in [27]). The quantitative calculations were done in the framework of a three-layer model of the photosynthetic membrane. This model takes into account the different dielectric properties of the membrane's low-polar inner core (hydrocarbon chains of lipids, α-helical segments of the hydrophobic intramembrane proteins; $\varepsilon_S \approx 3$ - 4) and of the membrane's outer layers of intermediate polarity (lipid polar heads, hydrophilic segments of proteins; $\varepsilon_S \approx$ 10-20). [97, 98] The different dielectric constants of inner and outer parts of a photosynthetic membrane reveal themselves in the magnitudes of electrogenic effects in these membranes. [99] The results of these calculations for the charge separation between the special pair (P) and bacteriopheophytin (Bph), *i.e.*, P + Bph = P^+ + Bph^-, are presented at Fig. 6.

Figure 6: Reorganization energy for charge separation P* - Bph calculated in the framework of the three-layer model as a function of the effective static dielectric constant of the membrane low-polar inner core. The distance between the center of the special pair P and the nearest boundary of the layers of low and intermediate polarity was taken as 0.45 nm (curve 1), 0.5 nm (2), and 0.6 nm (3). The arrows show the limits of the energy gap characteristic for the time intervals of picoseconds, 10 ns, and 100 μs.

We see that the effective static dielectric constants corresponding to the different time spans vary substantially. They are about 3.4 ÷ 3.5 at 100 μs, 3.2-3.3 at 10 ns and 2.8-2.9 at few picoseconds. In picosecond time interval, the inertial polarization has almost not developed, and the dielectric constant remains close to its optical (better to say quantum) value, and even after 10 ns it is far from the static one.

With the stationary value of $\varepsilon_S = 4$, the reorganization energy of the charge separation was found to be 0.31±0.01 eV. This is close enough to the stationary value of ΔG which is about −0.3 eV, the sum of these two quantities is practically zero, *i.e.*, as follows from Eq. 1, the reaction is activationless (practically activationless behavior corresponds to condition $\Delta G \ll kT$, which is met here).

It should be stressed that if the charge separation process reveals to be activationless at any value of ε_S it will remain activationless at any other ε_S. The reason is that the electrostatic part of the charge separation free energy, involving the energies of charging of two ions and of their Coulombic interaction, strictly equals, with the opposite sign, to the corresponding ε_S - dependent term in the expression for the reorganization energy (this result has been obtained in [26] for homogeneous medium, and for any arbitrary inhomogeneous one in [27]). Therefore, the sum of ε_S - dependent contributions to ΔG and λ_S is always zero.

The activationless character of the primary charge separation step of photosynthesis is a well established fact (for review see, *e.g.*, [100, 101]). Therefore, in many theoretical analyses of this process, one adjusted the value of λ to make it equal to the experimentally found ΔG. [1, 89, 102, 103] The consideration given above shows how this "adjustment" was made by Nature: the redox-active molecules have a negligibly small intramolecular reorganization energy, and their size and disposition ensure the medium reorganization energy equal to the free energy gap. [26, 27, 104]

The same result, namely the activationless charge separation, can be obtained not only by the semi-continuum approach but also by a microscopic way. This ultrafast process occurs in so short time spans that it can be analyzed directly by the molecular dynamics method. Such simulations have been performed in [55-58, 61, 62] for the time spans of few tens of picoseconds (up to 60 ps [62]). They have obtained reorganization energies of 0.19-0.22 eV. These quantities match quite well the energy ΔG = −(0.18÷0.19) eV characteristic of this time interval. We see that for this system the conclusions obtained by semi-continuum and microscopic methods are in a full agreement.

Besides the one-step charge separation process discussed above an alternative mechanism has been proposed, namely that the process proceeds in two steps with an intermediate formation of $Bchl^-$. [105, 106] Strong arguments in favor of this mechanism were given at this meeting by M. Michel-Beyerle. Our calculations [27] have shown that both these steps can be activationless if the energy level of $Bchl^-$ is intermediate between those of the excited special pair P^* and Bph^-. It was a revelation to get know at this meeting that such an estimate of its position was obtained experimentally by M. Michel-Beyerle and her coworkers.

The subsequent electron transfer to quinone was also computed to be practically activationless which agrees well with the experimental data. This result was obtained by the semi-continuum approach taking account of the reasonable range of ε_S typical of the times of the order of few hundreds picoseconds (as estimated from Fig. 6). [27] The same result follows also from molecular dynamic simulations. [56, 58]

The primary charge separation takes place at rather low free energy gap, hence the probability of the reverse reaction is not low enough. To ensure the irreversibility of the trans-membrane electron transfer, a larger energy gap between the initial and final electron levels is necessary. This additional substantial energy drop occurs at the transition from bacteriopheophitin to quinone A. To make this process ultrafast, and hence to impart to it a decisive preference over competing back reactions, it is desirable to make it activationless, *i.e.* to increase its reorganization energy up to the increased value of $-\Delta G$. This goal is achieved by using the final acceptor - quinone - having a smaller effective radius and, therefore, a higher reorganization energy. The same principle is used in other photosynthetic organisms: the primary acceptor (a compound of the chlorophyll series) is rather close in energy to the primary donor (a photoexcited pigment), the second step is accompanied by a substantially larger energy drop, and the secondary acceptor (quinone, iron-sulfur cluster) has a substantially smaller effective radius than chlorophyll. [27, 104]

The time-dependent values of ε_S were used also to calculate the activation energies for other electron transfers. In accordance with the experiment, practically activationless processes were found for recombination of the primary radical-ion pair with formation of a neutral triplet, and of recombination of the special pair cation with quinone anion.

The considerations given above show the importance of the time evolution of the protein's dielectric response for the electron transfer kinetics. These data are not the single argument in favor of this effect. Quite recently, a very wide, in time scale, temporal evolution of electrochromic shifts in RC have been reported, [107] which can be considered as an independent proof of the time evolution of the effective dielectric constant. Different dielectric relaxation times (*e.g.*, picoseconds to nanoseconds) were revealed also by time-resolved fluorescence spectra [108] and by the lifetime dependence of the fluorescence maxima [109] of dyes bound to protein.

5 Conclusion

In this paper, we have considered some consequences of the theory of charge transfer processes in polar media as applied to enzymatic reactions. In this context, the specificity of the systems lies in the specific features of proteins as polar media substantially different from low-molecular solvents. Proteins comprise a high concentration of highly-polar peptide groups being a part of a well-defined and rather stable spatial structure; therefore, the orientational mobility of these dipoles is strongly restricted. This structural principle has two interconnected important consequences: a rather weak dielectric response of this polar system to an applied external electric field, and the existence of a permanent time-averaged non-zero electric field of the system of dipoles, the field which is present in absence of any free charge. These properties can be briefly summarized in a definition of proteins as low-dielectric highly-polar media (this may be true also for other macromolecules constructed on the same principles).

Low static dielectric constant of proteins (*i.e.*, low mobility of their dipoles) results in low reorganization energies of the enzymatic charge-transfer reactions; this is a factor substantially decreasing, as a rule, the activation energies of reactions. At the same time, low dielectric constant is unfavorable for transfer of charges from, say, aqueous medium into protein, and hence this factor should strongly retard intraprotein charge transfer as compared to the same reaction in a high-dielectric solvent. However, this effect is compensated by the "medium preorganization", *i.e.* the existence of an intraprotein permanent electric field favoring the charged states of reactants at some definite positions inside the protein structure. It is just the combined action of these two effects, *viz.* the low reorganization energy and the medium preorganization, which ensures the high catalytic activity of enzymes.

A low reorganization energy of proteins has been confirmed by the method independent of the data on the enzymatic kinetics, namely by the spectroscopic study. The Stokes shift of the fluorescence spectrum of a dye proflavine related directly to the medium reorganization energy has been studied in different solvents and in the dye complex with α-chymotrypsin. Comparison of these data enables us to conclude that the reorganization energy of the protein corresponds to characteristics of a low-polar medium.

The dielectric constant reflects the total effect of small shifts of a large number of dipoles, and hence this factor is typical of practically all proteins irrelevant to details of their structure. On the contrary, the intraprotein electric field is highly structure-sensitive. Therefore, its favorable effect can be operative only for a definite charge-transfer reaction in the corresponding class of enzymes evolutionary-adjusted to carry out this process. This circumstance should be kept in mind, say, in constructing artificial enzymes: the same catalytic entities introduced by gene engineering at different positions inside a protein globule can have a quite different activity.

The third consequence of the complex and relatively rigid structure of proteins (in contrast to liquid solvents) is a very wide interval of times characteristic for different kinds of motion. This results in the time-evolution of the dielectric response. This effect seems to be especially important for very fast electron transfer processes as it was shown by an analysis of experimental data on different electron transfer processes in photosynthetic reaction center.

Acknowledgement

The research described here has been supported in part by grants from the ISF (MD 9000), NIH Fogarty International Center (TW00063), INTAS (93-2852) and RFBR (96-04-48374). E.L.M. was supported by A.N. Frumkin fellowship.

References

1. R.A. Marcus and N. Sutin, *Biochim. Biophys. Acta* **811**, 265 (1985).
2. R.R. Dogonadze and A.M. Kuznetsov, *Progress in Surface Science* **6**, 1 (1975).
3. J. Ulstrup, *Charge Transfer Processes in Condensed Media, Lecture Notes in Chemistry* (Springer-Verlag, Berlin etc., 1979).
4. L.I. Krishtalik, *Charge Transfer Reactions in Electrochemical and Chemical Processes* (Plenum, New York, 1986).
5. A.M.Kuznetsov, *Charge transfer in physics, chemistry and biology* (Gordon and Breach Publ., Amsterdam etc., 1995).
6. J. Applequist and T.G. Mahr, *J. Amer. Chem. Soc.* **88**, 5419 (1966).
7. A. Wada, in *Polyamino acids, polypeptides and proteins, Proc. Int. Symp.*, Univ.Visconsin Press, Madison, p.131 (1962).
8. L.I. Krishtalik and V.V. Topolev, *Molekulyarnaya Biologiya (Moscow)* **17**, 1034 (1983).
9. P.Th. van Duijnen and B.T. Thole, *Biopolymers* **21**, 1749 (1982).
10. S. Bailey, *Trans. Faraday Soc.* **47**, 509 (1951).
11. S. Maricic, G. Pifat and V. Pravdic, *Biochim. Biophys. Acta* **79**, 293 (1964).
12. S. Takashima and H.P. Schwan, *J. Phys. Chem.* **69**, 4176 (1965).
13. A. Tanaka and Y. Ishida, *J. Polymer Sci. Phys. Ed.* **11**, 1117 (1973).
14. R. Pethig, *Dielectric and Electronic Properties of Biological Materials*, (Wiley, Chichester 1979).
15. M.K. Gilson and B.H. Honig, *Biopolymers* **25**,2097 (1986).
16. H. Nakamura, T. Sakamoto and A. Wada, *Prot. Eng.* **2**, 177 (1988).
17. T. Simonson, D. Perahia and G. Bricogne, *J. Mol. Biol.* **218**, 1859 (1991).
18. T. Simonson, D. Perahia and A.T. Brünger, *Biophys. J.* **59**, 670 (1991).
19. G. King, F.S. Lee and A. Warshel, *J. Chem. Phys.* **95**, 4366 (1991).
20. P. E. Smith, R.M. Brunne, A.E. Mark and van W.F. Gunsteren, *J. Phys. Chem.* **97**, 2009 (1993).
21. L.I.Krishtalik, *Molekulyarnaya Biologiya (Moscow)* **8**, 91 (1974).

22. F.R.N. Gurd and T.M. Rothgeb, *Adv. Prot. Chem.* **33**, 74 (1979).
23. M. Karplus and J.A. McCammon, *CRC Crit.Rev.Biochem.* **9**, 293 (1991).
24. H. Frauenfelder, S.G. Sligar and P.G. Wolynes, *Science* **254**, 1598 (1991).
25. C.L. Brooks III, M. Karplus and B.M. Pettitt, *Proteins, A Theoretical Perspective of Dynamics, Structure, and Thermodynamics* (Wiley, New York, 1988).
26. L.I. Krishtalik, *Biochim. Biophys. Acta* **1228**, 58 (1995).
27. L.I. Krishtalik, *Biochim. Biophys. Acta*, **1273**, 139 (1996).
28. G. Johannin and N. Kellersohn, *Biochem. Biophys. Res. Commun.* **49**, 321 (1972).
29. A. Warshel and M.J. Levitt, *J. Mol. Biol.* **103**, 227 (1976).
30. A. Warshel, *Proc. Natl. Acad. Sci. USA* **75**, 5250 (1978).
31. W.G.J. Hol, P.T. Van Duijnen and H.J.C. Berendsen, *Nature, Lond.* **273**, 443 (1978).
32. P.T. Van Duijnen, B.T. Thole and W.G.J. Hol, *Biophys. Chem.* **9**, 273 (1979).
33. R.P. Sheridan and L.C. Allen, *Biophys. Chem.* **11**, 133 (1980).
34. B.H. Honig, W.L. Hubbell and R.F. Flewelling, *Annu. Rev. Biophys. Biophys. Chem.*, **15**, 163 (1986).
35. N.K. Rogers, *Progr. Biophys. Mol. Biol.* **48**, 37 (1986).
37. S.C. Harvey, *Proteins* **5**, 78 (1989).
38. K.A. Sharp and B.H. Honig, *Annu. Rev. Biophys. Biophys. Chem.*, **19**, 301 (1990).
39. M.E. Davis and J.A. McCammon, *Chem. Rev.*, **90**, 509 (1990).
40. A. Warshel and J. Åqvist, *Annu. Rev. Biophys. Biophys. Chem.*, **20**, 267 (1991).
41. A. Warshel, *Current Opin. in Struct. Biology*, **2**, 230 (1992).
42. D. Bashford, *Current Opin. in Struct. Biology* **1**, 175 (1991).
43. P.A. Kollman, *Current Opin. in Struct. Biology*, **2**, 765 (1992).
44. A.-G. Yang and B. Honig, *Current Opin. in Struct. Biology*, **2**, 40 (1992).
45. K.A. Sharp, *Current Opin. in Struct. Biology* **4**, 234 (1994).
46. P.A. Kollman, *Current Opin. in Struct. Biology* , **4**, 240 (1994).
47. M.K. Gilson, *Current Opin. in Struct. Biology*, **5**, 216 (1995).
48. M.K. Gilson, A. Rashin, R. Fine and B. Honig. *J. Mol. Biol.* **183,** 503 (1985).
49. L.I. Krishtalik, *J. Theor. Biol.* **139**, 143 (1989).
50. L.I. Krishtalik, A.M. Kuznetsov and E.L. Mertz, *Proteins: Structure, Function, Genetics,* in press [abstract published in *Biophys. J.* **70**, A225 (1996)].

51. A.M. Kuznetsov, J. Ulstrup and M.A. Vorotyntsev, in *The Chemical Physics of Solvation.* editors R.R. Dogonadze, E. Kálmán, A.A. Kornyshev and J. Ulstrup. (Elsevier, Amsterdam. Part C, 163, 1988).
52. L.I. Krishtalik, *Molekulyarnaya Biologiya (Moscow)* **13**, 577 (1979).
53. L.I. Krishtalik, *J. theor. Biol.* **86**, 757 (1980).
54. A.K. Churg, R.M. Weiss, A. Warshel and T. Takano, *J. Phys. Chem.* **87**, 1683 (1983).
55. S. Creighton, J.K. Hwang, A. Warshel, W.W. Parson and Norris J., *Biochemistry* **27**, 774 (1988).
56. A. Warshel, Z.T Chu. and W.W. Parson, *Science* **246**, 112 (1989).
57. W.W. Parson, Z.-T. Chu and A. Warshel, *Biochim. Biophys. Acta* **1017**, 251 (1990).
58. K. Schulten and M. Tesch, *Chem. Phys.* **158**, 421 (1991).
59. C. Zheng, J.A. McCammon and P.G. Wolynes, *Chem. Phys.* **158**, 261 (1991).
60. A. Yadav, R.M. Jackson, J.J. Holbrook and A. Warshel, *J. Amer. Chem. Soc.* **113**, 4800 (1991).
61. M. Treutlein, K. Schulten, A.T. Brünger, M. Karplus, J. Deisenhofer and H. Michel, *Proc. Natl. Acad. Sci. U.S.A.* **89**, 75 (1992).
62. M. Marchi, J.N. Gehlen, D. Chandler and M. Newton, *J. Amer. Chem. Soc.* **115**, 4178 (1993).
63. J. Åqvist, M. Fothergill and A. Warshel, *J. Amer. Chem. Soc.* **115**, 631 (1993).
64. Y-P. Liu and M.D. Newton, *J. Phys. Chem.* **98**, 7162 (1994).
65. R.A. Marcus, *J. Phys. Chem.* **98**, 7170 (1994).
66. A.M. Kuznetsov and I.G. Medvedev, *J. Phys. Chem* **100**, 5721 (1996).
67. L.I. Krishtalik and Yu.I. Kharkatz, *Biofizika* **29**, 19 (1984).
68. Yu.I. Kharkatz and L.I. Krishtalik, *J. theor. Biol.* **112**, 221 (1985).
69. J.A. Stratton, *Electromagnetic Theory*, (McGraw-Hill, New York, 1941).
70. A.V. Finkel'shtein, *Molekulyarnaya Biologiya (Mosc.)* **11**, 627 (1978).
71. L.I., Krishtalik and V.V. Topolev, in preparation.
72. L.I.Krishtalik, in *The Chemical Physics of Solvation* eds. R.R. Dogonadze, E. Kálmán, A.A. Kornyshev and J. Ulstrup (Elsevier, Amsterdam, part C, p.707, 1988).
73. L.I. Krishtalik, *Molekulyarnaya Biologiya (Mosc.)*, **15**, 290 (1981).
74. L.I. Krishtalik, *J. theor. Biol.* **112**, 251 (1985).
75. L.I. Krishtalik, in preparation.
76. T.R. Cech, D. Herschlag, J.A. Piccirilli and A.M. Pyle, *J. Biological Chemistry* **267**, 17479 (1992).
77. F. Michel and E. Westhof, *J. Mol. Biol.* **216**, 585 (1990).

78. E.L. Mertz and L.I. Krishtalik, in preparation.
79. R.A. Marcus, *J. Chem. Phys.* **38**, 1858 (1963).
80. N.S. Hush, *Progr. Inorg. Chem.* **8**, 391 (1967).
81. E.L. Mertz, *Chem. Phys. Lett.*, in press (1996).
82. A.N. Glaser, *Proc. Nat. Acad. Sci. USA* **54**, 171 (1965).
83. S.A. Bernhard, B.F. Lee and Z.H. Tashian, *J. Mol. Biol.* **18**, 405 (1966).
84. E.L. Mertz and L.I. Krishtalik, *Biofizika*, **41**, 320 (1996).
85. E.L. Mertz, E.D. German, A.M. Kuznetsov, *Chem. Phys.*, in press (1996).
86. G.E. Chudinov, D.E. Napolov and M.V. Basilevsky, *Chem. Phys.* **160**, 41 (1992).
87. S. Miertuš, E. Scrocco and J. Tomasi, *Chem. Phys.* **55**, 117 (1981).
88. E.L. Mertz, V.A. Tikhomirov and L.I. Krishtalik, *J. Phys. Chem.*, submitted.
89. M. Bixon and J. Jortner, *J. Phys. Chem.* **90**, 3795 (1986).
90. N.W.T. Woodbury and W.W Parson, *Biochim. Biophys. Acta* **767**, 345 (1984).
91. J.M. Peloquin, J.C. Williams, X. Lin, R.G. Alden, A.K.W. Taguchi, J.P. Allen and N.W. Woodbury, *Biochemistry* **33**, 8089 (1994).
92. N.W. Woodbury, J.M. Peloquin, R.G. Alden, X. Lin, S. Lin, A.K.W. Taguchi, J.C. Williams and J.P. Allen, *Biochemistry* **33**, 8101 (1994).
93. N.W.T. Woodbury, W.W. Parson, M.R. Gunner, R.C. Prince and P.L. Dutton, *Biochim. Biophys. Acta* **851**, 6 (1986).
94. R.A. Goldstein, L. Takiff and S.G. Boxer, *Biochim.Biophys.Acta* **934**, 253 (1988).
95. R.J. Shopes and C.A. Wraight, *Biochim. Biophys. Acta* **899**, 409 (1987).
96. A. Ogrodnik, W. Keupp, M. Volk, G. Aumeier and M.E. Michel-Beyerle, *J. Phys. Chem.* **98**, 3432 (1994).
97. D.A. Cherepanov and L.I. Krishtalik, *Bioelectrochemistry and Bioenergetics*, **24**, 113 (1990).
98. L.I. Krishtalik, *Molekulyarnaya Biologiya* **26**, 1377 (1992).
99. S.M. Dracheva, L.A. Drachev, A.A. Konstantinov, A.Y. Semenov, V.P. Skulachev, A.M. Arutjunjan, V.A. Shuvalov and S.M. Zaberezhnaya, *Eur. J. Biochem.* **174**, 253 (1988).
100. W.W. Parson and B. Ke, in *Phototsynthesis* Govindjee, ed., Vol. **1**, p. 331, (Academic Press, New York, 1982).
101. C. Kirmaier and D. Holten, *Photosynth. Res.* **13**, 225 (1987).
102. J. Jortner, *Biochim. Biophys. Acta* **594**, 193 (1980).
103. M. R. Gunner and P.L. Dutton, *J. Amer. Chem. Soc.* **111**, 3400 (1989).
104. L.I. Krishtalik, *Biochim. Biophys. Acta* **977**, 200 (1989).
105. V.A. Shuvalov and V.V. Klimov, *Biofizika* **32**, 814 (1987).

106. W. Holzapfel, U. Finkele, W. Kaiser, D. Oesterhelt, H. Scheer, H.U. Stilz and W. Zinth, *Proc. Natl. Acad. Sci. USA* **87**, 5168 (1990).
107. B. McMahon, J.D. Müller, B. Marino, P. Debrunner, C. Wraight and U. Nienhaus, *Biophys.J.*, **70**, A142 (1996).
108. D.W. Pierce and S.G. Boxer, *J. Phys. Chem.* **96**, 5560 (1992).
109. E.L. Mertz, in preparation.

PROTEINS AT METALLIC SURFACES INVESTIGATED BY CYCLIC VOLTAMMETRY AND *IN SITU* SCANNING PROBE MICROSCOPY

J.E.T. ANDERSEN, E.P. FRIIS, J.-J. KARLSSON,
K.G. OLESEN, M.H. THUESEN, J. ULSTRUP
Department of Chemistry, Technical University of Denmark, DK-2800 Lyngby, Denmark

L.L. MADSEN, P. MØLLER
Department of Manufacturing Engineering, Technical University of Denmark DK-2800 Lyngby, Denmark

M.F. NIELSEN
Department of Chemistry, Symbion Science Park, University of Copenhagen, Copenhagen Ø., Denmark

The organization and electron transfer (ET) dynamics of metalloproteins at metallic surfaces can be mapped in increasing detail by voltammetric methods. *In situ* scanning tunnelling (STM) and atomic force microscopy (AFM) of metalloproteins under potentiostatic control and directly in aqueous solution has also become feasible.

We overview briefly some recent results of ours in these areas. We discuss first the voltammetric pattern of the bacterial di-heme protein cytochrome c_4. Both heme groups can be brought to exchange electrons with the electrode and, interestingly, in spite of the close proximity of the heme groups, intramolecular ET is slow (ms range). We show next that the single-heme cyt *c* can be non-covalently and covalently immobilized on gold and platinum. Immobilized cyt *c* can be imaged by *in situ* STM to a resolution that qualifies for the characterization "molecular". The blue single-copper protein azurin can, finally be assembled on Au(111) by molecular chemisorption via the disulphide group. Chemisorbed azurin can be imaged with molecular resolution by *in situ* AFM and STM of which we show an interesting dynamic pattern in the former mode.

1. Introduction

Electron transfer (ET) patterns of metalloprotein systems in homogeneous solution have been mapped in considerable molecular detail[1-6]. Mapping extends to several fundamental ET features including particularly driving force and directional tunnel patterns through the bulk protein. Target systems have been modified small single-metal proteins such as cytochromes, blue copper proteins and iron-sulphur proteins. Composite multi-centre protein complexes such as the bacterial photosynthetic reaction centres in particular, have added new and untraditional features of organized long-range electron transport where electronic and vibrational coherence stand out conspicuously[7-11].

Protein function is also associated broadly with solid-liquid surfaces[12,13].

Biological liquids are, for example always in contact with solid surfaces such as membranes, bones, electrodes or implantates. Microbial corrosion is essentially induced and stimulated by interactions between proteins and metallic surfaces. In the overlapping areas of basic science and technology, bioelectrochemistry, biosensors, enzyme electrodes, controlled molecular desorption, and organized membrane-bound protein function are all examples of intimately interfacially controlled protein function.

Interfacial metalloprotein electrochemistry is a relatively recent area where protein dynamics at solid surfaces has been addressed in a way which approaches to qualify for the notion of quantitative detail[14,15]. Reversible promoter-induced voltammetry of small metalloproteins is, for example now straightforwardly accessible. In addition, voltammetry of multi-centre proteins such as metallo*enzymes* are disclosing interesting prospects for resolution of molecular mechanisms for composite electrochemical ET. At the same time environmental inhomogeneity and external electric fields strongly entangle theoretical approaches to interfacial metalloprotein ET. This is reenforced by the commonly different approaches to the electronic coupling which induces ET in homogeneous solution and interfacial systems. Molecular wave functions, rooted in quantum chemistry constitute the basic notion in the former class, while metallic overall electronic *densities* are commonly taken as the microscopic characterization of metallic interfacial systems[6,16].

In this report we overview the properties of some interfacial metalloprotein systems where molecular mechanistic detail has in some way been achieved. This applies, first to a composite but relatively simple, two-centre protein. As new perspectives we show, next some applications of *in situ* scanning tunnelling (STM) and atomic force (AFM) microscopy in the particular context of metalloproteins at metal/aqueous electrolyte interfaces. Such an integrated approach appears to offer new perspectives for experimental and theoretical characterization of metalloproteins at solid surfaces in contact with the natural aqueous medium for metalloprotein function.

2. The Molecular Mechanism of Electrochemical ET of a Two-Centre Metalloprotein: *Pseudomonas Stutzeri* Cytochrome c_4

2.1 Some Properties of Pseudomonas Stutzeri Cytochrome c_4

The cytochromes c_4 are a class of di-heme proteins isolated from both aerobic and anaerobic bacteria. They are believed to constitute part of repiratory chains but their exact function and position in the chains are not known. Physical chemical properties of cyt c_4 from *P. stutzeri* and *A. vinelandii* have been characterized in some detail. In the present context the following observations, specific to *P. stutzeri* cyt c_4 are important[17]:

A. The MW ≈ 19.000 Da.

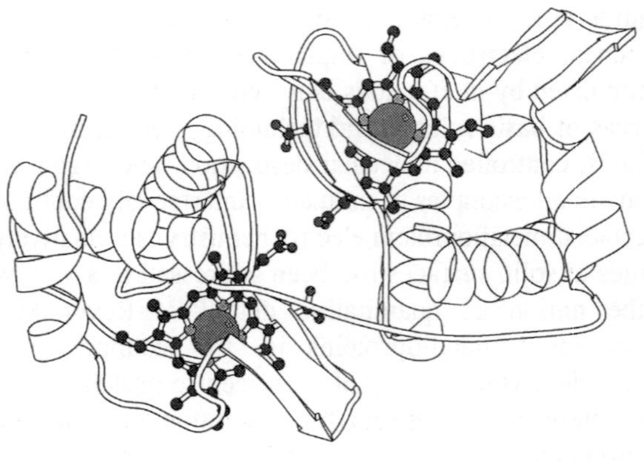

Figure 1: 2.2 Å X-ray crystal structure of *P. stutzeri* cyt c_4[18]. Presentation by the programme Molscript[19].

B. A 2.2 Å X-ray crystal structure has become available (fig. 1)[18]. The structure discloses:
two domains connected by a single peptide string. The heme groups are in H-bond contact via two propionates and, interestingly, tilted by ≈ 30 °. Other H-bond contacts are located between the two domains. His and Met are axial ligands in both hemes.

C. The protein is strongly dipolar (pH ≈7) with four excess negative charges in the N-terminal domain and one excess positive charge in the C-terminal domain (protein moiety). This holds interesting prospects for protein orientation at electrochemical interfaces by suitable choice of promoter (cf. below).

D. The protein has been characterized by UV/VIS, NMR, EPR, and resonance Raman spectroscopy. The most important outcome related to the present report is that the oxidized form is a high-spin/low-spin equilibrium while the reduced form is pure low-spin. In addition to entangling the overall ET kinetics, such a spin conversion may induce considerable intramolecular nuclear reorganization. In spite of the close proximity of the heme groups, intramolecular ET is therefore not necessarily a fast process.

E. The kinetics of ET between cyt c_4 and several external reaction partners has been characterized. Intermolecular ET is multi-exponential and in the μs-ms time range. The macroscopic reduction potentials are ≈ 240 and ≈ 320 mV (NHE), with the lower reduction potential in the negatively charged N-terminal domain.

Interaction potentials between the heme groups are negative and in the 30-40 mV range.

The ET kinetics points to electron exchange at both heme groups. One question left open is, however, the importance of intramolecular ET between the heme groups. The kinetics could only be resolved to the point that intramolecular ET is either *slow* or *fast* on the stopped-flow time scale and *slow* on the NMR time range. This would locate the rate constants k_{int} (s^{-1}) in the ranges $k_{int} << 10^2\ s^{-1}$ *or* $10^5\ s^{-1} >> k_{int} >> 10^2\ s^{-1}$. In the following we show that interfacial electrochemical ET of cyt c_4 holds an additional clue to the intramolecular ET process.

2.2 Interfacial Electrochemical Kinetics of the Two-Centre Protein P. Stutzeri Cyt c_4

Metalloprotein voltammetry commonly requires the use of promoters, i.e. organic chemical adsorbates[14,15]. Their exact molecular function is not known except that they prevent the proteins from collapsing or adsorbing irreversibly at the surface and ensures that pleasant-looking reversible voltammetry can emerge. The common promoter structure is X-Y where X adsorbs to the electrode and Y links to the protein by H-bond or other contacts. It is an experience of metalloprotein electrochemistry that broad classes of promoters are specific *either* to positively charged *or* to negatively charged proteins. A small group of promoters are active towards both categories. Fig.2 shows the promoters used below, namely cystamine (negatively charged proteins), bis-4,4'-pyridyl-bisulphide (4,4'bpySS, positively charged proteins), and 3-pyridinylmethylene)-hydrazinecarbothiamide (3-PMHC, either protein group).

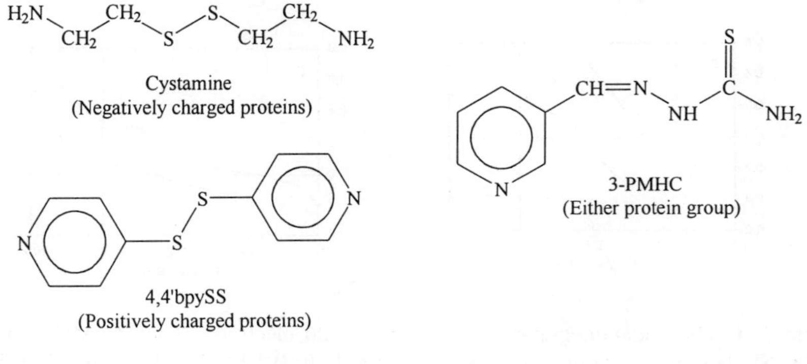

Figure 2: Promoters.

Cyt c_4 exhibits no voltammetric signals at gold electrodes in promoter-free solution. At high ionic strength (0.1 M NaCl), close to reversible voltammetry with two closely spaced voltammetric peaks is observed (fig.3)[20]. This is indicative that the partly electrostatically controlled promoter effect is screened and little selective to the protein surface orientation. On the other hand, a different character of the voltammograms emerges at low ionic strength (0.01 M NaCl). The voltammograms are now partly *kinetically* controlled. In the presence of the charge specific promoters 4,4'bpySS and cystamine they also show interesting asymmetry. The former (positively charged proteins) gives *two* peaks in the *cathodic* scan but a *single*, higher peak in the *anodic* scan. The latter (negatively charged proteins) gives a similar pattern but with *two* peaks in the *anodic* scan and a *single* higher peak in the *cathodic* scan[20].

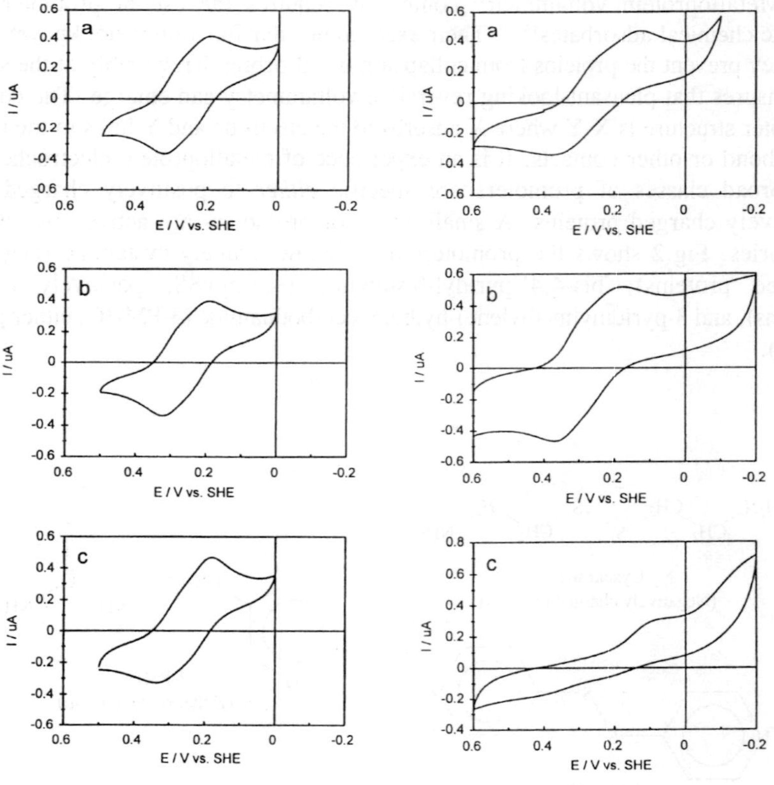

Figure 3: Cyclic voltammograms of cyt c_4 on a gold electrode. Top: 3-PMHC, middle: 4,4'bpySS, bottom: cystamine. Scanrate 10 mV/s. Left: 0.1 M phosphate, pH=7.5. Right: 0.01 M phosphate, pH=7.5.

All the observations are best in line with the view that cyt c_4 is adsorbed in a rather rigid, promoter-specific orientation at low ionic strength. The negatively charged N-terminal domain then points towards the electrode when cystamine is the promoter and the C-terminal domain points towards the electrode when 4,4'bpySS is the promoter. In the former case, as the potential is scanned anodically, the adjacent heme group is oxidized first, followed by oxidation of the remote group at a higher potential, *via* the adjacent group. As the scan is reversed the reduction of the remote, high-potential heme group is, however, hampered by the adjacent heme until the reduction potential of the latter is reached. Two merging single-ET steps then occur within a narrow potential range. The qualitative asymmetry of the voltammetry therefore straightaway follows from this view. It is also understandable that the detailed voltammetric shapes are determined by the intramolecular ET *rate* provided that the latter falls approximately in the time range of the potential scan rates.

Voltammograms according with this mechanism can be computer simulated by available procedures[20]. The reduction potentials, the interfacial electron exchange rate constants, *and* the intramolecular ET rate constant can subsequently be obtained with fair accuracy by comparison with the data. Details of this analysis are reported elsewhere[20] but we note the following:

A. The analysis is feasible because the selected protein is "simple" enough but as a two-centre protein it still holds the fundamental features of intramolecular ET functionality and cooperativity.

B. The intramolecular ET rate constants are 1600 s^{-1} (low- to high-potential heme) and 40 s^{-1} (high- to low-potential).

The values of the rate constants are first, low compared to a wealth of reported intramolecular ET rate constants in modified single-metal proteins[2,4,5]. The latter are, however, always strongly exothermic, photo-induced reactions. In addition they are not subject to the spin conversion as observed for cyt c_4. In this sense the values are therefore perfectly reasonable. On the other hand, they do not resolve the issue regarding the ET mechanism of cyt c_4 in homogeneous solution because the rate constants correspond to a time range where detection by the homogenous ET analysis should be feasible. This apparent discrepancy must presently be ascribed to the different external conditions in the two investigations (ionic strength, interfacial and homogeneous environment). The electrochemical investigation, however, still stands out as a quite powerful tool towards otherwise elusive ET dynamics in composite biological systems.

3. *In Situ* STM and AFM of the Single-Metalloproteins Cytochrome *c* and Azurin

STM and AFM have opened exciting new perspectives for mapping the structure and dynamics of solid surfaces and adsorbates in molecular detail. Reports of STM and AFM imaging of proteins at surfaces are available[21-23] but with a few exceptions *all* in the *ex situ* mode. Aqueous electrolyte solution is, however, the natural medium for protein function. In addition water is an integrated constituent of the molecular structure of all soluble proteins, either as solvation spheres or as internal protein structural elements. Mapping of protein dynamics by STM/AFM methodology should therefore ultimately rest on the *in situ* modes.

3.1 Some Theoretical Views on the Role of the Redox Metal Atom in the In Situ STM Imaging of Metalloproteins

Facile long-range ET through metalloproteins is strongly supported by broad experience from comprehensive investigations of metalloprotein ET in homogeneous solution[1-6]. *In situ* STM of metalloproteins can be expected to follow similar lines but attention to the particular three-level STM configuration is needed, illustrated by fig.4[24-26]. The redox level is strongly coupled to the protein and solvent environment. ET from the negatively potential biased electrode (fig.4, left) to the molecule induces an equilibrium shift in the nuclear coordinates. As the electron is transmitted to the positively biased electrode the original nuclear configuration is restored but the (Gibbs free) energy is now shifted by the bias potential. Theoretical approaches to the *in situ* STM patterns can therefore be brought to rest on system motion along the three-level

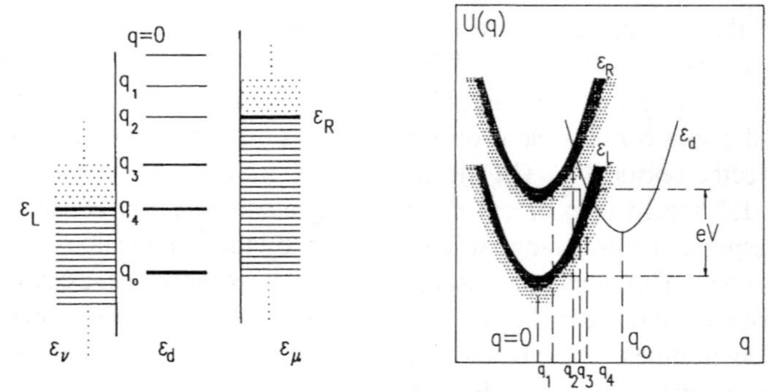

Figure 4: (Left) Energy diagram of three-level ET via the adsorbate level between two metallic continua, at different values of a fluctuating nuclear coordinate q. q=0 corresponds to equilibrium in the initial, vacant state, $q=q_0$ to equilibrium in a final, occupied state. (Right) Potential surface configuration corresponding to the level system to the left. The hatched regions represent continua of potential surfaces.

potential surface configuration shown in fig.4, right. The initially equilibrated local redox level is thus located well above the Fermi levels of both the substrate and tip. Environmental configurational fluctuations take the level to resonance with levels around the Fermi level of the negatively biased electrode where the first ET step occurs. Configurational relaxation continues subsequent to ET but before the relaxation is complete the electron is transmitted in a second step to the positively biased electrode. Two expectations regarding specifically the *in situ* STM mode are particularly interesting. One is the observation that *bias voltage* variation would give a smooth, rather structure-less current-voltage relation unless the local molecular transition metallic level crosses the Fermi level of one of the electrodes. The other observation is that *overvoltage* variation would give an interesting non-monotonous current variation as the local level crosses the region between the two Fermi levels. The latter expectation has received exciting support from current-voltage relations of iron protoporphyrin IX (the central prosthetic group in hemoglobins and other hemoproteins) at highly oriented pyrolytic graphite substrates[27].

The observations above illustrate the expectation that special inelastic and activated tunnel channels may apply to metalloproteins and large transition metal complexes. These channels further hold perspectives for the mapping of metalloprotein electronic *function* at metallic surfaces, illustrations of which are beginning to appear.

3.2 In Situ STM of Non-Covalently and Covalently Immobilized Cytochrome c at Gold, HOPG, and Platinum Substrates

Cytochrome *c* is a single-heme metalloprotein (MW ≈ 12 kDa) fairly universally central in respiratory ET chains. The reduction potential is quite high (265 mV (NHE), horse heart). ET patterns in homogeneous solution are well characterized, and promoter-induced reversible cyclic voltammetry straightforwardly accessible at high ionic strength.

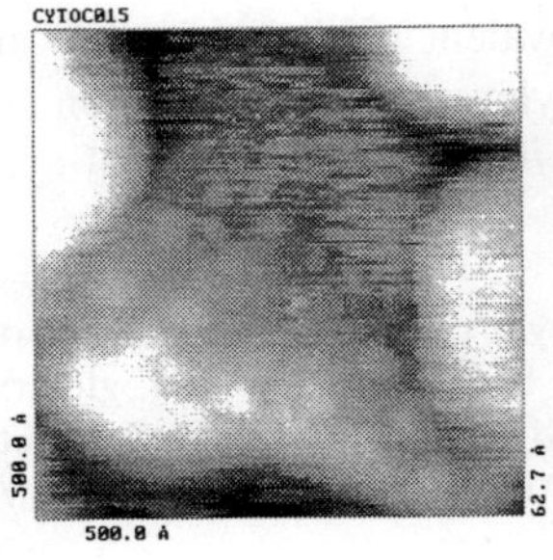

Figure 5: *In situ* STM image of non-covalently immobilized cyt *c* (low ionic strength)[28]. 20 mM MES buffer, pH=7, 10^{-3} M cyt *c*. Small, molecular-size structures between larger aggregates. z=6.3 nm. Substrate potential -100 mV (Cu reference). Bias voltage 50 mV. Tunnel current 1.25 nA.

Cyt *c* has been a target protein in recent *in situ* STM investigations[28,29]. Fig.5 shows common patterns for non-covalently immobilized assembled cyt *c* on flame annealed gold and on glassy carbon or pyrolytic graphite. The following is notable[28]:

A. Immobilization is achieved at low ionic strength only. No cyt *c* images are found at high ionic strength. The latter conditions are commonly used for electrochemistry but cyt *c* is also electrochemically active at low ionic strength.

B. Cyt *c* assembles in mesoscopic structures. By the lateral extension the structures hold a hundred or more individual molecules organized in a single or a few monolayers.

C. The mesostructures are mobile. Pinholes form, coalesce, and disappear. These structures may hold some prospects for investigations of heterophase fluctuations, phase transitions etc.

D. Depending on the time and patience available, small structures, of the size of individual cyt *c* molecules *between* the mesostructures can be observed. It appears as if the molecular-like structures adhere to the surface as the larger bodies are collectively - spontaneously or induced by the tip - shifted in space.

Self-assembled, non-covalently immobilized cyt *c* has thus been identified by *in situ* STM. By the small structures *between* the mesostructures it also appears that resolution has reached the molecular level. The balance between immobilization and aggregation is, however, fragile. Covalent immobilization holds alternative preparatory perspectives. Covalent linking would give more restricted molecular mobility with better prospects for molecular resolution. These conditions are, on the other hand less suitable for complementary electrochemical characterization of the same covalently immobilized monolayer samples.

Polycrystalline platinum surfaces are broadly composed of large, approximately atomically flat areas (up to several hundred nm). They contain enough surface oxides that silanization, glutaric aldehyde attachment, and covalent binding of cyt *c* are feasible. Cyt *c* can also be attached to Au(111) ny N-acetylcysteine chemisorption and subsequent carbodi-imide coupling to cyt *c*. Fig.6 shows *in situ* STM images of cyt *c*

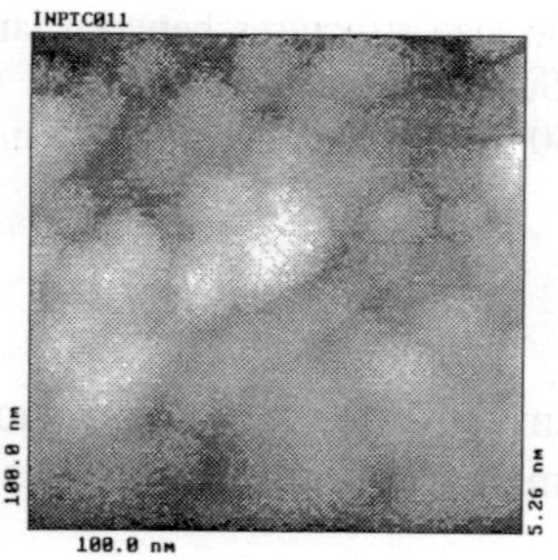

Figure 6: *In situ* STM of cyt *c* covalently immobilized on polycrystalline platinum by silanization and glutaric aldehyde coupling. 20 mM phosphate, pH=7. Substrate potential -500 mV (Au/AuO_x). Bias voltage 100 mV. Tunnel current 1.78 nA.

covalently immobilized on platinum by this procedure[29]. The character of the surface adsorbate is now quite different from both that of non-covalently adsorbed cyt *c*, from the free Pt-surface, and from the surface to which only the linker molecules have been attached. Particularly:

A. Spots are scattered all over the surface. The spot pattern is different from the free substrate pattern which can, however, be distinguished *underneath* the adsorbates.

B. The size range of the spots is ≈ 4-15 nm. The lower limit accords with the crystallographic size of cyt *c*. The larger structures could involve aggregation, convolution with the linker molecules or the tip, or inherent protein mobility.

Overall the covalent immobilization procedure is the more robust and gives systematic and better molecular resolution than the non-covalent immobilization procedure. Work towards optimization of the immobilization procedure and disclosure of immobilized protein function is now in progress.

3.3 In Situ AFM of the Single-Copper Redox Protein Azurin on Au(111)

Attempts to map the four-copper blue oxidase *P. versicolor* laccase by *in situ* STM on highly oriented pyrolytic graphite have only given low-resolution images of larger structures[30]. Molecular resolution was at best achieved only in the *ex situ* mode, probably due to incompatibility between the hydrophilic enzyme and the hydrophobic surface.

With a view on surface compatibility and adsorptive self-assembly, the bacterial respiratory single-Cu copper protein azurin (*P. aeruginosa*) was selected for STM and AFM characterization on gold substrates[31]. In this respect azurin offers several suitable properties: (a) There is a single Cu-atom with an accessible reduction potential (365 mV (NHE)); (b) the class of azurins are robust and structurally, electronically, and kinetically well characterized[32-34]; many mutant azurins are also available; (c) investigations of chemical ET of azurins in homogeneous solution indicate two sites for electron exchange, one directly at the Cu-atom, and another one a facile tunnel route *through* the protein[34]; (d) the protein shape is ellipsoidal with the copper atom in one end, suitably accessible for facile homogeneous and electrochemical ET; a disulphide group is, importantly, located in the opposite end; this group offers a convenient site for linking of the molecule to gold surfaces, in a configuration suitable for subsequent facile intramolecular ET through the molecule.

A final brief note of observation is that *P. aeruginosa* azurin shows stable, reversible, and unpromoted voltammetry at pyrolytic graphite electrodes[31]. Reversible voltammetry also appears at gold electrodes but the voltammograms here decay within a period of 10-20 min[31]. Approximately a monolayer of adsorbed azurin has been detected by XPS subsequent to exposure of the gold substrates to azurin solutions. The signal decay is in line with time dependent azurin adsorption via the disulphide bridge and concomitant attenuation of the interfacial ET rate by the need for electron tunnelling to the Cu-atom *through* the protein.

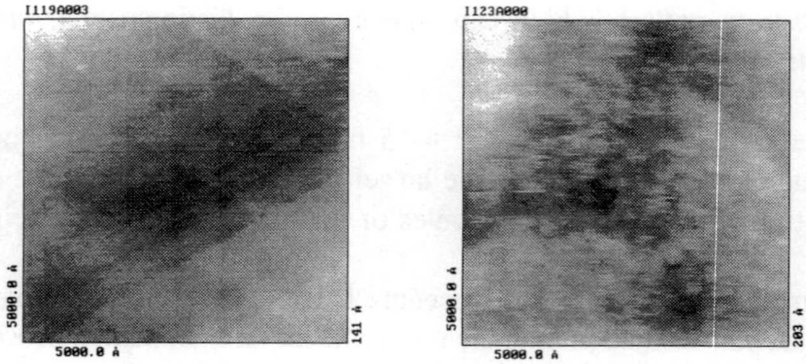

Figure 7: *In situ* (open circuit) AFM images of Au(111) (left) and Au(111) after pre-adsorption of *P. aeruginosa* azurin (right). 50 mM ammonium acetate solution, pH=4.65.

Fig.7 shows AFM images of oven-annealed Au(111) in contact with aqueous ammonium acetate buffer before and after exposure to azurin solution[31]. These images have been recorded under open circuit potential conditions. The triangular substrate surface structure is clearly visible, and an array of individual new structures is seen on the azurin-exposed samples. Their vertical height (≈ 5 nm) accords with the crystallographic molecular dimension while the lateral dimension is three to five times larger. This kind of discrepancy for soft materials is encountered elsewhere[23] and most likely due to convolution with the tip, the dimensions of which are much more crucial in AFM than in STM. Successive scanning gradually pushes off the azurin array, exposing again the free gold surface. This underlines the strong tip-sample interaction and the image convolution/deformation on the rather blunt contact mode used in the image recording in fig.7.

Better resolution and longer-time stability have been achieved using AFM in the tapping and phase contrast modes[35]. Other quite recent results are that high-resolution azurin array formation has been detected under full potentiostatic control at potentials negative of the reduction potential of azurin. These results will be reported elsewhere[35].

4. Concluding Remarks

The short overview above illustrates some recent developments and prospects for investigation of metalloprotein dynamics at metal-solution interfaces. Voltammetry and *in situ* STM/AFM are the two approaches addressed. Cyclic voltammetry of small proteins is now straightaway feasible but molecular detail such as adsorption patterns and promoter modes have remained elusive. Voltammetry of composite, multi-centre proteins including metallo*enzymes* is also feasible both in the direct and electrocatalytic modes, and molecular mechanisms can be addressed by strategic choice of suitably simple target proteins.

In situ STM/AFM in the natural aqueous medium for protein function, with molecular resolution, has also been achieved. Molecular and collective dynamics can be followed, and there are prospects for approaching molecular functionality and tunnel mechanisms of metalloprotein function at solid-liquid interfaces. It is understandable that sub-molecular resolution of these soft and internally mobile structures is not presently feasible. *In situ* STM/AFM of proteins is also fraught with interference caused by protein aggregation. Outstanding issues that need further *theoretical* attention are, finally associated with the contrast imaging and with large local current *densities* which would formally group the coherent two-step tunnel mechanisms close to the limit of environmental relaxational control.

Acknowledgement

Financial support from the Danish Natural Science and Technical Science Research Councils, The Carlsberg Foundation, Novo Nordisk Foundation, P.A. Fisker's Foundation, and the EU Human Capital and Mobility Network, Activation of Organic Molecules by Electron Transfer is acknowledged.

References

1. *Metal Ions in Biological Systems*, eds. H. Sigel and A. Sigel (M. Dekker, NY, 1991).
2. *Chem. Rev.* **92**, 365-490 (1992), ed. M.A. Fox.
3. *J. Photochem. Photobiol. A: Chem.* **82**, 1-232 (1994), ed. L.G. Arnaut.
4. For a list of many systems, see: G.W. Canters and C. Dennison, *Biochimie 77*, 506 (1995).
5. H.B, Gray and J.R. Winkler, *Chem. Rev.* **92**, 369 (1992).
6. G. Iversen *et al., Adv. Chem. Phys.*, in press.
7. *Chem. Phys.* **197**, 223-472 (1995), eds. M.E. Michel-Beyerle and G.J. Small.

8. *The Reaction Centre of Photosynthetic Bacteria*, ed. M.E. Michel-Beyerle (Springer-Verlag, Berlin, 1996).
9. Yu.I. Kharkats *et al., J. Phys. Chem.* **99**, 13545 (1995).
10. H. Sumi and T. Kakitani, *Chem. Phys. Lett.* **252**, 85 (1996).
11. M.H. Vos *et al., Nature* **363**, 320 (1993).
12. *Surface and Interfacial Aspects of Biomedical Polymers. Volume 2, Protein Adsorption*, ed. J.D. Andrade (Plenum, NY, 1985).
13. *Proteins at Interfaces. Physicochemical and Biochemical Studies*, eds. J.L. Brash and T.A. Horbett (ACS Symp. Ser., Washington DC, 1987).
14. F.A. Armstrong, *Struct. Bond.* **72** 137 (1990).
15. G. Huo and H.A.O. Hill, *Adv. Inorg. Chem.* **36**, 341 (1991).
16. A.A. Kornyshev, *Electrochim. Acta* **12**, 1829 (1989).
17. L.S. Conrad *et al., Eur. J. Biochem.* **231**, 133 (1995).
18. A. Kadziola and S. Larsen, *Struct.*, in press.
19. P. Kraulis, *J. Appl. Phys.* **24**, 946 (1991).
20. J.-J. Karlsson *et al., submitted.*
21. *Atomic Resolution Microscopy of the Solid-Liquid Interface*, eds. S.M. Lindsay and M.J. Weaver (Oxford University Press, NY, 1994).
22. *SPM and SFM in Biology*, eds. O. Marti and M. Armrein (Academic Press, San Diego, 1993).
23. S.N. Magonov and M.H. Whangbo, *Surface Analysis with STM and AFM* (Verlag Chemie, Weinheim, 1996).
24. A.M. Kuznetsov *et al., Surf. Sci.* **275**, 52 (1992).
25. A.M. Kuznetsov and J. Ulstrup, *Surf. Coat. Technol.* **67**, 193 (1994).
26. J.E.T. Andersen *et al., Electrochim. Acta*, in press.
27. N.J. Tao, *Phys. Rev. Lett.* **76**, 4066 (1996).
28. J.E.T. Andersen *et al., Surf. Sci.* **325**, 193 (1995).
29. J.E.T. Andersen *et al.*, *submitted.*
30. J.E.T. Andersen *et al., Electrochim. Acta*, **41**, 2005 (1996).
31. E.P. Friis *et al., Electrochim. Acta*, in press.
32. H. Nar *et al., J. Mol. Biol.*, **221**, 765 (1991).
33. A.G. Sykes, *Adv. Inorg. Chem.* **36**, 377 (1991).
34. O. Farver *et al., Biochemistry* **32**, 7317 (1993).
35. E.P. Friis *et al., in preparation.*